ADVANCES IN
GEOPHYSICS

VOLUME 17

Contributors to This Volume

FRANS DE MEYER
FRANCIS E. FENDELL
V. B. KOMAROV
A. QUINET
B. V. SHILIN

Advances in
GEOPHYSICS

Edited by
H. E. LANDSBERG

*Institute for Fluid Dynamics and Applied Mathematics
University of Maryland, College Park, Maryland*

J. VAN MIEGHEM

*Royal Belgian Meteorological Institute
Uccle, Belgium*

Editorial Advisory Committee

BERNARD HAURWITZ   R. STONELEY
ROGER REVELLE   URHO A. UOTILA

VOLUME 17

1974

Academic Press   •   New York   San Francisco   London
A Subsidiary of Harcourt Brace Jovanovich, Publishers

COPYRIGHT © 1974, BY ACADEMIC PRESS, INC.
ALL RIGHTS RESERVED.
NO PART OF THIS PUBLICATION MAY BE REPRODUCED OR
TRANSMITTED IN ANY FORM OR BY ANY MEANS, ELECTRONIC
OR MECHANICAL, INCLUDING PHOTOCOPY, RECORDING, OR ANY
INFORMATION STORAGE AND RETRIEVAL SYSTEM, WITHOUT
PERMISSION IN WRITING FROM THE PUBLISHER.

ACADEMIC PRESS, INC.
111 Fifth Avenue, New York, New York 10003

United Kingdom Edition published by
ACADEMIC PRESS, INC. (LONDON) LTD.
24/28 Oval Road, London NW1

LIBRARY OF CONGRESS CATALOG CARD NUMBER: 52-12266

ISBN 0-12-018817-1

PRINTED IN THE UNITED STATES OF AMERICA

CONTENTS

LIST OF CONTRIBUTORS vii

Tropical Cyclones
FRANCIS E. FENDELL

1. Introduction.. 2
2. Aspects of Tropical Meteorology... 17
3. Models of a Mature Tropical Cyclone.. 29
4. Theory of Tropical Cyclone Intensification................................. 75
5. Concluding Remarks... 86
Appendix A. Estimating the Kinetic Energy and Water Content
 of Hurricanes... 87
Appendix B. The Moist Adiabat... 88
Partial List of Symbols.. 92
References... 93
Note Added in Proof... 100

A Numerical Study of Vacillation
A. QUINET

1. Introduction... 101
2. The Laboratory Simulation of Large-Scale Atmospheric Flow 102
3. A Model Atmosphere for the Study of Vacillation........................ 113
4. The Spectral Dynamics and Energetics of the Model 131
5. The Numerical Study of Vacillation.. 149
6. Vacillation in the Atmosphere... 178
List of Symbols... 182
References.. 184

Filter Techniques in Gravity Interpretation
FRANS DE MEYER

1. Introduction... 187
2. Convolution Filtering 189
3. Upward and Downward Continuation of the Surface Gravity
 Effect... 203
4. Frequency Filtering.. 226
5. Calculation of Derivatives of Higher Order 248
Appendices.. 250
List of Symbols... 254
References.. 256

Aerial Methods in Geological–Geographical Explorations

B. V. Shilin and V. B. Komarov

1. Introduction	263
2. Radar Aerial Survey	265
3. Infrared Aerial Survey	282
4. Aerogeochemical Survey: Remote Sensing of Gases and Vapors	304
5. Conclusions	321
References	322
Subject Index	323

LIST OF CONTRIBUTORS

Numbers in parentheses indicate the pages on which the authors' contributions begin.

FRANS DE MEYER, *Royal Belgian Meteorological Institute, Uccle, Belgium* (187)

FRANCIS E. FENDELL, *Engineering Sciences Laboratory, TRW Systems Group, Redondo Beach, California* (1)

V. B. KOMAROV, *Laboratory of Aeromethods, Ministry of Geology, Leningrad, U.S.S.R.* (263)

A. QUINET, *Institut Royal Météorologique de Belgique, Bruxelles, Belgium* (101)

B. V. SHILIN, *Laboratory of Aeromethods, Ministry of Geology, Leningrad, U.S.S.R.* (263)

TROPICAL CYCLONES

Francis E. Fendell

Engineering Sciences Laboratory
TRW Systems Group, Redondo Beach, California

1. Introduction		2
	1.1. The Dangers and Benefits of Hurricanes	2
	1.2. Some Observational Facts on Hurricanes	8
	1.3. Tropical Cyclone Generation	3
	1.4. Properties of Mature Hurricanes	5
	1.5. Path Prediction	12
	1.6. The Importance of Tropical Cyclones in the Global Circulation	13
2. Aspects of Tropical Meteorology		17
	2.1. Stability in the Tropical Atmosphere	17
	2.2. Tropical Cumulonimbi	22
	2.3. CISK	26
3. Models of a Mature Tropical Cyclone		29
	3.1. Introduction	29
	3.2. The Carrier Model	29
	3.3. Critique of the Carrier Model	33
	3.4. Maximum Swirl Speed Estimate According to the Carrier Model	35
	3.5. The Swirl-Divergence Relation for the Frictional Boundary Layer	45
	3.6. The Energetics of the Frictional Boundary Layer and Throughput Supply	50
	3.7. The Riehl–Malkus Postulate of an Oceanic Heat Source	53
	3.8. The Intensity of a Tropical Cyclone and the Underlying Sea Surface Temperature	55
	3.9. Critique of the Riehl–Malkus Model	59
	3.10. Numerical Simulation of Hurricanes on Digital Computers	66
	3.11. Implications of Hurricane Models on Seeding	70
4. Theory of Tropical Cyclone Intensification		75
	4.1. Carrier's Outline of Intensification	76
	4.2. Critique of the Carrier Model of Intensification	78
	4.3. The Distribution of Cumulonimbi during Intensification	81
	4.4. The Time-Dependent Flowfield during Intensification	82
5. Concluding Remarks		86
	Appendix A. Estimating the Kinetic Energy and Water Content of Hurricanes	87
	Appendix B. The Moist Adiabat	88
	Partial List of Symbols	92
	References	93
	Note Added in Proof	100

1. Introduction

1.1. The Dangers and Benefits of Hurricanes

In an average year the Atlantic and Gulf coastal states of America suffer over $100 million damage and 50–100 fatalities owing to hurricanes; in a severe year damage will exceed $1 billion (Meyer, 1971a). Hurricane Agnes from June 19 to 26, 1972 caused damage from Florida to Maine that has been estimated at $3 billion (Anonymous, 1972), and Hurricane Camille in August 1969 cost 258 lives (White, 1972). The threat consists of high winds [over 200 mph is known (Riehl, 1972a)]; in rainfall [up to 27 in. in 24 hr (Schwarz, 1970)—since 1886 hurricanes have caused over 60 floods in the United States (Alaka, 1968), and Hurricane Agnes is estimated to have rained *in toto* 28.1 trillion gallons (Anonymous, 1972)]; in storm surges, especially along concavely curving coasts [coastal ocean levels can rise 15–20 ft, with Hurricane Camille setting the American record at 24 ft along the Gulf coast (White, 1972)];[1] and in ocean wave heights [wave heights of 45 ft are known (Riehl, 1954, p. 298)]. On the positive side, on a long-term basis the hurricane-associated rainfall over the Eastern seaboard states is about one-third of the total annual rainfall (Penner, 1972); without hurricanes, droughts seem inevitable on the west coast of Mexico, in southeast Asia, and elsewhere. About one to three hurricanes cross coastlines of the United States annually.

Globally, there are about eighty tropical storms annually, of which about fifty intensify into hurricanes; hurricanes occur in all oceans except the South

---

[1] For the East Coast of the United States, the hurricane-generated storm surge depends on storm direction, but can be roughly estimated from $s = 2.90$–$9.70 \times 10^{-2} (v)_{\max} + 1.33 \times 10^{-3} (v)_{\max}^2$, $74 < (v)_{\max} < 149$, where $(v)_{\max}$ is the maximum wind speed in miles per hour and s is the surge above normal in feet [based on Saffir (1973)]. However, the crudity of such simple empirical relations becomes evident when one recognizes that the storm surge (which may precede, accompany, or lag the center of the hurricane) is sensitive to the storm translation speed and size, as well as direction relative to the coast; the shelf slope and bathymetry; the radius and magnitude of the maximum wind speed, and the general radial profile of the swirl; and the central pressure deficit. In addition to onshore wind-driven waves, with the arrival of a hurricane, and persisting often for about a day after its passing along a coast with a sloping bottom (say, of inclination angle β), arise so-called edge waves. These waves travel parallel to the coast; their crests are normal to the coast. If the component of the hurricane translational velocity parallel to the coast is denoted U, the period of the waves is $(2\pi U/g \sin \beta)$ and the wavelength is $(2\pi U^2/g \sin \beta)$; typically for the East Coast of the United States the period is 5–7 hr, and the wavelength is 400 km. The amplitude vanishes rapidly from the shore seaward, and is negligible at the distance of one wavelength; the amplitude is roughly given by the inverse barometer rule (one centimeter water rise per millibar of atmospheric pressure drop) so heights of two to three feet are typical (Munk *et al.*, 1956) Sometimes waves generated because of the rate of along-shore hurricane movement relative to shelf depth cause more flooding than onshore waves.

Atlantic (Atkinson, 1971). Under the current state of the art for path predictions for either military or civilian use, three times the area actually hit by a hurricane is typically placed under hurricane warnings (Meyer, 1971a; Malone and Leimer, 1971). Anderson and Burnham (1973) note that a swarth 650 n miles wide will typically be placed under hurricane warnings; destruction usually covers about a 250 n miles wide path, but 200 n miles to each side must be placed under warnings owing to the current 122 n miles mean position error in 24 hr forecasts of hurricane landfall. If one assumes 10 landfalls a season, then Anderson and Burnham estimate a $15.2 million savings by the 20 % of U.S. Gulf Coast residents who take protective measures, the first year after the mean position error of landfall is reduced to 70 n miles. In the greater Miami area, current estimates suggest that at least $2 million are spent on preparations whenever hurricane warnings are issued (Simpson, 1971). *Unnecessary* preparation by the United States Department of Defense installations owing to false hurricane and typhoon warnings costs $8.3 million annually, exclusive of diversion-of-manpower costs. Thus, the mere threat of hurricanes incurs expenditure of prodigious sums.

1.2. Some Observational Facts on Hurricanes

"A tropical cyclone starts out as a *tropical disturbance* in which there is a slight surface circulation and perhaps one closed isobar. When the wind increases to about 20 knots and there is more than one closed isobar around the center, it is called a *tropical depression*. When the wind rises to more than 34 knots, and there are several closed isobars, it becomes known as a *tropical storm*. If the winds exceed 64 knots (74 mph), it is classified as a *hurricane* or *typhoon* or *cyclone* (depending on location)" (Day, 1966, p. 187). Thus any low-latitude low-pressure circulation is technically a tropical cyclone; however, probably to emphasize the physical similarity of all very intense tropical lows despite the plethora of local appellations, in practice *tropical cyclone* often refers to the hurricane stage only. Following this practice, the name tropical cyclone will be mainly used to refer categorically to the hurricane stage; hurricane will also be used as a synonym for variation. The other local names will be used mainly when geographically appropriate. The local designations for tropical cyclones (Fig. 1) are hurricanes (North Atlantic Ocean), typhoons (western North Pacific Ocean), papagallos (eastern North Pacific Ocean), baguios (Philippine Islands), cyclones (North Indian Ocean), trovadoes (near Madagascar), and willy-willies [near Australia—a term in use in the early twentieth century, although now it is reserved for dust devils (Spark, 1971; Vollsprecht, 1972)].

Annually, especially in the late summer and early fall, some disturbances over tropical oceans warmed by solar radiation (usually at least 26°–27°C, often 28°C and higher) intensify into tropical cyclones, which are typically

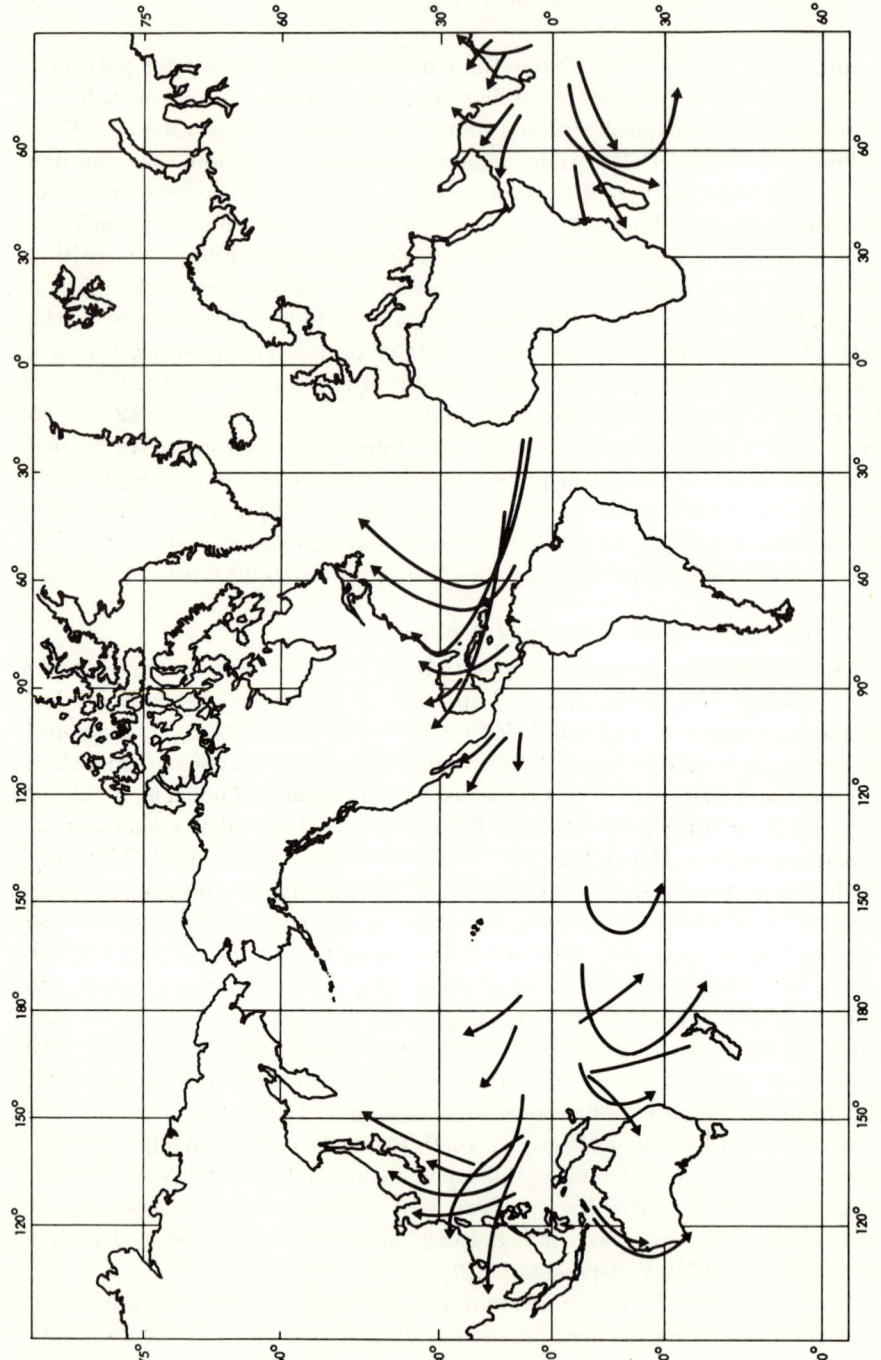

FIG. 1. Typical paths of tropical storms and cyclones.

at least a thousand miles in diameter and ten miles in height (i.e. they extend from sea level to tropical tropopause). Not only do most typhoons form in the autumn, but also the most intense ones occur then (Brand, 1970a). These vortical storms are cyclonic in the Northern Hemisphere and anticyclonic (North Pole reference) in the Southern Hemisphere, and take many days to intensify—indicating that the small Coriolis force is the source of angular momentum and explaining why intensification within 5° of the geographic equator is very rare. In fact, conservation of angular momentum in itself indicates that a fluid particle in the tropics drawn in about five hundred miles will swirl at several hundred miles per hour.

With satellite photography, inspection of broad ocean expanses has improved, and some former estimates about the frequency of tropical cyclones have had to be revised. (Large environmental-data-gathering ocean buoys, of which two experimental forerunners are currently deployed by the U.S. Department of Commerce in the Gulf of Mexico, may eventually play a large role in hurricane detection.) About three-quarters of the annual global total of fifty tropical cyclones occur in the Northern Hemisphere. In the North Atlantic about 60% of the nine tropical storms that typically occur annually intensify into tropical cyclones; in the eastern North Pacific, 33 % of 14; and in the western North Pacific, 66 % of 30 (Atkinson, 1971). Since there are daily disturbances in the tropics in autumn, a weak disturbance has a poor chance (~ 10 %) of becoming a tropical storm, but any disturbance that does manage to become a tropical storm has a good chance to become a hurricane (Palmén and Newton, 1969, p. 503). Since globally just over 60 % of tropical storms become hurricanes, there appears to be no criterion such that once a developing depression exceeds it, the depression will definitely become a hurricane. No one today can infallibly predict which tropical disturbances will intensify into hurricanes.

1.3. Tropical Cyclone Generation

Hurricanes form where there is sustained local convective activity over warm tropical seas, with surface temperature 27°C or higher (so that low-level air lifted on a moist adiabat remains convectively unstable relative to the undisturbed ambient up to 12 km); where there is enhanced cyclonic shear (as occurs when the Intertropical Convergence Zone lies at a considerable distance from the geographical equator); and where there is weak vertical shear of the zonal wind. Each of these three criteria warrants elaboration.

The first criterion regarding stability tends to be fulfilled only in the fall, and is discussed in detail in the discussion of tropical-atmosphere stability (Section 2.1). With regard to the need for increased cyclonic shear as well as warm ocean temperatures, it is noteworthy that, globally, 65 % of the disturbances that later become tropical storms were first detected between

10° and 20° of the equator, with only 13 % poleward of this region and only 22 % equatorward (Atkinson, 1971). The absence-of-a-vertical-wind-shear criterion may explain the anomalous cyclone season for the northern Indian Ocean (in the Arabian Sea, the Bay of Bengal, and even the South China Sea); in these areas, rather than a single autumnal peak, twin peaks in cyclone frequency occur in fall and spring, with a relatively uneventful summer season (Palmén and Newton, 1969). (The displacement during the summer of the low-level trough northward over land so westerlies generally cover most, if not all, of the North Indian Ocean, is probably another contributing factor for the lull between the greater peak of activity in the fall and the lesser peak of activity in the spring.) Gray (1968) suggests that unless cumulonimbi retain vertical integrity, the atmospheric lightening associated with them is dissipated by being advected in different directions at different altitudes ("ventilated"). Climatologically, regional differences in vertical wind shear within the tropics become most evident in the upper troposphere. Gray (1968) also asserts that the up- and down-drafts in cumulonimbi themselves help suppress enhancement of vertical wind shear as baroclinicity increases during intensification of tropical depressions. Vertical wind shear at upper tropospheric levels prevents the higher structure of some typhoons from getting fully organized, although the structure may be well defined in the lower troposphere; such so-called shallow typhoons are invariably of minimal hurricane intensity (Varga, 1971).

Particular features of the North Atlantic hurricane season are now enumerated. As the peak of the hurricane season approaches, the region where tropical storms reach hurricane intensity moves eastward from the Gulf of Mexico and the Caribbean to the Cape Verde Islands; as the peak hurricane season passes, the spawning ground moves westward again to the Caribbean (Meyer, 1971a). Coincidentally, there is a latitudinal movement northward in the first half of the season, then a retreat equatorward (Riehl, 1954, p. 323). Cyclogenesis poleward of 20°N is a particular characteristic of the North Atlantic; in fact, for the past four years the tropical North Atlantic has been mostly free of hurricanes, which have been forming at 25°N and higher (Simpson and Frank, 1972). In 1972, for example, there were only eight hurricane days in the North Atlantic Ocean (second lowest total since 1930), while there were 33 hurricane days in the eastern North Pacific Ocean, and 121 typhoon days in the western North Pacific (21 more typhoon days than average, and the highest number since 1959). Lower than normal sea-surface temperature and greater than normal vertical wind shear (owing to strong high-level westerlies) have been advanced as possible explanations for the recent decrease in hurricane activity in the North Atlantic.

At one time discussion of tropical cyclogenesis inevitably evoked discussion of waves in the easterlies (Riehl, 1954). But in recent years attention has focused on (1) the role of the ICTZ and (2) twice-weekly autumnal disturbances that begin as large cyclonic sandstorms of 1000 n miles extent over Africa and that drift westward at up to 10 knots. There may well be two different sources of hurricane seedlings, and if so, the relative roles of latent and sensible heat in each may differ (Garstang, 1972).

1. Tropical cyclones tend to form on the poleward side of the equatorial trough; in fact, 80–85 % of synoptic scale tropical disturbances form within 2°–4° of the equatorial trough on the poleward side. Disturbances are rarer, smaller, and weaker when the trough is closest to the equator. The trough is maintained by the CISK process (Section 2.3), since low-level meridional moist inflow sustains the persistent cloudiness of this perennial low-pressure region. Further, there is low-level cyclonic shear from interaction of easterlies poleward of the trough and westerlies equatorward of the trough. On a zonally averaged basis around the globe, the equatorial trough annually migrates from 15°N to 5°S, lagging the solar zenith by about two months as it does so (Riehl, 1972a). Byers (1944) attributes the absence of hurricanes in the South Atlantic largely to the failure of the ITCZ to become displaced south of the equator, even in February, at longitudes extending from the eastern Pacific to western Africa. Agee (1972) has presented an interesting sequence of satellite photographs documenting a case of tropical cyclogenesis in the vicinity of the ITCZ in late July, 1972.

2. If it is possible to correlate tropical North Atlantic disturbances with the ITCZ, it is also possible to correlate them with disturbances first formed over the mountainous east African bulge, that migrate westward (Carlson, 1969). Half the disturbances over the tropical North Atlantic can be so traced; further, half of these disturbances can then be traced across Central America to the eastern Pacific. Actually 75 % of eastern Pacific storms originate east of Central America. A significant percentage of midseason hurricanes (August and September) have African origins; analyses of dust samples taken on Caribbean isles after hurricane passage reportedly confirm the African origin of the storm systems (Jennings, 1970). The sometimes turbid disturbances migrating westward from the arid Sahara may not adjust to the maritime environment until they are over mid Atlantic. An interesting case documented by Denny (1972) is an African seedling identified on September 7, 1971, which became a depression on September 11 and Atlantic Hurricane Irene on September 18. The system crossed southern Nicaragua on September 20 and regenerated to eastern Pacific Hurricane Olivia (948 mb central pressure) before dying on September 28 near Baja, California.

There are many other correlations of tropical cyclone formation that may be attempted; for example Carpenter et al. (1972) have recently suggested a major peak in hurricane formation near a new moon, a minor peak at full moon, and minima at last quarter and several days after first quarter.

However, the correlations with the ITCZ and African seedlings just discussed have the advantage of inherent plausible physical mechanisms. Discussion of modeling of tropical cyclogenesis will be taken up in Section 4.

1.4. Properties of Mature Hurricanes

Tropical cyclones have a structure characterized in the mature stage by a relatively cloud-free calm eye (winds usually well below 15 mph) of about 10–20 miles radius. The eye is characterized by low pressure at sea level (often below 960 mb, infrequently below 910 mb) and high temperatures aloft (often 10°C above ambient at the same altitude). Such pressure drops are particularly spectacular in the tropics, where surface pressure usually varies by little more than 0.3 % (Riehl, 1954, Chapter 11). The eye is surrounded by an eyewall, an approximately ten mile wide annulus of intense convection, torrential rainfall, and deep thick cloudiness (marked by large numbers of cumulonimbi). Outside the eyewall are convective rainbands that appear like pinwheels or logarithmic spirals in some satellite photographs and/or radar displays taken from above the storm (Figs. 2 and 3).

The principal velocity component in much of the storm is azimuthal (or tangential). Hurricane force winds begin at the eyewall and extend outward 50 to 70 miles; winds usually fall to moderate gale level (~ 35 mph) at distances of 100 to 150 miles from the center. The vertical component of velocity is largest in the eyewall. In the Northern Hemisphere, there is low-level cyclonic inflow (below 3 km) and high-level outflow (above 8 km, with the maximum outflow near 12 km); toward the outer regions (between, say, 200 to 275 miles from the axis) the outflow turns from cyclonic to anticyclonic (relative to an observer rotating with the earth). If one subtracts off the symmetric flow from the total flow, the outflow layer possesses a horizontal anticyclonic eddy to the right of the path vector and a cyclonic eddy to the left; these are seemingly due to ambient streaming around the high-level outflow (Black and Anthes, 1971). In the midtroposphere (say, 3 to 8 km above sea level) there is little radial flow. While there is much spray, the lowest few hundred feet (at least) of the inflow layer remain cloud-free in as far as the eyewall (Riehl, 1954).

It is often agreed that there is slow downward motion in the eye and in the outer regions of the storm to balance the strong updrafts in the eyewall; that radiational cooling of about 1.5°C/day attends sinking in the outer, less cloudy regions; that, either directly or indirectly, the release of the latent heat or condensation reduces the density in the eyewall to establish a large radial pressure deficit relative to ambient conditions, from hydrostatic considerations; that a still further pressure deficit from ambient occurs in the eye owing to roughly dry adiabatic recompression of air that has arisen along

Fig. 2. NASA photograph of Hurricane Gladys west of Naples, Florida taken from Apollo 7 on October 7, 1968. The maximum wind speed was 65 knots at this time, and the tropopause was at 54,000 ft.

the near moist adiabat vertical sequence of states that characterize an appreciable portion of the eyewall; and that a cyclostrophic balance (balance of radial pressure gradient and centrifugal force), together with an algebraic decay of the swirl with increasing radial distance from the axis, yields on the average a fairly good estimate of the maximum swirl speed.

Outside the eyewall, the precipitation rate falls off roughly linearly with distance from the center of the tropical cyclone. Cloud cells in the eyewall are typically 5 to 20 km thick and a few of the smaller ones can rain over 15 cm/hr. Spiral bands out to 150 km yield 1 cm/hr, and further out 0.25 cm/hr —though individual convective cells of 1 km horizontal scale can give much heavier rainfall. In the low rainfall area the precipitation is probably snow

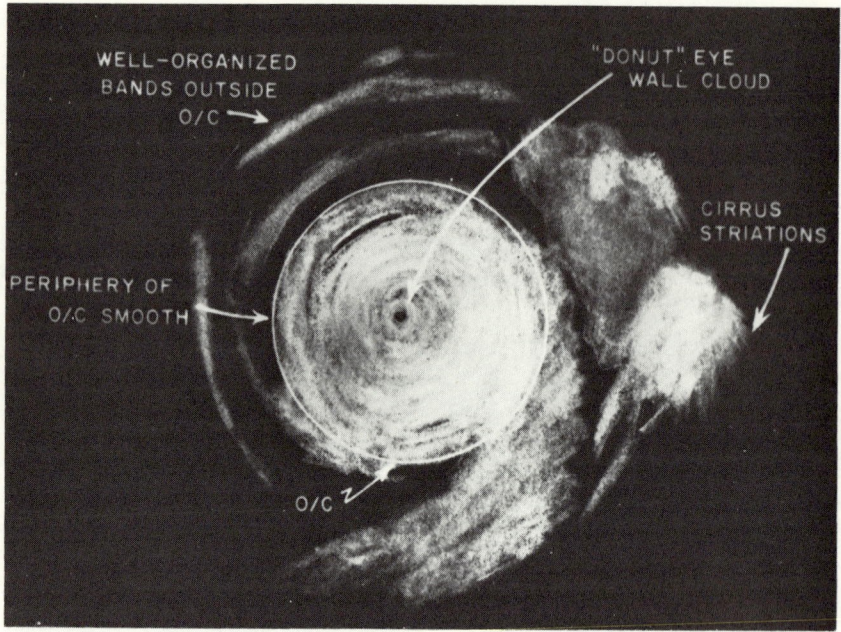

Fig. 3. NOAA photograph of mature hurricane revealing lower-level rainbands outside the region of overcast (O/C) and upper-level cirrus clouds of the outflow.

that turns to rain at the melting level (Meyer, 1971a,b). Incidentally, clouds in a hurricane are probably strained by the rapid swirl into the spiral band pattern seen on radar screens or in satellite photographs The spiral bands give visualization to parts of the strain pattern, rather than streamline pattern. A rainband persists typically for one to two hours (Gentry, 1964).

In addition to rainfall, storm surge, large waves, and high winds, tropical cyclones can spawn tornadoes and waterspouts of moderate intensity (Orton, 1970). These are usually reported for the region outside the domain of hurricane winds, but this *may* only mean that twisters are more easily discernible from the general vortical intensity in the outer portions. One would expect tornadoes to be most closely associated with the intense convection and large swirl of the inner rainbands. However, the fact remains that tornadoes have been observed mainly in the outer portions (i.e., where the surface pressure exceeds 1000 mb), soon after landfall of a formerly intense, rapidly decaying hurricane. Perhaps the intrusion (sometimes associated with the occurrence of tornadoes) into the storm of midtropospheric ambient dry air is significant. Most tornadoes are reported for the right forward quadrant, with respect to an observer looking along the direction of translation. This seems plausible because for North Atlantic hurricanes that quadrant is also the most severe with respect to rainfall, rainbands, and winds (Hawkins, 1971). As previously noted, only near the center are tropical cyclones axisymmetric to good approximation; near the outer edges there is asymmetry. The additive translational contribution to the azimuthal winds has been cited as one plausible source of the asymmetry in wind speeds (Riehl, 1954, p. 290). But Riehl also notes that the largest radial inflow occurs in the right forward quadrant once translational effects are subtracted out. It may be remarked that the so-called beta-plane effect (variation of the Coriolis force with latitude for fixed velocity) is not negligible over a system as large as a hurricane, which is normally *at least* of diameter 1000 miles.

The longest recorded lifetime for an Atlantic tropical cyclone was Ginger (September 5–October 5, 1972); of this 31-day lifespan, 20 were spent as a hurricane. Previously, Carrie (1957) had lasted 18 days as a hurricane and Faith (1966) had lasted 26 days as a system (Simpson and Frank, 1972). An unusually long-lived tropical cyclone in the western North Pacific was Rita (July 6–27, 1972); of this 22-day lifespan, about 18 days were spent as a typhoon.

Hurricanes tend to weaken moderately rapidly over land. The central pressure of Camille (1969) rose from 905 to 990 mb in about 13.5 hr after landfall (Bradbury, 1971).[2] Agnes (1971), a minimal hurricane with 986 mb central pressure and maximum winds of 75 mph with gusts to 95 mph over the Gulf of Mexico, was downgraded to a tropical storm eight hours after crossing the Florida panhandle (Anonymous, 1972). In an interesting report Grossman and Rodenhuis (1972) cite Hurricane Able (1952) and Hurricane Diane (1955) as examples of hurricanes which only weakly interacted with their environment and still maintained appreciable energy, circulation, and precipitation rates after passing inland. For example, 24 hr after Diane passed inland over the Carolinas on August 17, 1955 the central pressure had risen from about 985 mb to 1000 mb, and the core rainfall had decreased from 9.1 cm/day to half as much. Convective instability in the core decreased, and maximum rainfall occurred 75 miles from the center. However, in the next 24 hr the central pressure began to fall a few millibars and the central rainfall *at the center* of the storm began to increase; the convective instability in the core increased again. Hurricane Diane, although inland, encountered no major orographic changes during this time while moving northeastward.

---

[2] According to data collected on 30 typhoons for 1960–1970, tropical storms approaching the Philippines with maximum winds in excess of 90 knots leave with winds reduced by 40–50 %, while a storm with peak winds less than 90 knots experience only a 10–15 % reduction recovered within a day after leaving (and storms with peak winds under 60 knots typically undergo an intensification in crossing the islands). The average crossing time is 14.5 hr, with the weaker storms crossing more quickly (Brand and Bellock, 1972).

Grossman and Rodenhuis emphasize the uncertainty in numerical modeling of the role of latent heat release in complicated situations in which soundings indicate layers of stable and of convectively unstable air (Section 3.10). This lack of knowledge renders detailed analysis of how tropical cyclones die inland difficult.

Matano and Sekioka (1971), after examining several typhoons near Japan, seem to suggest that a tropical cyclone gradually decays without strong interaction with other atmospheric systems if it moves into extratropical regions where there is no marked midlatitude baroclinicity. If there is marked baroclinicity, the tropical cyclone will tend to interact with extratropical cyclones, either preexisting or else induced on fronts passing close to the typhoon core.

1.5. Path Prediction

Tropical cyclones move westerly in the trades, usually at 15 to 25 mph, before turning poleward, often at greater translational speeds. Except when the westerlies are furthest north in midsummer, one frequently observes a recurvature eastward in the midlatitudes along the western side of high-pressure cells (Riehl, 1972a); usually a decrease in intensity follows a recurvature (Riehl, 1972b). Not only does the latitude of recurvature (if it does occur) move poleward and then retreat equatorward as the peak typhoon season arrives and passes, but also the longitude of recurvature tends to move eastward and then retreat westward as autumn comes and goes; further, within two days after recurvature, the storm accelerates typically to two to three times its translational speed at the point of recurvature (Burroughs and Brand, 1973). However, there are so many special circumstances that many exceptions could be cited to virtually every generalization about path. For example, coexistent tropical cyclones of comparable size and intensity in the same hemisphere rotate about one another (Fujiwhara effect) (Brand, 1970b), while binary systems in different hemispheres tend to move parallel (Cox and Jager, 1969)—as suggested by classical hydrodynamic potential flow theory for line vortices. Prior passage, within a month, of a previous cyclone can also have an effect (Brand, 1971). Tropical cyclones can interact with extratropical cyclones; some typhoons are large enough to alter anticyclonic highs, the result on path being similar to that just cited for coexistent typhoons in different hemispheres (Palmén and Newton, 1969). Riehl (1954, p. 289) notes that hurricanes produce long waves that travel three times as fast as the storm and provide early warning. When hurricane swell arrives at a coastline, the normal wave frequency of 10 to 15 per minute, is reduced to 2 to 4; the wave direction can be interpreted to yield the path.

Tropical cyclone path prediction is in an imperfect state. Since forecast errors involving the intensity, rainfall, and movement of hurricanes can have serious consequences, the subject is pursued here briefly. For 1971, short term (12–36 hr) path forecasts by the National Hurricane Center were best made by the use of past climatological and analogue data. (What did previous hurricanes in a similar situation do?) Methods based on historical data are referred to as "objective." For 1971, for long term (48–72 hr), NHC found methods based on dynamical principles superior (R. Simpson, private communication). Today the average error for 24-hr predictions of hurricane movement is 129 n miles; the average landfall error for a 24-hr prediction is about 100 n miles. The reduction is due to the closer monitoring of tropical storms as they approach the Atlantic and Gulf coasts (Simpson, 1971). The less accurate forecasts often entail faster moving storms, and storms at latitudes poleward of the trades, where recurvature may occur.

Many dynamical techniques treat the tropical cyclone as a point vortex steered in a current, which has been smoothed to remove the influence of the circulation of the storm itself. The steering current may be the flow at a specific level, usually in the mid- or upper troposphere [Byers (1944, p. 447) cited 10,000 ft]. More recent work emphasizes the use of a steering layer; Riehl (1954, p. 345) advocated the use of a pressure weighted mean flow from the surface to 300 mb, over a band 8° latitude in width centered on the storm, to predict hurricane direction and speed. Today prognostic flow is sometimes used to predict path. Further, the barotropic model of Sanders and Burpee (1968) averages over the depth of the troposphere fron 100 to 1000 mb, and does *not* involve reduction of the tropical cyclone to a point vortex. Despite such improvements, the outlook for reducing path forecast errors by more sophisticated dynamical prediction models is not favorable:

> Numerical models for predicting the movement and development of hurricanes remain a frail source of guidance to the hurricane forecaster for three reasons. First, an error in direction of movement as small as 8 to 10 degrees—nominally an acceptable one in a 24-hour forecast for extratropical storm centers—can yield disastrous results in hurricane warnings if followed literally. Second, the performance of most hurricane prediction models depends significantly upon the initial direction of movement of the center, which in turn depends upn an exact knowledge of the current center position and the position 6 and 12 hours earlier. The average positioning error is more than 20 nm, and often leads to initial direction errors of 15–20 degrees. Finally the forecaster remains hard put to identify and diagnose the frailties of numerical prediction models for individual forecasts. All too often this has led to near abandonment of the guidance materials and the application of empirical and individual experience factors in decision making. (Simpson, 1971, p. 1.)

1.6. *The Importance of Tropical Cyclones in the Global Circulation*

Lorenz (1966, p. 409, 418) suggested that hurricanes were of secondary, rather than primary, importance in the global circulation of the atmosphere (Fig. 4). However, he noted that it was unclear how great a role hurricanes

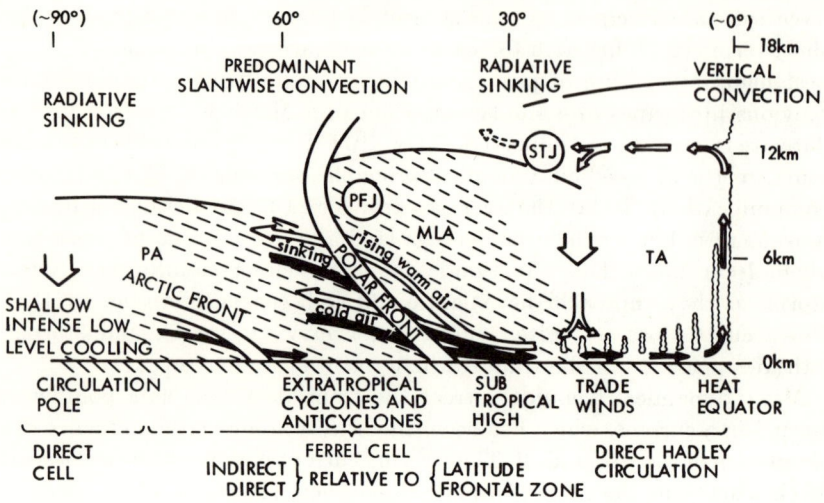

Fig. 4. Schematic features of the atmosphere in winter. Latent heat evaported from tropical oceans to the lower-level trades is carried toward the ITCZ by the sketched meridional flow; the air rises as the ITCZ is approached, and the latent heat ultimately is released as sensible heat and gravitational potential energy. Height of the local tropopause is marked. Portions of the general circulation are less well defined in summer. TA denotes tropical air; MLA, midlatitude air; PA, polar air; STJ, subtropical jet stream; and PFJ, polar front jet stream. (By permission from Palmén and Newton (1969, p. 569), Academic Press, New York).

play in maintaining the currents of larger scale. He also noted that hurricanes often do not appear in numerical simulations of the general circulation on advanced digital computers. Of course, appearance is hardly to be expected when restraints on computing time severely limit resolution; for example, the Mintz–Arakawa model is currently treated for a 5-deg longitudinal and a 4-deg latitudinal spatial grid with a 6-min time step, such that about 20 min of computer time is needed to simulate a day of climate (Rapp, 1970; Gates et al., 1971). The resolution for disturbances of dimension less than 1000 n miles is poor, but two weeks of computer time per day of simulated climate would be required on most computers if satisfactory resolution on a 100 n mile scale were sought. Meteorologists at the Rand Corporation in Santa Monica, California planned to program the two-level Mintz–Arakawa model of the general atmospheric circulation for the advanced ILLIAC IV computer. Even if such plans are realized, a grid no finer than $1° \times 1°$ seems practical; thus, if and when hurricanes are added to the global circulation model, their role may have to be introduced by parameterization guided by more fundamental studies (such as a subroutine with finer grid activated on appropriate occasions). After all, cumulus convection, radiational cooling, and turbulent mixing must currently be parameterized into computer simulations of tropical cyclones themselves.

Manabe et al. (1970), in their computer simulation of the global circulation, do observe the two-week intensification of tropical disturbances of scale 2000–3000 km near the ITCZ in regions where tropical cyclones are observed; further, there is low-level convergence and upper-level divergence. However, the resolution is not great, the pressure fell only to about 980 mb, and the disturbance appears more like subtropical cyclones (warm core above 800 mb, especially above 400 mb, and cold core below 800 mb). Appreciable doubt remains whether these disturbances are to be identified as hurricanes; Bates (1972) discusses possible inadequacies in the parameterization of cumulus convection adopted.

Some observational evidence has been presented that in certain regions during certain seasons hurricanes do play a role in the global circulation. From interpretation of satellite photographs of tropical storms and typhoons for 1967–1969, Erickson and Winston (1972) have suggested that substantial heat and moisture transfer from tropical cyclones to midlatitude westerlies plays a role in the autumnal build-up of the planetary scale circulation. A tropical storm is believed to transport energy from the lower to upper tropical atmosphere; the energy is then carried to fronts in the midlatitudes in broad extensive cloud bands extending northeastward over 20° latitude. The warming so conveyed to the extratropics intensifies the midlatitude zonal westerlies around the 300 mb level. Satellite photography suggests that unless upper-tropospheric westerlies favor a heat-exporting outflow from the cyclone toward the subtropics, as evidenced by an upper-level cloud plume emanating from the storm, intensification can be suppressed and the pressure deficit abruptly reduced. When the Hadley cell circulation is weak in late summer, perhaps hurricanes do serve locally as a substitute mechanism for conveying energy from the tropics to the extratropics.

That hurricanes could locally and seasonally play such a role seems at least possible. First, some models of hurricane structure require for self-consistency an export of heat from the storm to the rest of the atmosphere. For example, various models require an export of roughly 10^{26} ergs/day (Palmén and Riehl, 1957; Erickson and Winston, 1972); unless this energy (in the form of sensible heat), produced by transformations within the hurricane, is removed by other members of the general circulation via high-level advection on the periphery of the storm, the models break down (Riehl, 1954, p. 338; Palmén and Riehl, 1957, pp. 158–159). Anthes and Johnson (1968), by applying the theory of available potential energy, estimate that about 4×10^{24} erg/day are contributed by a hurricane to the global scale. Second, a hurricane has been estimated to possess a kinetic energy comparable to that of a hydrogen bomb, or 4×10^{23} erg (Battan, 1961; see also Appendix A), and one has rained over 95 in. on one location in four days (Silver Hill, Jamaica in November,

1909) (Alaka, 1968). The daily production of condensed water for precipitation in a mature hurricane has been estimated at 1.6×10^{16} gm (Ooyama, 1969, p. 29). A weather system with such energy and water content, that extends radially hundreds of miles and vertically from sea level to the tropical tropopause, and that persists for weeks, would seem no local accident. Third, hurricanes occur annually, mainly in the autumn after the long heating of the tropical oceans by solar radiation, and hurricanes do turn poleward after drifting westward in the trades. These facts invite the previously stated speculation that hurricanes are seasonally part of a substitute mechanism for relaxing energy poleward when the Hadley cell mechanism is not sufficient. If so, then the paradox discussed by Bates (1972, pp. 2, 14) is perhaps resolved: "The Hadley cell of the summer hemisphere is weak or non-existent.... There seems to be a paradox in the fact that while the Hadley cell is most intense in winter, the frequency of oceanic tropical disturbances, which one would expect to be an important contributor to its rising branch, is greatest in summer." Even if temporally and longitudinally varying eddy transfer associated with pressure troughs, rather than the zonally symmetric mean meridional (Hadley) cell, is the primary mechanism by which moisture and angular momentum are transported from the tropics to the midlatitudes (Riehl, 1954, Chapter 12; Riehl, 1969a), still, a significant portion of the eddy transfer occurs in the upper troposphere. The mode by which quantities to be transferred are convected to appreciable height in the tropics is cumulonimbus clouds, and, as discussed later, these occur in abundance in hurricanes. The basic suggestion remains: hurricanes may play some role in the fall in the export of energy from the tropics to the midlatitudes by what they export themselves, and by what they convect to the upper tropospheric levels in the tropics for other mechanisms to export.

However, on an annual and global basis, vertical transfer in the tropics is carried out by smaller scale, but more numerous convective systems (Palmén and Newton, 1969, p. 572). In fact, it is now suggested that only about one percent of the total energy annually exported from the tropics to the midlatitudes is conveyed by hurricanes. For convenience, suppose in a year there are 100 hurricanes, each lasting 18.25 days and covering 10^6 miles². The surface area of the tropics (with 30° serving as the limit) is about 10^8 miles². The fraction of the total annual total static enthalpy H transferred by hurricanes is

(1) $\qquad R = (100)(18.25)10^6(\partial H/\partial z)_{0,\,h}/(365)10^8(\partial H/\partial z)_{0,\,a}$

where the numerically unassigned quantity in the denominator is the ambient sea level transfer rate, and that in the numerator is the *rate above ambient* associated with hurricanes (the transfer coefficients have been cancelled in numerator and denominator). According to Carrier et al. (1971), the ratio of the two unassigned quantities is about 0.2, so $R \simeq 0.01$. The value of R would be larger if the model of Malkus and Riehl (1960) were used. These authors assert that at the sea surface in a moderate intensity hurricane (966 mb central pressure) the Bowen ratio (ratio of sensible to latent heat transfer) is about 0.2, and that the latent heat transfer is augmented by a factor of 10–12 over the ambient level in the trades.

2. Aspects of Tropical Meteorology

2.1. Stability in the Tropical Atmosphere

Stability in the tropical ambient atmosphere introduces consideration of circumstances rare in the extratropics. For convenience these properties of the tropical atmosphere crucial to tropical cyclogenesis are introduced at this point.

Stability is conveniently described in terms of the total static enthalpy H where

$$(2) \qquad H \equiv c_p T + gz + LY$$

Here c_p is the heat capacity of the atmosphere (effectively that of dry air, and independent of temperature for present purposes); T is the static temperature; g the magnitude of the gravitational acceleration; z the height above sea level; L the relevant latent heat of phase transition for water substance; and Y the mass fraction of water vapor. Throughout this paper, whenever H is (loosely) described as a temperature, reference is to (H/c_p). Actually the total enthalpy of a fluid particle is the total stagnation enthalpy

$$(3) \qquad H_t \equiv c_p T + gz + LY + q^2/2$$

where q is the magnitude of the velocity vector (the wind speed).

The magnitude of the various contributions in the tropics to H and H_t are worth mentioning. As shown in Fig. 5, the static enthalpy decreases from a sea level maximum of about 350°K ($L =$ heat of condensation plus fusion) to a midtropospheric minimum of about 330° at about 650 mb (around 13,000 ft), before recovering the sea level value of 350° just below the tropical troposphere (that is, at almost 100 mb or 50,000 ft). This midtropospheric minimum is observed with monotonous regularity throughout the tropics in all seasons, although the local sea level and near-tropopause maxima are reduced from the 350° level in the equatorial trough to 340° near the subtropics. Over the lowest few tens of millibars there may be a layer of well-mixed air in which H is virtually constant, but this is of no major consequence here. The dynamic contribution of $(q^2/2)$ is always small, and usually negligible; for example, even in a 200 mph hurricane, $(q^2/2c_p) = 4°K$, or just over a 1 % contribution to (H/c_p) near sea level. For stability one can virtually always consider H,

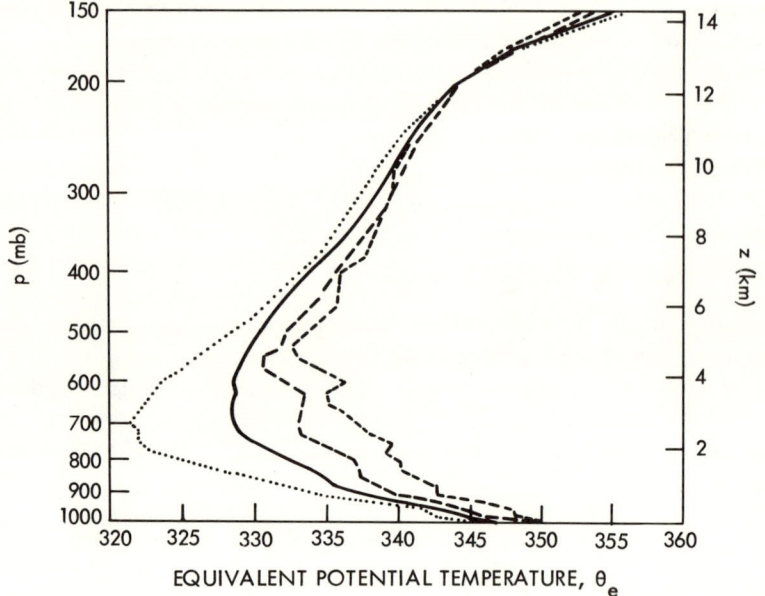

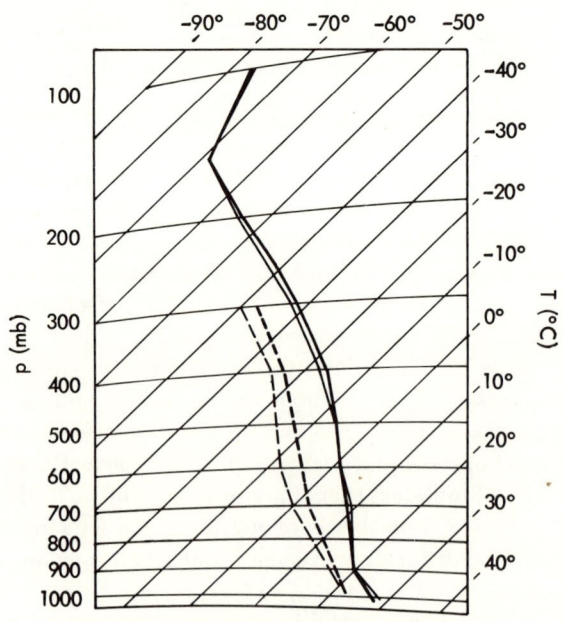

rather than H_t. At sea level, the static temperature contribution T is about 300°K near the trough, the latent heat contribution (LY/c_p) is about 50°K, and, of course, the gravitational potential energy contribution (gz/c_p) is zero. The moisture content of the tropical ambient falls off roughly exponentially with increasing height; the e-folding distance is crudely 650 mb. Whatever modest variation in H normally occurs from day to day over a locale in the tropics is more likely due to changes in water vapor content than to changes in temperature (Johnson, 1969, pp. 122-123); the lapse rate alters very little as one moves from cloudy areas to clear areas, but there is a drop in the dew point (Gray, 1972b, p. 14). In any case, certainly by 400 mb (or 25,000 ft) for current purposes $H \doteq \theta$, where the potential temperature-like quantity (actually an enthalpy) θ is defined by

(4) $$\theta \equiv c_p T + gz$$

A quantity describing the stagnation potential-like enthalpy may also be defined:

(5) $$\theta_t \equiv c_p T + gz + q^2/2$$

At 50,000 ft the gravitational contribution (gz/c_p) is about 150°K, or over 40% of θ (or H), and hence no longer negligible.

Clearly H plays a role much like the equivalent potential temperature used by most meteorologists, but H is preferred here. For one thing, what a dry-bulb and a wet-bulb thermometer measure (of relevance below) is readily understood in terms of H_t and θ_t. However, unless L is constant, H is not rigorously a thermodynamic state variable. However, L varies slowly with temperature and H remains a very useful concept.

The condition that convection is absent (mechanical equilibrium) is that the entropy increases with height. If air is approximated as a perfect gas, the unsaturated atmosphere is stable if $(\partial H/\partial z) > 0$, convectively unstable if $(\partial H/\partial z) < 0$. If the atmosphere is saturated, then the atmosphere is stable if $(\partial H/\partial z) > 0$, and statically unstable if $(\partial H/\partial z) < 0$. Convectively unstable means a perturbed particle will resume its former position unless lifted enough

---

FIG. 5. (*top*) The equivalent potential temperature θ_e as a function of height z and pressure p, from measurements near Barbados in the Lesser Antilles in July–August, 1968. Four characteristic weather types are represented: average (solid), suppressed convection (dotted), moderately enhanced convection (long dashes), and strongly enhanced convection (short dashes). (From Warsh *et al.* 1971, p. 127; with permission of the American Royal Meteorological Society.) (*bottom*) Soundings taken at Gan Island in the Indian Ocean (00°41′S; 73°09′E). The heavier curves (solid for static temperature, dashed for dew point) are the average for 42 wet days; the lighter curves are the average for 113 dry days. Measurements were made in Julys 1960–1964. (From Johnson, 1969, p. 122; with permission of the Royal Meteorological Society.)

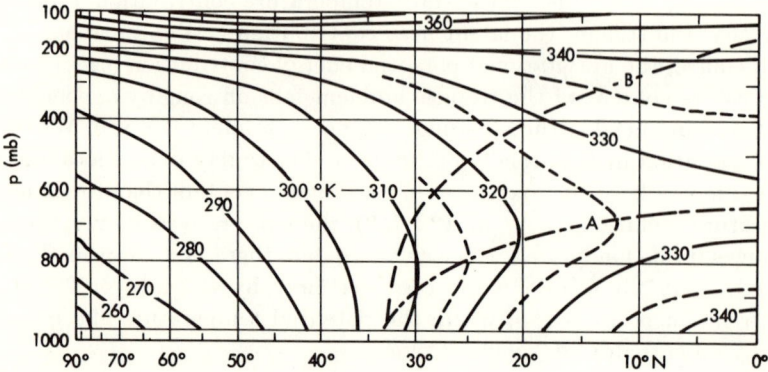

Fig. 6. The average total static temperature (H/c_p) for the Northern Hemisphere in winter. The curve marked A denotes the locus of the minimum of (H/c_p); free convection can most readily occur below level A, and only undilute ascent can continue much above A; undilute ascent continues to level B, where sea level values are recovered. (By permission from Palmén and Newton, 1969, p. 574. Academic Press, New York.)

for the onset of condensation, in which case it will rise until its density discrepancy relative to ambient is reduced to zero (i.e. rise to a new equilibrium position). Statically unstable means unstable without the requirement of a sufficiently large displacement. If the atmosphere is dry, or if for some reason one wants to consider stability excluding the role played by condensation (dry ascent as opposed to moist ascent), then the atmosphere is stable if $(\partial\theta/\partial z) > 0$, and unstable if $(\partial\theta/\partial z) < 0$.

In the extratropics, H and θ both increase monotonically with altitude usually, so the atmosphere is stable to both dry and moist ascent (Fig. 6). Only in exceptional circumstances, as in thunderstorms, does penetrative convection, with its vertical (as opposed to slantwise) ascent, occur. In the tropics θ increases monotonically with height, but H (as previously noted) has a midtropospheric minimum (Fig. 7); thus air is stable to dry ascent, but air *in the lower atmosphere* is unstable to moist ascent (convectively unstable). How turbulent mixing, cumulus convection, radiational cooling, cumulonimbi (see below), and large-scale circulation maintain this condition of convective instability deals with tropical meteorology in general, and lies largely outside the scope of this review of tropical cyclones (although such distinction will probably soon prove unsound with incipient progress in tropical cyclogenesis). Actually a full understanding of how the tropical ambient is maintained does not exist; the complexity is indicated by the fact that, as will now be explained, the H profile generates different types of clouds, which in turn sustain the H profile.

In the convectively unstable tropical atmosphere, the larger the lapse rate, the more suppressed is the cumulus activity. Only as the ambient lapse rate

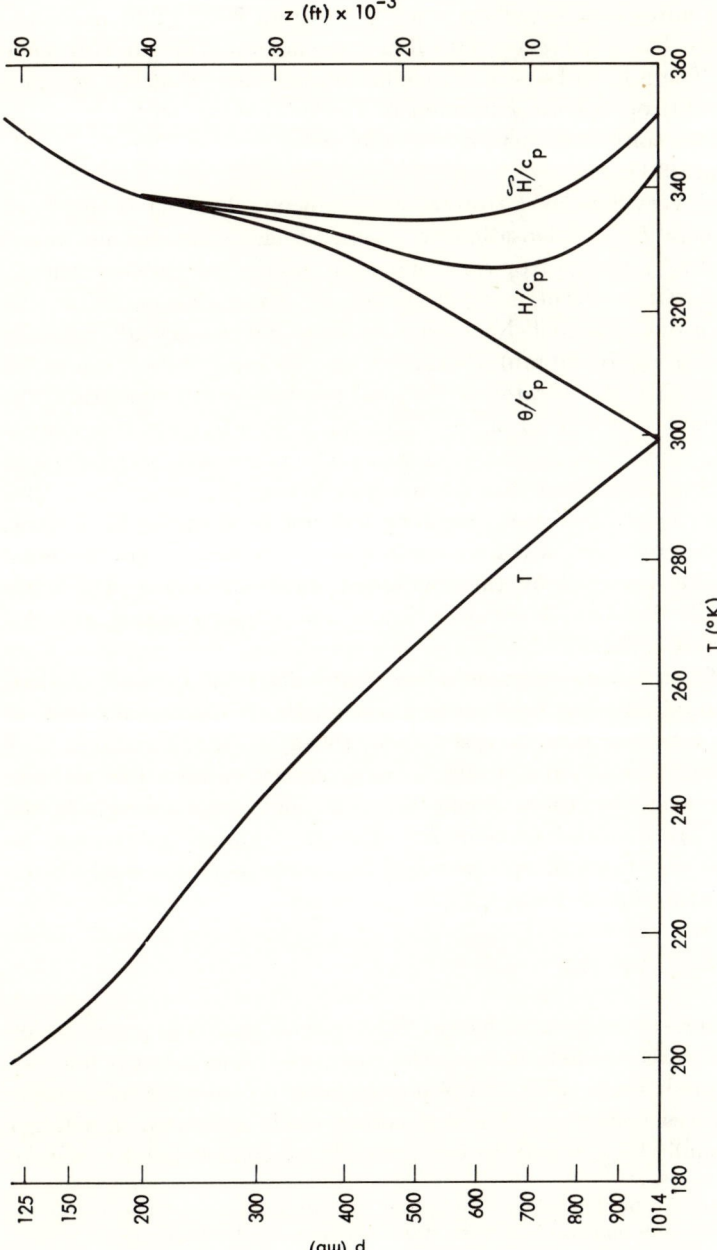

Fig. 7. Static temperature T, potential temperature-like measure (θ/c_p), total static temperature (H/c_p), and the total static temperature for a hypothetical atmosphere saturated at the actual local temperature and pressure $(\tilde{H}/c_p)$ vs. height and pressure for the West Indies ambient for September. Based on data given by Jordan (1957). A parcel at height z_1 will become buoyant at z_0 where (H/c_p) at z_1 exceeds $(\tilde{H}/c_p)$ at $z_0(>z_1)$; the inequality is satisfied for air in the planetary boundary layer, and little else.

approaches the moist adiabat is there penetrative convection. For example, Malkus (1960) notes that $-(\partial T/\partial z) \sim 6.6°C/km$ from 900 to 200 mb in the tropics normally, but $-(\partial T/\partial z) \sim 6.0°C/km$ in the inner rain area of Hurricane Daisy (1958). There have been many confirmations since, that the less pronounced the midtropospheric minimum of the total static enthalpy in the tropics, the more convective activity is likely to be present (Garstang et al., 1970; Aspliden, 1971).

If a glob of very low-level tropical air (buoyant element) is displaced vertically enough for condensation to occur, ascent will continue many kilometers virtually to the tropical tropopause, under the stability criteria just presented, if the ascent is rapid enough for no mixing or radiational cooling to occur during ascent. Near sea level air would rise dry adiabatically until saturated (roughly, 50 mb); thereafter, enough precipitation would fall out to leave the air just saturated at the local pressure and temperature, the air retaining the latent heat of phase transition. Such a locus of thermodynamic states is conventionally referred to as the moist adiabat; for rapid ascent in which ambient processes are too slow to act, $H \doteq \text{const.}$[3] (Air that has risen to its level of neutral buoyancy will not be unstable to descent, because any compressional heat from work done on the fluid by gravitational forces cannot be absorbed by the condensed water substance, which has precipitated out. However, descending flow will accompany ascent, and this will be addressed in Section 2.3.)

The core of a towering cumulonimbus is described by a moist adiabat. The tropical ambient must be close to moist adiabatic for cumulonimbi to be significant; this is in fact the case (Riehl, 1969a, b). If the ambient were moist adiabatic, cumulonimbi would be unnecessary because the ambient would be too unstable (there would be gross convective lifting). If the ambient were far removed from moist adiabatic, cumulonimbi would be nonexistent because the atmosphere would be too stable (there would be no lifting, just heating up of the air).

2.2. Tropical Cumulonimbi

About one percent of the area within the tropics is receiving precipitation on average; this area is likely to be in the equatorial trough, but is not uniformly distributed (Riehl, 1969b). The precipitation is concentrated in large-scale disturbances consisting of a few cumulonimbi immersed in a larger number of cumuli (Figs. 8 and 9); Gray (1972b) estimates that 10–20 % of

---

[3] The convective instability of the tropical ambient explains how cumulonimbi can occur, but cumulonimbi explain how high static enthalpy levels of the upper tropical troposphere can arise. This interdependence exemplifies the earlier remark on the link between the presence of clouds and the circumstances regarding stability.

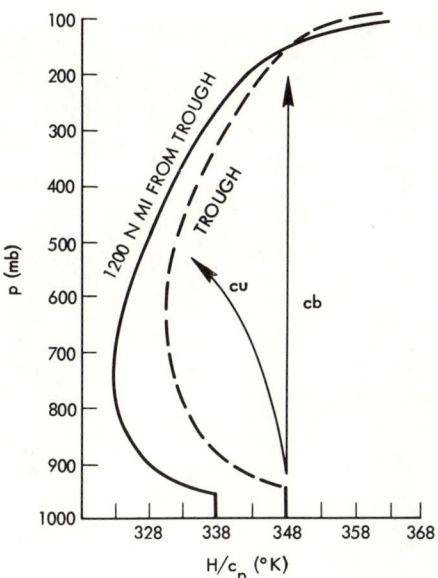

FIG. 8. The total static temperature (H/c_p) over the ocean in the equatorial trough and in the tropics near the subtropics, as a function of pressure. Both typical cumulus loci (cu) and also cumulonimbus loci (cb) are noted. (Based on, with permission, Palmén and Newton, 1969, p. 575. Academic Press, New York.)

the rain areas are covered by active cumulonimbi. Agglomerations of cumulonimbi are sometimes described as cloud clusters. Cumulonimbi tend to align themselves in east–west rows, or bands; the bands can be tens to hundreds of kilometers long (Kuettner, 1971), and are typically 20–50 km wide (Charney, 1971). The east–west alignment is consistent with the low-level wind field of large-scale disturbances in the tropics. The ITCZ, itself composed of one or more narrow bands of vigorous cumulonimbus convection, is typically of width 300 km (Charney, 1971). It is emphasized that the ITCZ, while a region of variable winds, appreciable cloudiness, large precipitation, and low pressure (Godshall, 1968), does not consist of a long unbroken band of heavy cloudiness. Rather, it consists of intermittent cloud clusters (with strong low-level convergence, appreciable vertical movement with precipitation, cyclonic vorticity, and upper-level outflow with possibly anticyclonic relative vorticity) with interstitial clear areas (with divergence, subsidence, and anticyclonic vorticity) (Holton et al., 1971).[4] Gray (1972b) also emphasizes this patchwork nature of the entire tropical belt, which consists of 20 % cloud area (which contains the cumulonimbi, than can rain at the rate of 2.5

---

[4] These authors emphasize that the ITCZ consists of westward propagating cloud clusters that pass at four- to five-day intervals.

Fig. 9. NASA photograph taken in June, 1965 from Gemini IV of cumulus, cumulonimbus, and cirrus clouds over the Pacific Ocean off the western coast of Central America; the view is northeast toward Mexico.

gm/cm^2-day), and 40 % variable cloud area and 40 % clear area (this 80 % of the area yields negligible precipitation). There tends to be moist ascent in the cloudy areas, and dry descent in the less cloudy areas, though by evaporation of liquid water and by diffusion of water vapor, water substance is transferred from the cloudy to less cloudy regions. If the tropics do consist of cloud clusters of 500 km scale separated by clear (or at least mostly cloud free) areas of 200–10,000 km scale, then one anticipates that there is a greater incidence of cumulonimbi relative to cumuli in those clusters near the ITCZ as opposed to those clusters in the trades (see below). Hence, there would be appreciable upper (lower) tropospheric detrainment from those clusters in the ITCZ (trades). If the clusters propagate westward at intervals of several days, then the clusters would be associated with thin wave troughs (i.e., low pressure regions) and the clear areas, with ridges.

These cloud systems are described here because in some still incompletely understood manner, they play a role in tropical cyclogeneisis (Palmén and Newton, 1969, p. 585). Although tropical cyclogenesis is marked by the exist-

synoptic scale motion, the water substance distribution in the summertime mean tropics (in which individual cumulonimbi can rain at 2.5 cm/day even though the evaporation rate at the sea surface is a uniform 0.5 cm/day). This large local recirculation may have implications for the momentum balance because it would seem to inhibit vertical wind shear development.

Malkus and Riehl (1960) described the tropical cumulonimbus as a hot tower because sensible heat from condensation was supposedly available to warm the ambient surrounding air. Lopez (1972b), however, asserts that *cooling tower* would be more apropos because the condensational heat is primarily used to raise the air in the column to its level of neutral stability (i.e., the heat goes into potential energy). The air that is detrained from the cloud serves to *cool* the environment because the heat extracted from the ambient to reevaporate condensed moisture in the detrained air exceeds any sensible heat transferred to the surrounding environment. Measurements around cumulus activity indicate that the static temperature often falls (e.g., Riehl, 1969, p. 588). Of course, the ascent in a cloud cluster does indirectly engender drying and warming through the subsidence of surrounding air that inevitably accompanies ascent in the cloud; this heating counteracts heat loss to detrainment and radiation. Thus on a large scale cumulonimbus activity generates a net warming, but the effect on the locally surrounding air is cooling (Gray, 1972b). Most models parameterizing the role of cumulus convection in large-scale disturbances have the direct sensible heat transfer model, rather than the indirect warming-by-subsidence mechanism, in mind (see Section 1.10). The Riehl–Malkus hot-tower model is inconsistent with the existence of large local recirculatory motion (Gray, 1972b).

2.3. CISK

Among the most important and fruitful concepts introduced in tropical meteorology in several decades is an idea due to Charney (Charney and Eliassen, 1964; Charney, 1971); the concept of CISK (conditional instability of the second kind), in addition to whatever contribution it may make to extratropical meteorology, seems of great utility in describing tropical phenomena of widely different scales, ranging from a single cumulonimbus to hurricane rainbands to cloud clusters to tropical cyclones to the ITCZ (the intertropical convergence zone is, for present purposes, taken as identical with the equatorial trough, the doldrums, and the trade confluence). In all these phenomena there is low-level convergence with large moisture content in the presence of appreciable Coriolis force. Phillips (1970) notes that while the physical processes involved in CISK can be qualitatively described, much remains to be done with respect to detailed quantitative description (Geissler, 1972).

ence of relatively little variation of the zonal wind with height, cloud clusters in which the easterly wind speed aloft exceeds that near the surface tend to spawn storms. Cumulonimbi are given particular attention because although they cover one one-thousandth (Palmén and Newton, 1969, p. 440) to possibly four-thousandths (Gray, 1972b, p. 33) of the area of an equatorial belt extending 10° latitude on each side of the equatorial trough, they cover one-tenth of the area of the trough itself (Bates, 1972, p. 2). Tropical cyclogenesis often occurs near the equatorial trough, and cumulonimbi occur in concentration in the hurricane eyewall (Section 4.3). In fact, the absence of cumulonimbi in *subtropical* cyclones occurring off western India, over the Pacific, and over the Atlantic serves as one means of distinguishing these storms (in which maximum intensity is reached in midtroposphere, and in which a warm core occurs at high levels but a cold core at low levels) from *tropical* cyclones (Walker, 1972). [Simpson and Pelisser (1971) discuss various so-called hybrid storms in and near the tropics, that may have hurricane force winds at least transiently, but do not have conventional hurricane structure.]⁵

A cumulonimbus (cb) is typified by a radius of 2 km, a lifespan of 30 to 40 min, a height of 14 km, a peak water substance content of 4 gm/cm³, maximum updraft speed of 25 m/sec, a thermal anomaly of 6°C, and a total rainfall of 2.4 cm. In contrast, a cumulus (cu) is typified by a radius of 1 km or less, a lifetime of 10 to 25 min, a height of 6 km, a peak water content of 1 gm/cm³, maximum updraft speed of 6 m/sec, a thermal anomaly of 4°C, and a total rainfall of 0.3 cm (Lopez, 1972a). Cumulonimbi are observed to overshoot their neutral stability level and then oscillate about it; further, only the *core* of a cloud rises undilute because only the larger clouds reach upper tropospheric levels (Malkus, 1960). Riehl and Malkus (1961) estimate that a cumulonimbus conveys about 6×10^{10} gm/sec and that a fluid particle rises the full vertical extent in about 30 min.

In the summer tropics there is far more local ascent and descent than might be suggested by the mean synoptic scale motion (Johnson, 1969; Gray, 1972b); in fact, if air in the equatorial trough rose only with the mean vertical flow, the ascent would be so slow that radiational cooling alone would stop it in midtroposphere (Palmén and Newton, 1969) and cumulonimbi would not exist. Gray (1972b) suggests that there is ten to twenty times the mean synoptic convergence in the lower summertime tropical atmosphere, and this is the source of the initial forced vertical displacement that leads to condensation and permits free convection to continue undiluted ascent in the unsaturated conditionally unstable ambient conditions. Without one or two orders of magnitude more up and down recirculatory local motion than suggested by

---

⁵ Further discussion of hybrid (or semitropical) storms, which rely for energy source not only on latent heat release but also on baroclinicity associated with positioning of warm and cold air masses, is given by Spiegler (1972).

The CISK process is the feeding of convective activity in a swirling flow by frictionally-induced inflow in a surface boundary layer. The idea is that rotating flows like that in a local low over the ocean tend to suppress convergence except in frictional boundary layers (about a kilometer or two thick and contiguous to the ocean). In this frictional layer there is radial inflow because the Coriolis and centrifugal forces (which balance the radial pressure gradient above the boundary layer) are diminished by friction; in the frictional layer radial advective transport responds to the "partially unbalanced" pressure gradient. The radial inflow in the boundary layer is furnished by slight sinking in the swirling flow above the boundary layer, in those regions where the circulation relative to the earth decreases with radial distance from the center of the low. In the tropics the converging boundary layer flow is warm and moist, because of transfer from the warm ocean during inflow and because the boundary layer air is warm and moist at the outset. The boundary layer flow erupts near the center of the disturbance in the region where the circulation relative to the earth of the fluid above the boundary layer increases with radial distance. In the conditionally unstable tropical atmosphere the core of the erupting, rotating column can rise vertically without dilution to the tropical tropopause. [Buoyant elements from higher in the boundary layer may have somewhat lower total static enthalpy, and may rise to a level of neutral buoyancy somewhat lower than that of sea level air. In fact, a buoyant element may contain air from a wide range of the boundary layer thickness; the core of the element, which does not mix with outside air, may not mix internally either, so there may well be shedding off of fractional pieces at various upper tropospheric levels. This refinement need not be pursued here. But it should be noted that there is good laboratory evidence that entrainment into a buoyant plume is reduced appreciably with increasing rotation rate of the environment, for all but very small rotation rates (Emmons and Ying, 1967).] The hydrostatically computed weight of the convective column, which thermodynamically is described by a moist adiabatic locus of states, is taken to be less than the weight of a column of ambient air. The difference in weight of the two columns implies a radial pressure gradient, which (by the conservation of radial momentum) in turn implies swirling. The swirling leads to further downflux into the surface frictional layer, further inflow, further moist adiabatic ascent, and hence maintenance (or even augmentation) of the radial pressure gradient. A crucial additional point about the CISK process is that it furnishes the basis of explaining how, instead of competing, cumulus scale activity can cooperatively interact with cyclone scale activity to their mutual enhancement.

The emphasis in CISK on both Coriolis force and conditional instability suggests that both the ITCZ and tropical cyclogenesis occur near, but not at, the equator. The critical point of the CISK process is that the weight of the convective column is lighter than that of the ambient gas. From a facile view, the release of the appreciable heat of condensation during moist adiabatic ascent can only lighten the column, and the existence of a radial pressure gradient is obvious. However, Gray (1972b) has recently viewed this typical summary of the CISK mechanism as an incomplete description of certain tropical phenomena, because it fails to emphasize the actual indirect heating mechanism associated with cumulus activity and consequently fails to emphasize the large local up and down recirculation above the mean synoptic motion (Section 2.2).

Another problem with quantitative analysis of the CISK process is that the surface frictional layer over the oceans may not be adequately described even in time average by a linear quasi-Ekman layer theory, and particularly not at low latitudes. For instance, the zonal translational speed, latitude, and scale of some tropical disturbances are such that the frictional layer does not have time to equilibrate, and steady linear Ekman theory is inappropriate. Gray (1972a) notes that dissipation of kinetic energy in the tropical layer so exceeds production that indeed there must be mesoscale and synoptic scale sinking of new momentum-carrying fluid to sustain the boundary layer. However, Gray claims that, contrary to Ekman's theory, measured planetary boundary layer thickness does not increase with decreasing latitude; he suggests vertical stability and turbulence scales, rather than rotation, determine the boundary layer thickness. Gray's general remarks, it may be noted, are not addressed to the exceptional circumstances of a tropical cyclone. Mahrt (1972a,b) has attempted theoretically to demonstrate the sensitivity of tropical disturbance development to the details of the subcloud level moisture convergence; especially equatorward of the ITCZ where the Coriolis parameter is small and its latitudinal variation is large, nonlinear advective acceleration may so alter the boundary layer flow that spiraling may be in the opposite sense to that predicted by Ekman's theory. The thickness, latitudinal variation, and strength of cross-isobar flow may deviate from what would be predicted by linear analysis. Also, whether the amount of fluid emerging from a quasi-linear Ekman layer is fully adequate to describe the growth of finite-sized tropical disturbances is a point of current contention; perhaps at least some entrainment in excess of the Ekman pumping into the cloud base will be found necessary for the early-stage growth of tropical disturbances. For example, alternatives to CISK (barotropic instability owing to north-south wind shear) and modifications to CISK (low-level convergence by a frictionless, internal-wave-type mechanism) have been proposed to describe early-stage growth. Finally, Petrosyants (1972) comments that the ITCZ in the Atlantic Ocean is most clearly evident in the eastern part, where a trade inversion prevents the development of deep layer tropical convection, and mainly trade cumulus clouds are found; Petrosyants questions compatibility of this observation with the deep convective penetration concept of CISK as an explanation of the ITCZ.

3. Models of a Mature Tropical Cyclone

3.1. Introduction

In sorting out the thermohydrodynamics of a tropical cyclone, one is faced with understanding the interaction of two scales of phenomena: the larger cyclone scale (~ 500 to 1000 miles), and the smaller cumulus scale (~ 1 mile). The cyclone must feed the cumulus scale, which in turn sustains the cyclone scale, in a cooperative interdependence. Such ideas suggest that the CISK process is not only relevant to the maintenance of an individual cumulonimbus cloud, but also (properly scaled) to the maintenance of the tropical cyclone in the large (Section 2.3).

The cyclone scale thermohydrodynamics of the *mature stage*, taken to be axisymmetric and quasi-steady for most purposes to a lowest order of approximation, will be set forth in terms of a model introduced in the last three years by Carrier and his co-workers (Carrier, 1970, 1971a,b; Dergarabedian and Fendell, 1970, 1972b; Carrier *et al.*, 1971; McWilliams, 1971). These articles delineate *overall* dynamics and thermodynamics, and imply (in contrast to other models to be introduced later) no major augmentation of ambient sensible and latent heat transfer is needed to explain hurricanes. These articles emphasize that many details and refinements in structure still remain to be quantitatively treated.

3.2. The Carrier Model

Subsequently (Section 4) transient analysis will be devoted to describing how a severe vortical storm may be generated in the tropics. Here, the quasi-steady mature hurricane is studied.

The Carrier model, on the basis of subdividing the tropical cyclone into segments where different processes and scales predominate, is a four-part analysis. The four regions, indicated in Fig. 10, are the throughput supply I, the frictional boundary layer II, the eyewall and efflux region III, and the eye IV. Some of this subdivision is conventional, some not. Besides clarifying locally dominant physical processes, subdividing permits retention of the minimal number of terms in locally valid quantitative formulation; this procedure simplifies the mathematical solution in a manner unavailable to any direct finite-differencing of uniformly valid equations.

The Carrier model is closed for convenience—there is no very significant amount of mass convected across any boundary, although mass may be diffused across any boundary. The cylindrical-like volume encompassing the entire storm has the sea surface for its bottom; its sides lie far enough from the center (about 500 to 1000 miles) so that the winds are virtually reduced to ambient, and the swirl relative to the earth is taken as zero for compatibility with the Ekman condition for no radial inflow across the outer boundary. The top of the storm is taken to be that height (~ 150 mb) at which sea level air in the outer part of the storm, if lifted rapidly so that the total enthalpy of a fluid particle remained constant because relatively slow ambient-maintaining processes would not have time to act, would no longer be unstable relative to the local ambient air. Such rapid lifting is, of course, the moist adiabatic ascent discussed in Section 2.1. This "instability lid" lies at so great a height that there is virtually negligible swirl, as explained below; the ambient pressure and temperature at this height are taken to describe all radial positions at this height, from the center to the outer edge. Thus, the top of the storm is an isothermal, isobaric, constant altitude lid with no water vapor content for current purposes.

Discussion now turns to describing each of the four regions comprising the tropical cyclone in some detail.

In region I there is warm moist air typical in stratification of the ambient atmosphere in which the hurricane was generated. This air spun up under conservation of angular momentum as it moved in toward the axis of symmetry during the formative stage. As the mature stage was approached, a gradient wind balance of pressure, Coriolis and centrifugal forces choked off any further inflow; the inflow is only enough to prevent the eyewall III from diffusing outward, and that requires only an exceedingly small radial flow. The air in I, then, is rapidly swirling, the azimuthal velocity component greatly increasing and the pressure greatly decreasing from the edge to the center. Under such a radial profile for the swirl, there is a small downflux from the throughput supply I into the frictional boundary layer II. The small downflux leads to a large net mass flux into the frictional layer because of the large area involved. Furthermore, the downdraft is only a gross temporal and azimuthal average because locally and transiently there is intense convective activity by which clouds form and rain falls.

In the frictional boundary layer II, the only region in which angular momentum is not conserved but is partially lost to the sea, there is appreciable influx. In fact, the azimuthal and radial velocity components are of comparable magnitude; typically, for fixed radial position, the maximum inflow speed at any axial position in the boundary layer is about one-third the maximum azimuthal speed. [This fraction is close to the one reported by Hughes (1952) from flight penetration of hurricanes at altitudes of 1000 ft or less.] The vertical velocity component is *much* smaller. The reason for the inflow is, as previously discussed, that the no-slip boundary condition reduces the centrifugal acceleration, and a relatively uncompensated pressure gradient drives the fluid toward the axis of symmetry (so-called "tea cup effect"). Far from the axis in II the classical balance of the linear Ekman layer (friction, pressure, and Coriolis forces) suffices; since the downdraft from I to II is probably fairly independent of radial position (especially far from the eyewall) for swirl distributions of practical interest, three-quarters of the flux inward through II comes from downflux across the interface between II and I where $(r_0/2) < r < r_0$, where r_0 is the radial extent of the storm. Closer in to the axis the nonlinear accelerations, especially the radial acceleration of radial and tangential momentum and also centrifugal acceleration, must enter.

As the pressure gradient in I lets more air sink into II, the influx into II drives boundary layer air moderately rapidly up a cloudy eyewall III. In the eyewall hydrostatic and cyclostrophic approximations hold; the locus of thermodynamic states is the moist adiabat based on sea level conditions in the eyewall (this particular point of the Carrier model will be examined in the critique to follow). The swirl is reduced in III owing to boundary layer friction. Near the top of III the flow moves further away from the axis of symmetry such that, in the outflow, the air seems to an observer on earth to be rotating opposite in sense to the rotation in I and II (which is cyclonic). The air in III slips over the air in I with no interaction; there is no large radial pressure gradient in much of III, unlike I.

At low altitudes the eyewall flushes moist air out of IV, and at high altitude rained-out air is entrained into IV. In time the eye becomes drier and better defined; it is the central core in which relatively dry air sinks, is warmed by compression, and is entrained out into the eyewall or recirculated within the eye (stagnation at the base of the eye would permit transport processes to cool the eye). The relatively light eye permits much greater pressure deficits from ambient and hence supports much higher swirling speeds. At a fixed altitude, the density in the eye is less than that in the eyewall, which in turn is less than that of the ambient gas at the storm edge. Since (except in cumulonimbi) the hydrostatic approximation is uniformly valid, spatially and temporally, in a hurricane, the sea surface eye pressure is less than the sea surface eyewall pressure, which is less than the sea surface ambient pressure. While Carrier has not explicitly written so, it would seem that as a general trend, the smaller the eye radius, the more spin-up under conservation of angular momentum. Hence the more intense storm might generally have a relatively small eye, the eye broadening as the storm weakens.[6]

---

[6] Data on this matter appears contradictory. Using all recorded data for Atlantic hurricanes for 1961–1968, Sheets (1972) found no correlation between the size of the radar eye and storm intensity indicators (such as maximum wind speed or minimum sea level pressure); there was reportedly a slight tendency toward smaller eye diameters at lower latitudes. Questions concerning the use of radar by itself to determine the eye radius will be discussed in Section 3.4. In contrast, the eye diameter increased typically from 20 n miles to 33 n miles, while the intensity typically decreased by 33 %, according to data for typhoons crossing the Philippines during 1960–1970; furthermore, for typhoons near the Philippines the eye diameters for intense typhoons (maximum wind speed over 90 knots) are typically 13% smaller than for weaker typhoons (maximum wind speeds

Carrier's model thus pictures the tropical cyclone as a once-through process in which a "fuel supply"—the warm moist air in I, which is part of the tropical cyclone at its inception and is convected with the storm—is slowly exhausted. The storm weakens because the air drifting down into the frictional layer toward the end is typical of the higher tropical environment and hence of lower total static enthalpy. Eventually the fuel supply is exhausted, and the boundary between III and I sinks toward II. Some models (e.g., Eliassen and Kleinschmidt, 1957) picture a recirculation through the storm of outflow air; the storm does not survive long enough for this, nor could such recycled air maintain the storm.

There is one important omission to the foregoing description that has been intentionally deferred: the energetics of the surface frictional layer and associated questions of air/sea transfer of latent and sensible heat. The relevant quantity to consider is the total stagnation enthalpy (the sum of static enthalpy the heat associated with condensible moisture, gravitational potential energy, and kinetic energy contributions). This quantity is conserved at roughly its ambient stratification throughout regions I and II; therefore, it is described by a profile that decreases with height from 1000 mb to 650 mb, and then increases with height, as mentioned above. The implication is that the heat and mass transfer from the ocean to the atmosphere is *about the same* within the hurricane as in the ambient. This transfer helps compensate for the rain-out in the spiral bands and helps maintain the warm, moist nature of the air in I. [Occasionally the Carrier model is still grossly misrepresented as proposing adiabatic conditions (constant total stagnation enthalpy in II so the net heat and mass transfer from sea to air is zero); such a solution cannot possibly satisfy the parabolic boundary value problem describing the energetics of the frictional boundary layer because it obviously violates the initial condition at $r = r_0$, the outer edge. In fact, if the supplemental flux from the ocean is entirely eliminated, as from passage over land, the spin-down time is $O(a/v^{1/2}\Omega^{1/2})$ where the eddy viscosity $v \doteq 10^{-2}$ miles^2 hr^{-1}, the normal component of the rotation of the earth $\Omega \doteq 2 \times 10^{-1}$ hr^{-1}, and the height of the throughput supply $a \doteq 1$ mile—so the spin-down time is half a day to two days.] The model of total stagnation enthalpy fixed at its ambient stratification breaks down in the eyewall III; there the vertical velocity component is at least one, probably two orders of magnitude larger than the relatively small downdraft into the boundary layer; the result is

---

under 90 knots). Incidentally, if the size is determined by the average diameter of the closed surface isobar, the size of a storm decreases by 17% in area in crossing the Philippines, and near the Philippines the intense typhoon has a mean outer circulation diameter 60–150 n miles greater than that of a weaker typhoon (Brand and Blellock, 1972).

that convection dominates the slow ambient-sustaining processes, so for at least portions of the eyewall the total stagnation enthalpy is virtually constant at its local sea level value, which (as just explained) is roughly its *ambient* sea level value.

3.3. *Critique of the Carrier Model*

A closed system seems naturally definable since a single hurricane probably does not interact significantly with the entire planetary atmosphere, especially when structure rather than path is under consideration. However, the closed system is considered optional by Carrier because the same conclusions would also hold for an open system, although the arguments would be far more difficult to construct. Clearly treatment of asymmetric effects should be a future goal of the Carrier model since complete elucidation of the spiral-band phenomena and other properties are probably unattainable otherwise. Initial emphasis on axisymmetry has been the traditional path of development for tropical cyclone models. Nevertheless, examination of the modification of the axisymmetric theoretical solution by accounting for the beta-plane effect (variation of the Coriolis parameter with latitude) may well prove interesting. The disruption of axisymmetry might prove appreciable, and further, particular intensity might be ascribed by theory to the outer western edge of the storm with respect to an observer looking along the direction of translation; if the height of the tropopause over the storm increased with latitude, this effect would not be expected, and it is in fact not observed. As noted earlier, what is observed is that the storm remains basically axisymmetric, with particular intensity occurring in the inner portions of the northeast quadrant.

The sketch in Fig. 10 is schematic, but it will appear below that in lieu of a greatly augmented oceanic sea-to-air latent and sensible heat transfer, the Carrier model relies upon the pressure deficit achievable by dry adiabatic recompression in the eye to be available to maintain dynamic equilibrium in a system with wind speeds known to reach 200 mph. Thus, some small but finite outward sloping of the eyewall at least down to midtropospheric levels, if not lower, is required for internal consistency in the Carrier model. Such ideas were present in the work of Haurwitz (1935) and in the early papers of Palmén (1948).

The particular emphasis placed on careful scrutiny of the dynamics and energetics of the frictional inflow layer is a major contribution of the Carrier model. [Incidentally, the thickness of this layer is too great to let increased frictional drag explain the accelerated decay after landfall, unless the coastline is mountainous. Even then, the orographic effects are more likely to be felt in deleterious premature rain-out owing to lifting of I, which would result in relatively dry air descending into the boundary layer and running inward toward the eyewall, so that moist adiabatic ascent would no longer be possible. The rate of decay of intensity owing to sea-to-air enthalpy transfer reduction

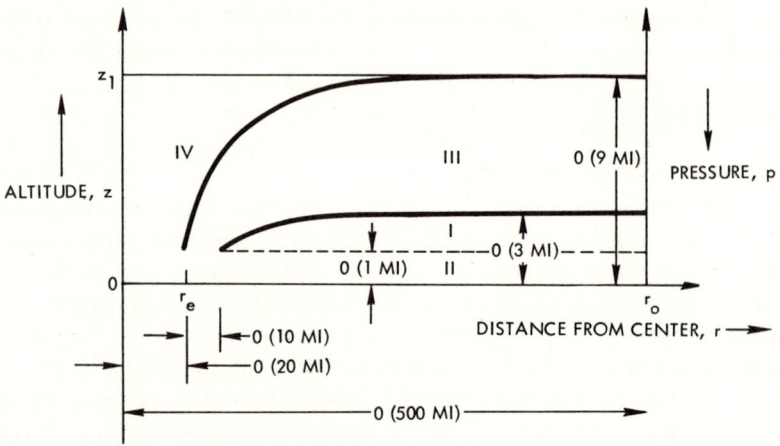

Fig. 10. This conjectured configuration of a mature hurricane with rough order of magnitude dimensions is not drawn to scale. The subdomains are: I, throughput supply, a region of rapid swirl and very slow downdrift; II, frictional boundary layer; III, eyewall; and IV, eye. Across the boundary layer there is about a 100 mb drop, and across I, a further drop of about 200 mb; the pressure at the top of the hurricane is about 150 mb, i.e. the top is near the tropopause. The eye–eyewall interface is taken to slope outward, though the effect may not be so pronounced as sketched.

after landfall, according to the Carrier model, was discussed earlier. The fact that friction effects the low-level moisture convergence suggests small increases in friction may even cause intensification.] However, the most novel contribution of the Carrier model is identification of the "fuel supply" in I, which requires only ambient level sea-to-air total stagnation enthalpy transfer for maintenance. That the fluid in I does sink down into II is suggested by the observation by Gentry (1964, p. 64) that at lower altitudes the temperatures are lower in the outer rainbands than in the surrounding air. There is little undiluted ascent from the inflow to the outflow layer. Rather, even the air which ascends in the outer rainbands also descends again in the storm area, and is *not* immediately carried away through the outflow layer in the upper troposphere.

The grossest feature of the Carrier model is the absence of any refinement to the eyewall structure. While for decades modelers have presented mean soundings in the eyewall that suggest moist adiabatic ascent from the surface layers (Palmén and Newton, 1969, pp. 477–482), Shea (1972) emphasizes that the ascent is limited to cumulonimbi that cover only 10 to 20 % of the eyewall area. Subsidence occurs in the regions of the eyewall outside cumulonimbi. Using flight data that include intensifying and decaying as well as mature tropical cyclones, Shea asserts that the relative humidity probably is not 100 % throughout the eyewall, and there is a small midtropospheric minimum in mean vertical profiles of the equivalent potential temperature (far less pronounced than the ambient minimum). In fact, wet-bulb effects may have led to spuriously low temperature measurements at midtropospheric heights (Gentry, 1964). In any case, the failure to delineate this structure does not have significant repercussions for Carrier's work on the mature hurricane, but does have important implications on his work on intensification, to be discussed later (Section 4). Shea also suggests that the gradient wind balance does not hold to good approximation in the eyewall; if so, this is probably due to a contribution from the transient partial derivative of the radial velocity component, and is not taken to modify analyses of a mature hurricane appreciably.

In the following sections three specific analyses among the many carried out by Carrier and his co-workers to help substantiate the model will be briefly reviewed. The three analyses involve maximum swirl speed estimation, the dynamics of a nonlinear Ekman layer, and the energetics of the surface frictional layer. The quantitative results achieved to date concerning the Carrier model have been attained without large-scale digital computation. The emphasis on subdivisional investigation of the four regions of the storm, with interfacial compatibility, permits a substantially analytic approach. It should be noted that although certain linearizations are sometimes adopted for tractability by Carrier and his co-workers in the course of their analysis, Carrier's model for the mature hurricane is definitely nonlinear, and solutions derived by such linearizations are acceptable only if, *a posteriori*, they can be demonstrated to satisfy the original *nonlinear* boundary value problem with acceptably small error. Without prior proof of internal self-consistency by subdivisional analyses, a full numerical treatment of the basic boundary value problem would seem premature.

3.4. *Maximum Swirl Speed Estimate According to the Carrier Model*

An upper and lower bound on the central pressure deficit achievable in a known spawning atmosphere will now be set forth using the Carrier hurricane model, hydrostatics, and the thermodynamics of moist and dry air. Specifically, the weights of various columns of air in the storm will be determined in the light of different moisture content and thermodynamic processes nvolved. The bounds on the central pressure deficit can then be translated into an estimate of bounds on the maximum swirl speed through dynamics (the radial momentum equation). Fletcher (1955) had suggested the use of the cyclostrophic balance once pressure deficits were known, and Malkus (1958) had suggested that pressure deficits could be calculated from moist adiabatic considerations for the eyewall and dry adiabatic considerations for the eye. Here the concepts are combined to achieve quantitative bounds, but just as important, to demonstrate that *hurricane speeds could be achieved without requiring any enthalpy transfer from the ocean greatly in excess of the ambient transfer, provided the eyewall is not perfectly vertical.* Miller (1958) performed somewhat similar calculations to those to be described and also noted that the Riehl–Malkus postulate of large oceanic latent and sensible heat transfer to air flowing in through the frictional layer was not required to explain low central pressure.

The first step is to neglect the frictional boundary layer II, which is relatively thin and across which, except for hydrostatic variations, the pressure does not change according to lowest order boundary layer theory.

The variation of pressure p, density ρ, and the temperature T with height above the ocean z, for any ambient tropical atmosphere in which a hurricane forms, may be computed from

(6) $\quad p_a = \rho_a R_a T \quad\quad$ (a = dry air)

(7) $\quad p_v = \rho_v R_a T/\sigma \quad\quad$ (v = water vapor; $\sigma = 0.622$)

(8) $\quad p = p_a + p_v, \rho = \rho_a + \rho_v, p_v = P(T)(\text{RH})$

(9) $\quad dp/dz = -\rho g$

(10) $\quad T = f(p), \text{RH} = g(p)$

where the temperature profile $f(p)$ and the relative humidity (RH) profile $g(p)$ are taken as known from measurement. The saturation pressure $P(T)$ is well tabulated for vapor and liquid phases above freezing, and vapor and solid below freezing (Keenan and Keyes, 1936); a convenient and accurate expression for $P(T)$ in millibars was given by Tetens (Murray, 1967):

(11a) $\quad P(T) = 6.1078 \exp[a(T - 273.16)/(T - b)]$

(11b) $\quad \begin{matrix} a = 21.8745584 \\ b = 7.66 \end{matrix} \Big\}$ over ice $\quad \begin{matrix} a = 17.2693882 \\ b = 35.86 \end{matrix} \Big\}$ over water

where T is in °K. The integration proceeds from the sea level upward in altitude z; data typically extend from about 1000 mb to 150 mb (Fig. 11). The top of the storm is normally taken as the height at which the ambient total stagnation enthalpy (for which the kinetic energy contribution is negligible) recovers its sea level value, as noted earlier; here, however, a slightly different procedure explained below will be used. The sea level ambient state is henceforth denoted by subscript s.

In a fully developed storm the air rising up the eyewall ascends in cumulonimbus clouds, and thus follows moist adiabats. The initial states to be used for the moist adiabats are not necessarily known *a priori*. *If* one believes the total stagnation enthalpy is constant along a streamtube in the surface frictional layer from the ambient to the eyewall, then the sea level tropical

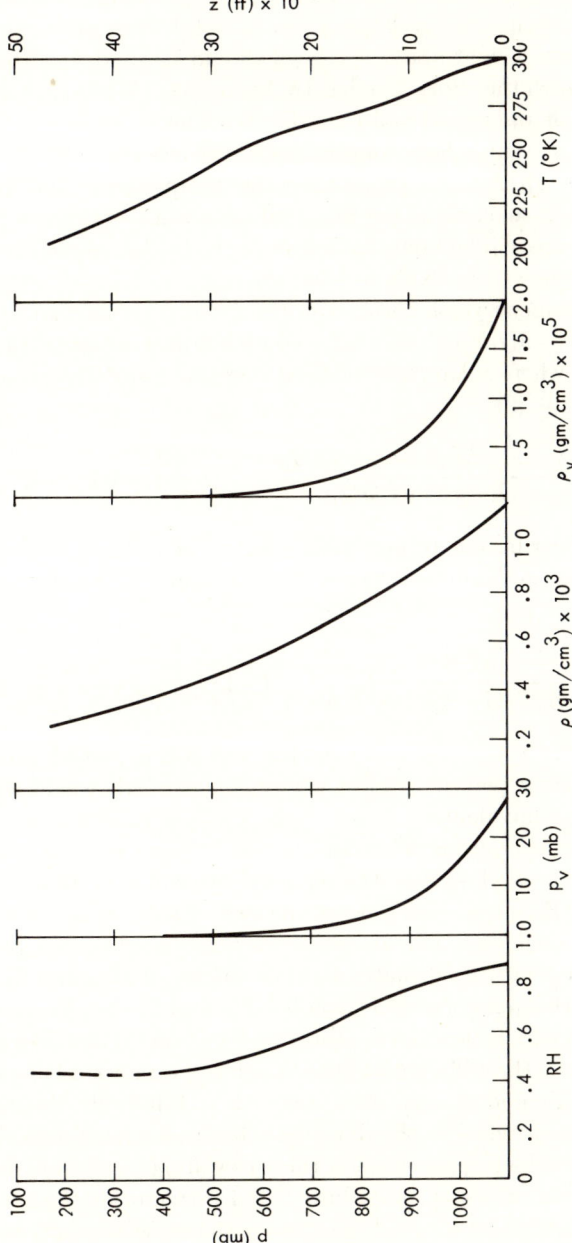

FIG. 11. Typical vertical profiles of thermodynamic state variables for the West Indies in September, based on data by Jordan (1957). The relative humidity RH, the water vapor pressure p_v, the total gas density ρ, the water vapor density ρ_v, and temperature T are plotted. Jordan's data for the relative humidity stop at 400 mb.

ambient state may characterize the total stagnation enthalpy of the streamtube rising in the eyewall closest to the storm center (procedure A). [Actually, mixing inevitably occurs so perhaps a lower total stagnation enthalpy characteristic of some height above sea level should be used, but the small distinction is not worth the effort (Carrier, 1971b, p. 158).] While such a choice for the initial state of an eyewall moist adiabat is sometimes made, and while such results will be presented here, another *preferable* procedure B will also be developed. In this alternative procedure, the temperature and relative humidity of the sea level state of the moist adiabat will be taken as known, but the initial pressure will be taken as unknown (to be determined by iteration for self-consistency to be explained below).

There is also a comment worth noting concerning the equation that describes the moist adiabat. If one simply takes $dH_t = 0$ with L held constant (Charney, 1971, pp. 357–358), where again H_t is the total stagnation enthalpy, then after manipulation with (6)–(9):

$$(12) \quad \frac{dT}{dp} = \frac{(1/\rho) + (L\sigma/p^2 x^2)P - d(q^2/2)/dp}{c_p + (L\sigma/px^2)(dP/dT)}, \quad x = 1 - [(1-\sigma)P/p]$$

The $[d(q^2/2)/dp]$ contribution is negligible and is henceforth dropped. A more careful derivation from basic principles (Appendix B) yields a slightly different expression

$$(13) \quad \frac{dT}{dP} = \left(1 + \frac{\sigma LP}{R_a Tp[1-(P/p)]}\right) \bigg/ \frac{\gamma}{\gamma - 1}\left(\frac{p}{T}\right)\left[1 + \frac{\sigma L}{c_p p[1-(P/p)]}\frac{dp}{dT}\right]$$

However, if one takes $p \doteq p_a$ [so $x = 1$ in (12)] and $\rho \doteq \rho_a$, which incurs an error of only about three percent at most at any point in the flowfield, then the two expressions are equivalent.

The procedure for computing the moist adiabat is to use the dry adiabatic relation $T \sim p^{\gamma/(\gamma-1)}$, $\gamma = 1.4$, from the sea level eyewall conditions to that pressure at which $RH = 1$, at which point one switches to the moist adiabat and continues the calculation. The integration is terminated at that pressure p_1 for which the temperature calculated from the moist adiabat and from the ambient are equal; this temperature is denoted T_1, and the height above sea level at which T_1 occurs is denoted z_1 (the "lid" on the cyclone). One then integrates (6)–(9) and the differential equation for the moist adiabat from $z = z_1$ (where $T = T_1$ and $p = p_1$) to $z = 0$; $RH = 1$ initially during this integration, but one switches to the dry adiabat where appropriate. Under procedure A the sea level pressure for the eyewall $p(z = 0) \equiv p_e$ is that computed. Under the superior procedure B, an estimate of the sea level eyewall pressure p_e had to be adopted to compute the moist adiabat, and this value must be recovered for convergence. If no eye existed in the vortex—as seems to be the case for some tornadoes and waterspouts—then the just calculated $p(z=0) \equiv p_e$, $\rho(z=0) \equiv \rho_e$ would characterize conditions at the center of the vortex.

The lid on the storm as calculated here will be higher in procedure B than in procedure A, for a given ambient. The taller the storm, the more intense it is according to calculations developed here. The same relation holds observationally (Riehl, 1972a, p. 248).

In a mature hurricane a pressure deficit in excess of $(p_s - p_e)$ is achieved by having rained-out air entrained from the eyewall sink in a relatively dry eye under adiabatic recompression. Thus in a hurricane $(p_s - p_e)$ is a lower bound on the central pressure deficit. For an upper bound on the deficit that may be achieved, one may adopt the idealized model that the eye is completely dry (so no compressional heat is lost to reevaporation) and that the air entrained into the eye is drawn from the top of the eyewall (or, in any case, has $T = T_1$, $p = p_1$, $Y = 0$ at $z = z_1$). The relevant equations are (6)–(9), RH $= 0$, and $T \sim p^{(\gamma-1)/\gamma}$; integration in the direction of decreasing z yields $p(z=0) \equiv p_c (< p_e < p_s)$ and $\rho(z=0) \equiv \rho_c (< \rho_e < \rho_s)$—the density discrepancies so calculated are at most 25% and Fletcher (1955) estimates the density does not vary by even 15%, so the density may be held constant at its ambient value throughout the dynamical calculations now discussed.

If one adopts the cyclostrophic balance, holds ρ constant at (say) ρ_s, and (since the core is observed not to rotate) lets

(14) $$v(r) = 0 \qquad 0 \leq r \leq R$$
$$= (v)_{\max}(R/r)^n \qquad R \leq r \leq \infty$$

then

(15) $$(v)_{\max} = [2n(p_s - p_c)/\rho_s]^{1/2}$$

First, for a one-cell vortex (when there is no eye so the moist adiabat calculation is appropriate all the way to the axis), a rigid body-like rotation lies near the core so then

(16) $$v(r) = (v)_{\max}(r/R) \qquad 0 \leq r \leq R$$
$$= (v)_{\max}(R/r)^n \qquad R \leq r \leq \infty$$

and from the cyclostrophic balance

(17) $$(v)_{\max} = [(2n/(n+1))(p_s - p_e)/\rho_s]^{1/2}$$

Clearly, (15) holds for the mature hurricane; the maximum swirl is larger in (15) than in (17) by the factor $(n+1)^{1/2}$ because none of the pressure deficit from ambient needs to be expended to maintain rotation of the central column. Next, although power law decays of swirl with radial distance are frequently adopted and suffice for current purposes, it will become evident that other forms are at least as plausible, and more convenient, for $r \geq R$. In any case, Miller (1967) suggested from limited data that $0.5 < n < 0.65$ usually suffices, and Riehl (1963a) had chosen $n = 0.5$; earlier, Byers (1944, p. 435) had also recommended $n \doteq 0.5$ and Hughes (1952) considered $n \doteq 0.6$–0.7 suitable for an average hurricane. The upshot is that estimates will be made on the limits $n = 0.5$ and $n = 1.0$, although $n \doteq 0.5$ seems the more realistic. Finally, the gradient wind equation would probably be more appropriate than the cyclostrophic equation, but the more complicated formula would give maximum speeds reduced by only five percent from those obtained from the simple forms (15) and (17).

First, results using procedure A (i.e. taking the total static enthalpy H constant throughout the radial inflow as well as eyewall region) will be given. If H is held constant at sea level ambient conditions [$p_s = 1014$ mb, $T_s = 299.4°$K, $(RH)_s = 0.84$], then use of (12) for the moist adiabat gives condensation at 972 mb ($T \doteq 295°$K, $z \doteq 180$ ft). One finds $z_1 \doteq 48{,}150$ ft, $p_1 = 138$ mb, $T_1 = 203°$K. The eyewall pressure at sea level $p_e \doteq 978$ mb [with $T_e \doteq 296°$K, $(RH)_e \doteq 0.975$]. The eye pressure at sea level $p_c = 894$ mb (with $T_c = 346°$K); this upper bound reflects, it is reiterated, the idealization of a completely moisture-free eye. As a variant, if sea level ambient conditions were revised slightly for computing the moist adiabat only [$p_s = 1014$ mb, $T_s = 299.4°$K, $(RH)_s = 1.0$], then use of (12) gives $z_1 = 52{,}500$ ft, $p_1 = 110$ mb, $T_1 = 196.6°$K. Also, $p_e \doteq 945$ mb, $T_e = 297°$K; $p_c = 850$ mb, $T_c = 352.7°$K.

As an illustration of the change attendant upon using (13) in place of (12), one finds for H referenced to the unsaturated sea level ambient [$p_s = 1014$ mb, $T_s = 299.4°$K, $(RH)_s = 0.84$], saturation again occurs at 972 mb, $z \doteq 450$ ft; however, $z_1 = 46{,}500$ ft, $p_1 = 150$ mb, $T_1 = 205.4°$K; further, $p_e = 988$ mb, $T_e = 297.2°$K, $(RH)_e = 0.94$ (Figs. 12 and 13). If the sea level ambient state is taken as saturated for computing the moist adiabat only, $z_1 = 50{,}530$ ft, $p_1 = 122$ mb, $T_1 = 199.6°$K; whence, $p_e = 958$ mb, $T_e = 297.5°$K.

These results, to reiterate, have followed procedure A [moist adiabat based on sea level ambient, as given by Jordan (1957) or modified to be saturated at the same temperature and pressure]. The results have been presented here because such calculations have been performed in the past, and because the size of the variances resulting from the use of (13) in place of (12) seemed worth investigating.

For the *superior* procedure B (iteration to determine the consistent pressure at the base of the eyewall moist adiabat), only (13) will be used. If one adopts $T_e = 299.4°$K, $(RH)_e = 0.84$, one finds convergence for $p_e = 958$ mb (Figs. 14 and 15); incidentally, $z_1 = 50{,}100$ ft, $p_1 \doteq 125$ mb, $T_1 \doteq 200°$K; further, saturation occurs at 918 mb, $z \doteq 1300$ ft. If L is given the value of the heat of condensation only, $(H/c_p)_s = 343.4°$K (sea level ambient), and $(H/c_p)_e \doteq 346.4°$K (sea level under the moist adiabat). [Under (13), (H/c_p) is not pre-

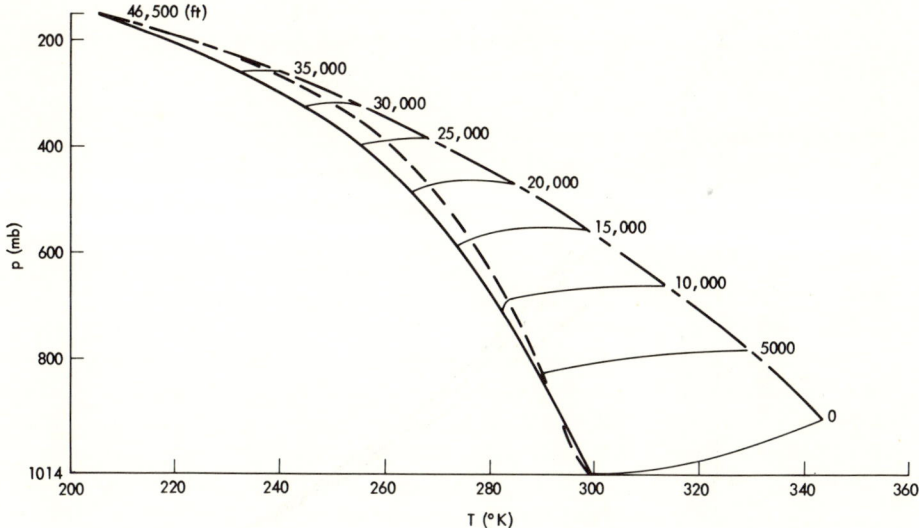

Fig. 12. The ambient pressure–temperature curve is based on data for the Caribbean in September given by Jordan (1957). The moist adiabat is based on having sea level ambient air undergo dry adiabatic expansion until saturation, then moist adiabatic ascent; the resulting sea level pressure deficit from ambient gives a lower bound on the central pressure deficit under the adopted model. An upper bound on the deficit is furnished by having the air that rose on the so-called moist adiabat, recompressed dry adiabatically back down to sea level. Altitudes are associated with the thermodynamic states by use of hydrostatics and the equations of state for dry air and water vapor. ——— Ambient, — — — moist adiabat, – – – dry adiabat.

cisely constant on a moist adiabat.] Also, $p_c = 875$ mb, $T_c = 349°$K. Thus it would seem that the sea level total static temperature needs only to increase about 3°K, with relative humidity and sea surface temperature held fixed, for an appreciably reduced pressure under eyewall cumulonimbi cores. According to (15), for the deficit $(p_s - p_e) = 56$ mb, $(v)_{max} = 155$ mph for $n = 0.5$ and $(v)_{max} \doteq 219$ mph for $n = 1.0$. Any eyewall bending would permit appreciably higher speeds; in fact, since $(p_s - p_c) = 139$ mb, theoretically sufficient bending could explain any speed up to 244 mph for $n = 0.5$ and 345 mph for $n = 1.0$. If one again takes $T_e = 299.4°$K, but now lets RH $= 1.0$, one finds convergence for $p_e = 890$ mb. Here $z_1 = 57{,}500$ ft, $p_1 = 84$ mb, $T_1 = 191.8°$K. In addition, $(H/c_p)_e = 359.4°$K; so an increase of about 16°K in the total static temperature could explain any recorded hurricane wind intensity without any eyewall bending *if* the sea level eyewall air were saturated, *and if* a moist adiabat were taken to be an adequate approximation to the eyewall sounding. In fact, increases above 16°K appear impossible under these conditions as long as the sea surface is taken as an isothermal

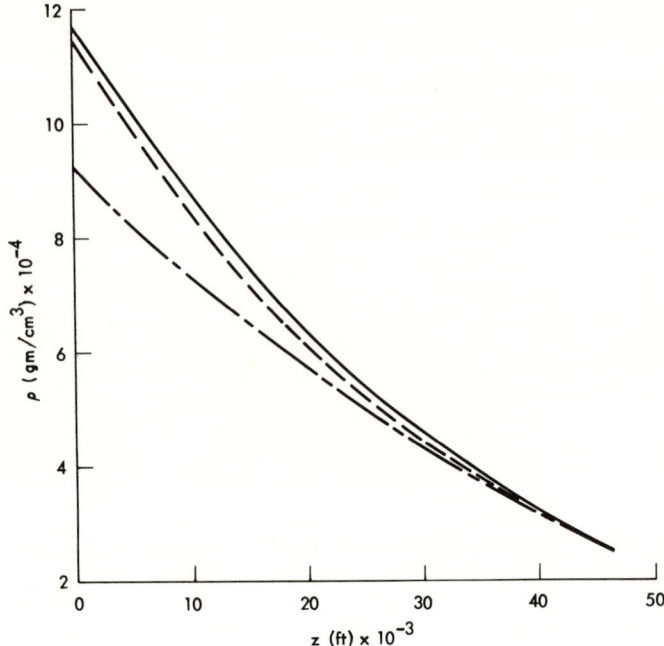

Fig. 13. The density versus altitude relation for each of the three thermodynamic loci of Fig. 12 are presented. The density variations from ambient within the storm are less than 30 %, and speeds are far below sonic everywhere. —— Ambient, — — — moist adiabat, —·—·— dry adiabat.

surface. Of course, use of the moist adiabat for the entire eyewall and the dry adiabat for the eye involves idealizations yielding upper estimates, but the large magnitude of the winds so calculated suggests the ideas behind the estimates are correct.

At this point a brief summary of the calculated results and their implication seems apropos. Even though cumulonimbi occupy only 10 to 20 % of the eyewall (Charney, 1971, pp. 358–359), results have been presented for moist adiabatic ascent and hydrostatics in the eyewall which suggest that significant pressure deficits relative to ambient may be achieved by moist adiabatic ascent of air whose total static temperature (H/c_p) is increased by only 1 % over ambient, and *very* intense storms may be achieved by ascent of air with a 5 % increase over ambient. The 5 % increase involves arbitrarily increasing the eyewall relative humidity at shipboard height to 100 % from an ambient level of 84 %, and it would seem more likely that the means by which the winds of very intense hurricanes are achieved is that the eye–eyewall interface, as well as it can be defined, slopes at least slightly outward from the top

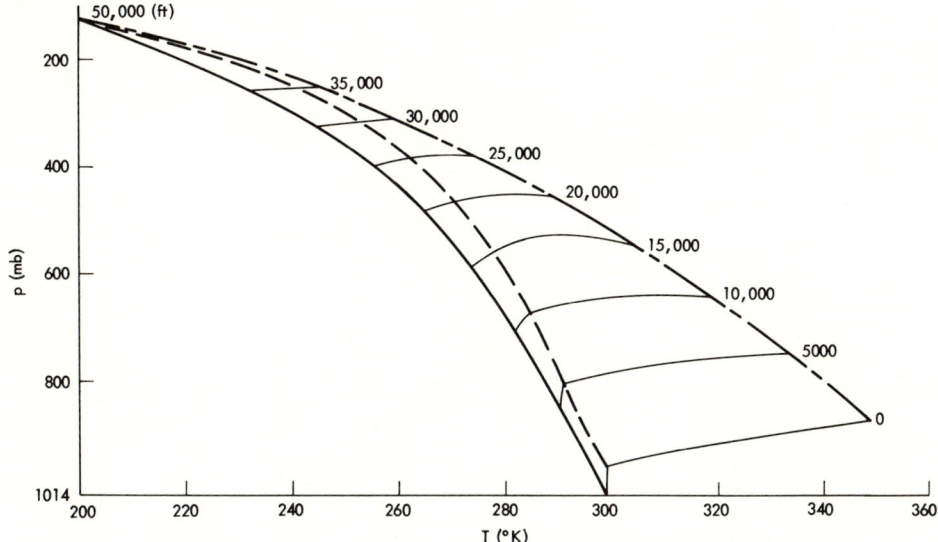

Fig. 14. The three curves are analogous to those of Fig. 12, except that the moist adiabat, while still taken to be based on ambient sea level values for relative humidity and temperature, is now computed from an iteratively determined sea level pressure, consistent with the definition of the top of the storm as an isobaric, isothermal surface of constant altitude. ——— Ambient, ——— moist adiabat, ——— dry adiabat.

of the inflow layer to the top of the storm. The calculation of wind speed for a storm with an eye implicitly assumes such sloping. The additional dry adiabatic compression-associated pressure deficit available to sustain swirling is reduced if the eye–eyewall interface is vertical to a certain height, and only then slopes outward; the additional pressure deficit is unavailable if the eye–eyewall interface is perfectly vertical virtually to the top of the storm. (Whether or not the eye–eyewall interface slopes outward does not alter the conceptual thermodynamic stratification of the eye nor the sea level eye pressure deficit from ambient.) Dissenting publications will now be discussed.

Malkus and Riehl (1960) find that moist adiabatic ascent of ambient sea level air to the level of neutral buoyancy produces a sea level pressure of 1000 mb; insufficient detail is presented to ascertain what precise calculation was conducted. In any case, such a pressure drop is insufficient to sustain even a moderate hurricane, and since these authors believe the eye–eyewall interface to be virtually perfectly vertical, another means of achieving and sustaining winds known to arise in hurricanes was required. The means ultimately postulated is latent and sensible heat transfer from sea to air greatly in excess of ambient transfer rates; the air ascending in the eyewall would then possess large enough equivalent potential temperature to achieve the pressure deficits necessary to sustain even the highest known hurricane winds. The Riehl–Malkus

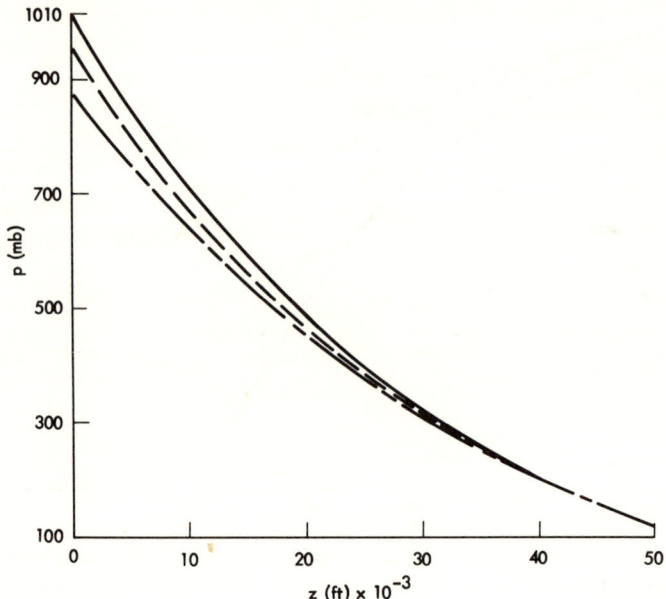

Fig. 15. In this replotting of Fig. 14, explicit pressure versus altitude curves for the ambient, the moist adiabat taken to characterize the eyewall, and the dry adiabat recompression taken to characterize the eye are presented. The moist adiabat is computed by the iterative procedure described in the caption to Fig. 14. ——— Ambient, ——— moist adiabat, ——— dry adiabat.

extra oceanic heat source postulate will be examined below (Section 3.7–3.9). Here it will be noted that even extrapolation of low-speed transfer coefficients to extreme hurricane conditions could not justify the equivalent potential temperature increases required by Riehl and Malkus; thus for very intense hurricanes (only) an outward sloping of the eyewall was conceded by these modelers. Such a procedure directly contradicts the conclusions of Shea (1972), based on consideration of flight radar data; Shea finds that the eye–eyewall interface slopes modestly outward for moderate hurricanes, but questions whether it does so for intense hurricanes. However, radar data are based on returns from precipitation and err on the side of verticality because the strongest returns will come from the torrential rains of cumulonimbi, rather than from weaker or decaying cumuli (Palmén and Newton, 1969, pp. 487–491). Direct observational distinction between an interfacial slope of a few degrees and no slope at all, especially when some cloudiness may occur in the eye, is difficult, and observational evidence can be cited to support either position. For instance, Palmén (1948) sketches an eyewall with appreciable outward slope down to nearly inflow-layer heights, based on observations of a September 1947 hurricane off Tampa, Florida. Palmén and Newton (1969) construct a schematic diagram of Hurricane Helene, based on observational data of September 26, 1958, that indicates an outward eyewall slope commencing a few kilometers above the sea surface. On the other hand, Riehl (1954, pp. 312–313) cites reports published in 1945 that described an observed eye–eyewall interface that did not have appreciable outward slope until 30,000 ft. Definitive proof concerning whether or not dry adiabatic recompression contributes to the pressure difference available to sustain hurricane winds appears nonexistent.

This section is concluded by noting that maximum wind speeds in hurricanes are estimated, in the absence of aircraft reconnaissance, by means of previously correlated formulas involving the area of the overcast " circle " on satellite photographs (Hubert and Timchalk, 1969) (Fig. 3).

3.5. *The Swirl-Divergence Relation for the Frictional Boundary Layer*

Because the maximum speed achieved in a tropical cyclone is rarely much over 200 mph, the Mach number rarely reaches even 0.3. Hence, when examining the dynamics (as opposed to the energetics), an incompressible constant-property model suffices.

A steady axisymmetric flow of an incompressible fluid is now studied to confirm the crucial point that, under rapid swirling, there is downflux from region I to region II, and sufficient downflux enters the surface frictional layer to account for the mass flux up the eyewall. The analysis will be carried out in a noninertial coordinate system rotating at the constant speed of that component of the rotation of the earth which is normal to the local tangent plane. Because the boundary layer divergence under an impressed swirl (the major constraint furnished by the boundary layer on the inviscid flow above it) is relatively small in magnitude, careful formulation and solution of the coupled quasilinear parabolic partial differential equations and boundary conditions describing the layer are required.

The relevant equations are

(18) $$\nabla \cdot \mathbf{q} = 0$$

(19) $$\nabla(q^2/2) + (\nabla \times \mathbf{q}) \times \mathbf{q} + 2\mathbf{\Omega}_e \times \mathbf{q} = -\nabla \tilde{p} - \nu \nabla \times (\nabla \times \mathbf{q})$$

where $\tilde{p} = (p/\rho) + (\mathbf{\Omega}_e \times \mathbf{r})^2/2 + gz$, the gravitational acceleration $\mathbf{g} = -g\hat{z}$, the velocity in noninertial coordinates $\mathbf{v} = \mathbf{\Omega}_e \times \mathbf{r} + \mathbf{q}$, the component of the rotation of the earth normal to the local tangent plane $\mathbf{\Omega}_e = \Omega \hat{z}$, and the kinematic viscosity (later given eddy-diffusivity values) is ν.

Nondimensionalization is effected by letting $\mathbf{q}' = \mathbf{q}/(\Psi_0 \Omega)^{1/2}$, $p' = \tilde{p}/(\Psi_0 \Omega)$, $\mathbf{r}' = \mathbf{r}/(\Psi_0/\Omega)^{1/2}$, and $E = \nu/\Psi_0$ where the Ekman number $E \ll 1$

and Ψ_0 characterizes the circulation away from the boundary (such as the maximum swirl speed times the radius at which it occurs). Dropping primes, one has

(20) $$\nabla \cdot \mathbf{q}$$

(21) $$\nabla(q^2/2) + (\nabla \times \mathbf{q}) \times \mathbf{q} + 2\hat{z} \times \mathbf{q} = -\nabla p - E\nabla \times (\nabla \times \mathbf{q})$$

These equations are studied in axisymmetric cylindrical polar coordinates

(22) $$\mathbf{q} = u\hat{r} + v\hat{\theta} + w\hat{z}, \quad \mathbf{r} = r\hat{r} + z\hat{z}$$

Away from the boundary (i.e. in region I) the following expansions are adopted:

(23) $$p = \pi(r, z) + \cdots, \quad v = V(r, z) + \cdots,$$
$$w = E^{1/2} W(r, z) + \cdots, \quad u = o(E^{1/2})$$

Substitution of (23) in (20) and (21) gives the gradient wind equation:

(24) $$\pi_z = 0, \quad W_z = 0, \quad \pi_r = 2V + V^2/r$$

Subscripts r and z here (and x and ζ below) denote partial differentiation. The axially invariant solution is complete when $\pi(r)$ or $V(r)$ is specified [here $V(r)$ will be given]; $W(r)$ is found by matching the solution to (24) to the solution for the frictional layer II, and in this sense $W(r)$ is determined by the boundary layer dynamics.

If $\zeta = zE^{-1/2}$ [which implies that the frictional layer is $O(E^{1/2})$ in thickness] and if near the boundary

(25) $$u = u_b(r, \zeta) + \cdots, \quad v = v_b(r, \zeta) + \cdots,$$
$$w = E^{1/2} w_b(r, \zeta) + \cdots, \quad p = p_b(r, \zeta) + \cdots,$$

then the axial component of the momentum conservation equation degenerates to $(\partial p_b/\partial \zeta) = 0$ in conventional fashion, so the pressure field in the boundary layer is known from (24). If

(26)
$$\psi = rv_b, \quad \Psi = rV, \quad \phi = ru_b, \quad \tilde{w} = 2^{-1/2} w_b, \quad x = r^2, \quad \tilde{\zeta} = 2^{1/2}\zeta,$$

then in terms of dimensional quantities $\phi = ru/\Psi_0$, $\psi = rv/\Psi_0$, $\tilde{w} = w/(2\Omega\nu)^{1/2}$, $\tilde{\zeta} = z/(\nu/2\Omega)^{1/2}$, $x = \Omega r^2/\Psi_0$; the boundary layer thickness is $O(\nu/\Omega)^{1/2}$. In terms of quantities introduced in (26), one has from (20) and (21), upon dropping the tildes,

(27) $$\phi_x + w_\zeta = 0$$

(28) $$\phi\phi_x + w\phi_\zeta + (\Psi^2 - \psi^2 - \phi^2)/2x - (\psi - \Psi) - \phi_{\zeta\zeta} = 0$$

(29) $$\phi\psi_x + w\psi_\zeta + \phi - \psi_{\zeta\zeta} = 0$$

Matching of expansions gives

(30) $$\zeta \to \infty: \phi \to 0, \quad \psi \to \Psi'(x) \text{ given}$$

and at $\zeta = 0$ no-slip conditions are adopted:

(31) $$\zeta = 0: \phi = w = \psi = 0$$

Initial conditions are conveniently given by noting that at $x = x_0$, for x_0 large enough, the solution is given by discarding all the nonlinear terms and retaining the linear equations treated by Ekman, in which x enters parameterically only. The solution to the balance of Coriolis, pressure, and friction forces is well known:

(32) $$\phi = -[\Psi'(x)]\sin(2^{-1/2}\zeta)\exp(-2^{-1/2}\zeta)$$

(33) $$\psi = [\Psi'(x)][1 - \cos(2^{-1/2}\zeta)\exp(-2^{-1/2}\zeta)]$$

(34) $$w = 2^{-1/2}[\Psi'_x(x)]\{1 - [\sin(2^{-1/2}\zeta) + \cos(2^{-1/2}\zeta)][\exp(-2^{-1/2}\zeta)]\}$$

Specifically what is sought is $w(r, \zeta \to \infty) = W(r)$ for $\Psi'(r)$ of interest. For r large, from (34)

(35) $$w(r, \zeta \to \infty) = W(r) = 2^{-1/2}\Psi'_x(x)$$

Numerical integration by finite-difference methods is formidable because the flow component in the timelike direction, u, is, in successively thinner strips lying parallel to the boundary, alternately in the direction of integration (numerically stable) and opposite to it (unstable). Though the radial flow is, on net, in the direction of integration, the integration is marginally stable. The only finite-difference results *applicable to the hurricane problem* known to the author were given by Anthes (1971) and are discussed later (Section 3.10).

George (Carrier et al., 1971) and Dergarabedian and Fendell (1972b) independently but simultaneously applied the method of weighted residuals to the boundary value problem, and found that, except near the eyewall where the method was inadequate, the linear result given in (35) sufficed for the nonlinear problem as well (Fig. 16). Thus, for a form like $\Psi = A(1 - x/x_0)$ [for which $\phi(r_0, \zeta) = 0$ according to (32)—as required by the closed system model] $w(r, \zeta \to \infty) \to -(A/x_0)$, a small negative constant quantity (equivalent to about 0.005 mph downdraft for physically interesting values).

Incidentally, the form just mentioned for the rapid swirl in the inviscid flow above the frictional inflow layer is stable for parametric values of practical interest, according to Rayleigh's criterion (Greenspan, 1968, pp. 271–272).

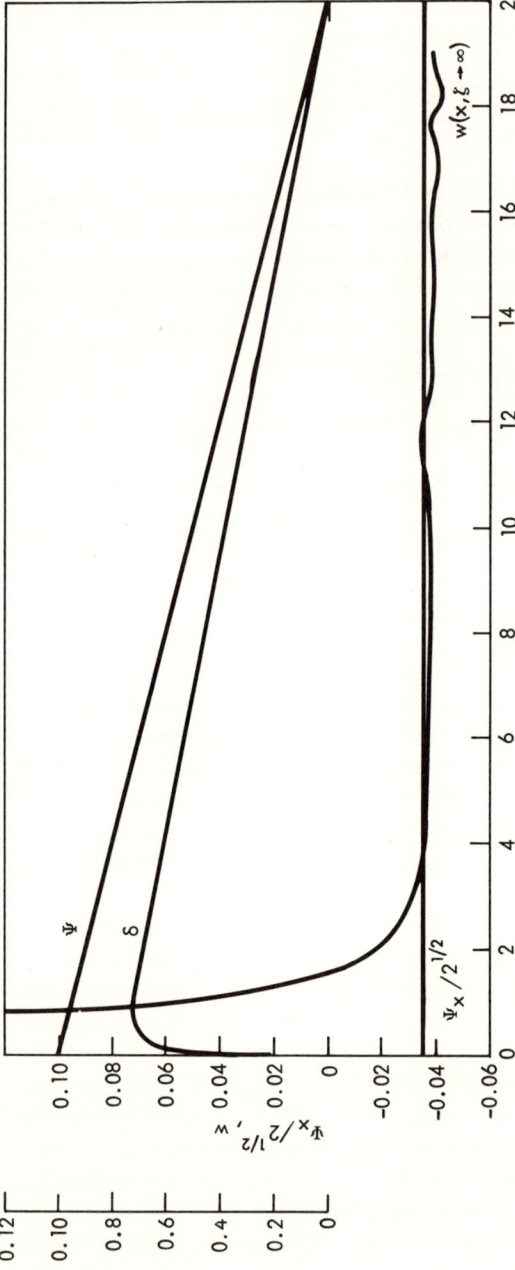

FIG. 16. Nondimensional results for the frictional boundary layer obtained by the method of weighted residuals, from Dergarabedian and Fendell (1972b). The divergence $w(x, \zeta \to \infty)$ and the volumetric flux $\delta = -\int_0^\infty \phi(x, \zeta) \, d\zeta$ are presented for the impressed swirl $\Psi = 1 - x/x_0$, $x_0 \doteq 20$, believed pertinent to a hurricane outside the eyewall. Except near the axis where nonlinear inertial effects dominate, the linear Ekman layer result, $w(x, \zeta \to \infty) \doteq \Psi_x/2^{1/2}$, is an excellent approximation to the numerical results. The volumetric flux $\delta(x)$ is thus linearly proportional to $(x_0 - x)$ to good approximation. Normalized residuals indicate large errors for $x < 3$, and discount the premature eruption as an artifact of the method. The solution by Carrier (1971a) for small x indicates the adequacy of the linear Ekman result for the divergence to within a factor of two. Since $\nu \doteq (1/75)$ miles^2/hr and $\Omega \doteq (1/16)$ rad/hr, dimensionally the results imply the boundary layer is of thickness $O(\nu/\Omega)^{1/2} = O(1 \text{ mile})$, the downflux into the boundary layer is $(2\nu\Omega)^{1/2}$ $w(x, \zeta \to \infty) = O(5 \times 10^{-3} \text{ miles/hr})$, and the volumetric flux erupting up the eyewall is $(2\pi^2\Psi_0{}^2\nu/\Omega)^{1/2}$ $\delta(x) = O(7 \times 10^3 \text{ miles}^3/\text{hr})$ where Ψ_0 characterizes the eyewall relative angular momentum per unit mass. The implication is that the fluid initially in the boundary layer sustains the hurricane for about a week, and the fluid lying above the boundary layer (with supplementary replenishment of moisture from the ocean) can readily sustain the hurricane for more than another week.

The reason weighted-residual calculations fail near the axis, as discovered by Carrier (1971a) and McWilliams (1971) by modified Oseen linearization and by Burggraf *et al.* (1971) by seminumerical analysis, is that the structure of the boundary layer so varies with radial distance that adequate representation in terms of one set of orthonormal polynomials is difficult. Far from the axis of symmetry, friction is important across the entire layer of thickness $O(\nu/\Omega)^{1/2}$; near the axis, friction is significant only in a small sublayer near the "wall" of thickness $O(r^2\nu/\Psi)^{1/2}$, and the remainder of the inflow layer of $O(\nu/\Omega)^{1/2}$ thickness is inviscidly controlled. However, for conditions of interest in hurricanes, (35) is everywhere correct to within a factor of two, and often far better.

Still further confirmation of the adequacy of the classical linear Ekman swirl-divergence relation for the nonlinear frictional inflow layer under a hurricane vortex has recently been given, in as yet unpublished work, by K. K. Tam of McGill University. Professor Tam has applied the theory of differential inequalities to the solution proposed by the Carrier group to the boundary value problem (27)–(31) and indicated the satisfactory accuracy of the result.

But while the solution may satisfy the boundary value problem, one may ask how well the boundary value problem reflects the actual situation, since a constant eddy viscosity has been adopted to model the turbulent transfer. It would seem premature to adopt a second-order (or field-closure) model that reduces the Reynolds time-averaged formulation to a determinate set by means of a differential relationship between the rate of change of the Reynolds stress and diffusion, production, and dissipation of turbulence; still, an eddy viscosity invariant with distance from the air–sea interface seems outdated, and further, some small slip might better model the boundary condition at the air–sea interface, taken planar.

It should be noted, however, that only one *quantitative* result from the nonlinear Ekman layer plays a role in the Carrier model, and that is the swirl-divergence relation appropriate at the extreme outer edge of the frictional inflow layer. Furthermore, the eddy viscosity is almost always taken as independent of distance from the "wall" over at least the outer ninety percent of the turbulent boundary layer thickness, aside from intermittency effects, which play very little role in altering the profiles of time-averaged macroscopic variables and are very often entirely neglected in calculations. In fact, one does not really know how to model slip and effective roughness effects in the hurricane viscous sublayer, so restoration of eddy-viscosity dependence on the normal coordinate would be a very speculative undertaking at best.

Other workers introduce a constant or wind-dependent drag coefficient to model frictional effects near the air–sea interface for computer solutions.

There seems to be no inherent reason why either of these empirical, phenomenological devices (drag coefficients, eddy viscosity coefficients) is superior to the other; either could be made to succeed by proper adjustment (curve-fitting). Both face the problem that empiricism developed for relatively small wind speeds and relatively smooth interfaces must be extrapolated to higher wind speeds and very rough seas (Ooyama, 1969, p. 19). There exist attempts to study the frictional inflow layer entailing both simpler and more detailed modeling than that just presented; for example, Riehl (1963a) invokes conservation of potential vorticity for the frictional inflow layer, and Riehl and Malkus (1961), who assert that dissipation of kinetic energy in the surface layer is independent of distance from the center, find that they must adopt exceptionally large eddy transfer coefficients to resolve problems that arise in computing a refined mechanical energy budget.

3.6. *The Energetics of the Frictional Boundary Layer and Throughput Supply*

For the frictional boundary layer II and throughput I, Carrier et al. (1971) take the following approximations as adequate for the quasi-steady mature phase: (1) the Prandtl and Schmidt numbers are unity; (2) the hydrostatic approximation holds; (3) the boundary layer approximation holds (derivatives normal to the boundary exceed those tangential to the boundary, but velocity components parallel to the boundary exceed those normal to the boundary); (4) the eddy transfer is adequately modeled by the laminar flux-gradient relations for diffusion of mass, momentum, and heat (as given by Fick, Newton, and Fourier, respectively), except that the augmented kinematic (eddy) viscosity may vary with radial position (but not with axial position); and (5) the mixture of dry air and water vapor may be taken as a perfect gas with constant heat capacity over the range of temperatures of interest here.

In view of the limited understanding of quantitative formulation of cumulus convection and turbulent transfer, on the cyclone scale of interest here these five approximations seem reasonable. Clearly the familiar Reynolds analogy, together with the constancy of the eddy viscosity over most of the boundary layer, is being adopted. It can then be shown that for L held constant, the following equation, a generalization of those given by Crocco in fluid dynamics and Shvab and Zel'dovich in combustion, holds in the meteorological context of interest here:

(36) $$u(\partial H/\partial r) + w(\partial H/\partial z) = \nu(\partial^2 H/\partial z^2) + D$$

where D denotes radiational loss, taken by Carrier et al. (1971) to be representable as

$$D = -f(z)H$$

with $f(z)$ (to be discussed below) known. The H used throughout this section is H_t as defined in (3); the subscript t is dropped to avoid double subscripts below. The boundary-initial conditions are taken to be

(37) $$z = 0: H = H_s; \qquad z = z_1: H = H_1$$

(38) $$r = r_0: H = H_a(z)$$

where, again, r_0 is the outer edge of the storm and z_1 is the top of the storm. The ambient profile $H_a(z)$ is given; $H_s = H_1 = H_a(z = z_1) = H_a(z = 0)$. Taking $H_s = H_a(z = 0)$ implies that for $r > R$ (where R is the radius of the eyewall), the ocean surface temperature is a constant and the water vapor mass fraction (which takes on its saturation value at the nominally plane sea surface $z = 0$) is taken independent of pressure (see below). Since the thermal conductivity of water greatly exceeds that of air, uniform sea surface temperature appears to be a good approximation. That $H_1 = H_a(z = 0)$ follows from the definition of the lid on the storm.

The terms, from left to right, in (36) represent radial advection; axial convection; turbulent diffusion and cumulus convection (both parameterized in ν, which will henceforth be treated as constant, though this is not necessary); and radiation loss. Throughout I, and in II at $r = r_0$, (36) may be approximated as

(39) $$w(\partial H/\partial z) - \nu(\partial^2 H/\partial z^2) = -f(z)H$$

this follows because $\phi(x_0) = 0$ [cf. (32)] and $u(\partial H/\partial r)$ is negligible in I [cf. (23)]. In view of the boundary conditions $f(z)$ must be chosen to permit $H(z) = H_a(z)$, where $H_a(z)$ is given (Jordan, 1957); Carrier et al., (1971) take $w(r_0, z) \equiv w_a(z) = 0$, but there appears to be no need to require this. Thus,

(40) $$w_a(z)(\partial H_a/\partial z) - \nu(\partial^2 H_a/\partial z^2) = -f(z)H_a$$

In fact, taking $H = H_a(z)$ throughout I and II satisfies the initial and boundary conditions, and renders $u(\partial H/\partial r) = 0$. Further, the function $w(r, z)$ is available from the nonlinear boundary layer dynamics just discussed; the dynamics reveals $(w)_{\max} = w(r, \zeta \to \infty) = W(r)$, the value at the outer edge of the boundary layer—and W is independent of r over a wide expanse. Only if W is appreciably different from w_a can H depart from H_a, and such a variance occurs only in the eyewall, where vertical convection dominates all other terms in (36) because (by simple continuity considerations) w is increased two orders of magnitude over its maximum magnitude in I or II. Hence, for at least portions of the eyewall,

(41) $$w \, \partial H/\partial z = 0$$

and the relevant boundary data for this hyperbolic suboperator is that given at $z = 0$ by (37). All this discussion, more carefully argued in Carrier et al. (1971), leads to two significant results:

1. Throughout the frictional boundary layer and the throughput supply, outside the eyewall the total stagnation enthalpy is approximately fixed at its ambient stratification. Furthermore, the small correction owing to inertial effects in a hurricane is readily seen to be a decrease of H with z such that the enthalpy gradient at $z = 0$ is increased slightly (about ten percent or so).

2. In portions of the eyewall, the total stagnation enthalpy is, to good approximation, constant at its sea level value, i.e. columns of air are very nearly rising on a moist adiabat.

Numerical values of interest are the ambient net sea/air enthalpy transfer and the eddy viscosity (Carrier et al., 1971):

(42)
$$-\rho\nu[\partial H(r, z = 0)/\partial z] \doteq -\rho\nu[\partial H_a(r, z = 0)/\partial z]$$
$$= 1.8 \times 10^5 \text{ ergs cm}^{-2} \text{ sec}^{-1}$$
$$\nu = 2.7 \times 10^5 \text{ cm}^2 \text{ sec}^{-1}$$

Because this result is not compatible with many existing hurricane models, it becomes necessary to identify the reasons for disagreement. Hence in the following sections the Riehl–Malkus theory is scrutinized.

It may be worth emphasizing that (36) yields a steady solution for the frictional inflow layer with all inertial terms considered. It is *not* based on obviously inadequate models of the inflow layer, such as frictionless or purely horizontal flow.

However, this result is really not quite so accurate as suggested by Carrier, Hammond, and George, owing to the omission of compressibility effects. A decrease in gas density attends the decrease in gas pressure (of up to 12 %) as one moves radially inward in the frictional inflow layer from the outer edge toward the eyewall. This decrease in air density is negligible for the dynamics, but not the energetics. If the gas density decreases as one goes radially inward, but the total stagnation enthalpy sea to air transfer remains approximately at the ambient rate (Carrier, 1971b, p. 158 footnote), then the vapor mass fraction (ratio of vapor density to gas density) increases because there is less mass to accept the same transfer. Carrier et al. (1971, p. 162) acknowledge the effect, but fail to account for it in their calculations. It seems more appropriate to study the vapor density, which at sea level depends only on sea surface temperature, rather than the vapor mass fraction. A rough first estimate is that perhaps a 20 % increase over tropical ambient levels in latent and sensible heat transfer from the ocean to the atmosphere occurs near the eyewall of hurricanes owing to this density reduction effect.

3.7. The Riehl–Malkus Postulate of an Oceanic Heat Source

While the exposition of the mature tropical cyclone has been given here in terms of the Carrier model, most published works adopt the framework of the Riehl–Malkus theory alluded to earlier (Riehl, 1954, 1963a; Palmén and Riehl, 1957; Malkus, 1958, 1962; Malkus and Riehl, 1960; Riehl and Malkus, 1961). The cornerstones of this theory have been summarized by Malkus (1962, p. 232):

> The new model relates core maintenance to mechanism, namely, cumulonimbus convection and sea–air exchange. *To sustain the required pressure gradients, two coupled processes are necessary: first, a greatly magnified oceanic imput of sensible and latent heat, and secondly, the undilute release of the latter in concentrated hot tower ascent, so that air of high heat content is pumped rapidly into the upper troposphere.* The quantitative establishment of these crucial relationships was carried out in a joint analytic and observational framework (Malkus and Riehl, 1960).

The important role of large numbers of cumulonimbi in the eyewall has been confirmed, although the Riehl–Malkus concept of a direct hot-tower heating mechanism has since been challenged by Lopez (1972b), as discussed in Section 2.2. Here the other major contention ["... it is postulated that lowering of surface pressures in hurricanes arises mainly through an 'extra' oceanic heat source in the storm's interior" (Malkus and Riehl, 1960, p. 12)] is questioned. There is undoubtedly some augmentation of the ambient level sensible and latent heat transfer from sea to atmosphere over a wide ocean expanse under a hurricane, and the warm autumnal tropical seas are in a major way responsible for the atmospheric total static enthalpy profile to begin with; however, an augmentation of over an order of magnitude in oceanic heat transfer concentrated, according to Malkus and Riehl, largely in the region 30 to 90 km from the cyclone center seems unnecessary and unlikely (Section 3.8).

The Riehl–Malkus theory, as noted earlier in Section 3.4, takes the eye–eyewall interface to be effectively perfectly vertical so some mechanism other than dry adiabatic recompression must be found to sustain hurricane force winds. Why this mechanism is stated to be an "extra" oceanic heat source is understood by reconstructing their logic. It may be worth noting at this point that Riehl and Malkus permit the relative vorticity (rather than the relative velocity) to vanish at the outer edge of the storm; workers using their concepts tend to permit convective, but not diffusive, transport across the outer cylindrical boundary, while workers on the closed Carrier model permit diffusive, but not convective, transport across boundaries.

Malkus and Riehl (1960, pp. 3, 7) acknowledge the existence of a low-level frictional inflow layer, although in a sample calculation there is *efflux* from the boundary layer from the outer edge of the eye out to *500 km* at an average ascent rate of 0.7 mph. Thus air flows in through the boundary layer from the outer edge of an open system, rather than sinking down into the frictional layer as in Carrier's model. The boundary layer air, according to Malkus and Riehl, undergoes adiabatic expansion as it spirals inward toward the center, yet it remains isothermal. This requires a vast, rather localized increase in sea-to-air transfer between the ocean and the contiguous atmosphere. That the gradient normal to the air/sea interface of temperature and of water vapor mass fraction is large enough to be consistent with vastly increased air/sea transfer is accepted as possible:

> In the outskirts of a hurricane the temperature of the inflowing air drops slowly due to adiabatic expansion during (horizontal) motion toward lower pressure. It is one of the remarkable observations in hurricanes that this drop ceases at pressures of 990–1000 mb and that thereafter isothermal expansion takes place. Presumably, the temperature difference between sea and air attains a value large enough for the oceanic heat supply to take place at a sufficient rate to keep the temperature difference constant (Malkus and Riehl, 1960, p. 9).
>
> The actual transports [between sea and air], of course, are very large in the hurricane compared to the trades. Sensible heat pickup is 720 cal/cm²/day, an increase by a factor of 50 over the trades ...; latent heat pickup is 2420 cal/cm²/day, higher by a factor of 12–13 (Malkus and Riehl, 1960, p. 12).

Similar arguments are made elsewhere by Riehl (1954, pp. 286–287):

> Many published records, notably those by Deppermann ..., have proved that the surface temperature outside the eye is constant or decreases very slightly toward the center. The implications of this remarkable fact passed without notice until Byers ... drew attention to it. The temperature of the surface air spiraling toward a center should decrease if adiabatic expansion occurred during pressure reduction. For instance, air entering the circulation with the average properties of the mean tropical atmosphere should reach the 930 mb isobar with a temperature of 20.5°C and specific humidity of 17 gm/kg. Because of condensation, a dense fog should prevail at the ground inward from the 970 mb isobar. But this is never observed. It follows that *the potential temperature of the surface air increases along the inward trajectories.* We also know that the specific humidity increases and that the cloud bases remain between a few hundred and 1000 feet
>
> The surface air thus *acquires both latent and sensible heat during its travel toward lower pressure....*
>
> A source for the heat and moisture increment is obvious. The ocean is greatly agitated, and large amounts of water are thrown into the air in the form of spray. It is hard to say where the ocean ends and where the atmosphere begins! As the air moves toward lower pressure and begins to expand adiabatically, the temperature difference between ocean and air suddenly increases. Since the surface of contact between air and water increases to many times the horizontal area of the storm, rapid transfer of sensible and latent heat from ocean to air is made possible. In the outskirts, say beyond the 990 mb isobar, the turmoil is less and the process of heat transfer is not operative.[7]

[7] From "Tropical Meteorology" by Herbert Riehl. Copyright 1954 by McGraw-Hill Book Company. Used with permission.

The Riehl–Malkus theory that greatly augmented heat and mass transfer sustains the tropical cyclone has, in fact, been parameterized into all existing computer simulations (see Section 3.10). For example, Rosenthal (1971b) closely reflects the Riehl–Malkus theory and notes similar logic in the work of another computer modeler of tropical cyclones (Ooyama, 1969):

> Air–sea exchanges of sensible and latent heat have long been considered important ingredients in the development and maintenance of tropical storms. Palmén (1948) showed, on a climatological basis, that tropical storms form primarily over warm ocean waters ($T_{sea} > 26°C$). Malkus and Riehl (1960) showed that the deep central pressures associated with hurricanes could not be explained hydrostatically unless the equivalent potential temperature, θ_e, in the boundary layer was 10° to 15°K greater than that of the mean tropical atmosphere. Byers (1944) pointed out that the observed near-isothermal conditions for inward spiraling air in the hurricane boundary layer required a source of sensible heat to compensate for the cooling due to adiabatic expansion. ...
>
> Ooyama (1969) frouund drastic reductions in the strength of his model storm when the air–sea exchanges of sensible and latent heat were suppressed. He pointed out that at sufficiently large radii, the boundary layer is divergent (the so-called Ekman layer "sucking"). ... This subsidence tends to decrease the boundary layer θ_e since $\partial \theta_e / \partial z < 0$ in the lower troposphere. Ooyama argued that unless the energy supply from the ocean can again raise the θ_e of the boundary layer air to sufficiently large values before the inflowing air reaches the inner region, the convective activity will diminish in those regions and, hence, the storm will begin to weaken.
>
> Ooyama's line of reasoning can be extended to show that evaporation is far more important than sensible heat flux. The air sucked into the boundary layer has a higher potential temperature than the original boundary layer air. The subsiding air has a smaller θ_e only because it is relatively dry. (Rosenthal, 1971b, p. 772).

3.8. The Intensity of a Tropical Cyclone and the Underlying Sea Surface Temperature

Palmén and Riehl (1957, p. 156) state: "Evidently, a cyclone will decay rapidly if it encounters thermally unfavorable conditions. The generation term in [the equation for conservation of kinetic energy] will then decrease rapidly; it will become negative if the air ascending in the core is cooler than the surroundings in spite of release of latent heat. This would happen when a storm moves over a relatively cold ocean surface, over a continent, or when colder air masses invade the cyclone near the surface."

In discussing the Riehl–Malkus theory, one should distinguish between what these authors themselves assert, and what others citing their work have added. For example, Shuleykin (1970, 1972) goes beyond Riehl and Malkus and asserts that the "... power of the hurricane ... increases sharply with the temperature of the underlying water surface" (Shuleykin, 1972, p. 1). However, attempts to correlate central pressure deficit and sea surface temperature, without sufficient account of other factors, seems simplistic. This particular point will be now discussed.

First, ocean temperatures are difficult to determine accurately from currently available records and by currently available methods (Perlroth, 1967; Hidy, 1972, p. 1091). Next, Gentry (1969a, p. 406) presents data relating "... the maximum intensity of several tropical cyclones to the temperatures of the sea beneath them and shows that both severe and weak tropical cyclones occur when ocean temperatures are relatively high. This suggests that variations in parameters other than the transfer of heat from the ocean to the atmosphere also influence the storm's intensity...." Actually Gentry's data (Fig. 17) indicate that some intense tropical cyclones lie over relatively cold water ($<28.0°C$). While Brand (1971) cites a supposed correlation of central pressure deficit with sea surface temperature throughout the lifetime of Hurricane Ester of September 9–26, 1961, it may be noted that during one twelve-hour period of constant sea temperature, the central pressure rose 15 mb; for three days while the sea temperature hovered about 84°F, the central pressure nonmonotonically rose from 930 mb to 955 mb; and at various times when the sea temperature was at 86°F, the central pressure was at values as low as 927 mb and as high as 953 mb. Perlroth (1967) shows that for Hurricane Ginny of October 21–28, 1963, for five days while the sea temperature hovered near 80°F, the central pressure fell from 995 mb to 970 mb (Fig. 18). Perlroth (1969) cites the importance in tropical storm *generation* of considering not only high surface temperature (above 26°C) but also small oceanic temperature variation with depth (less than 7°C within 200 ft of the surface). Perlroth emphasizes that the oceanic environment is only one factor in tropical storm intensity, and the sea surface temperature is only one factor in the oceanic environment.

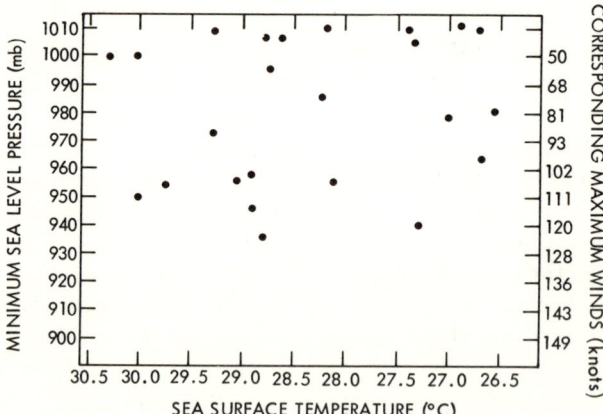

Fig. 17. Measurements of minimum sea level pressure versus local sea surface temperature for several tropical cyclones (from Gentry, 1969a, p. 406; reproduced with permission of the American Meteorological Society). Each dot represents relation for an individual storm selected from 1955–67 seasons.

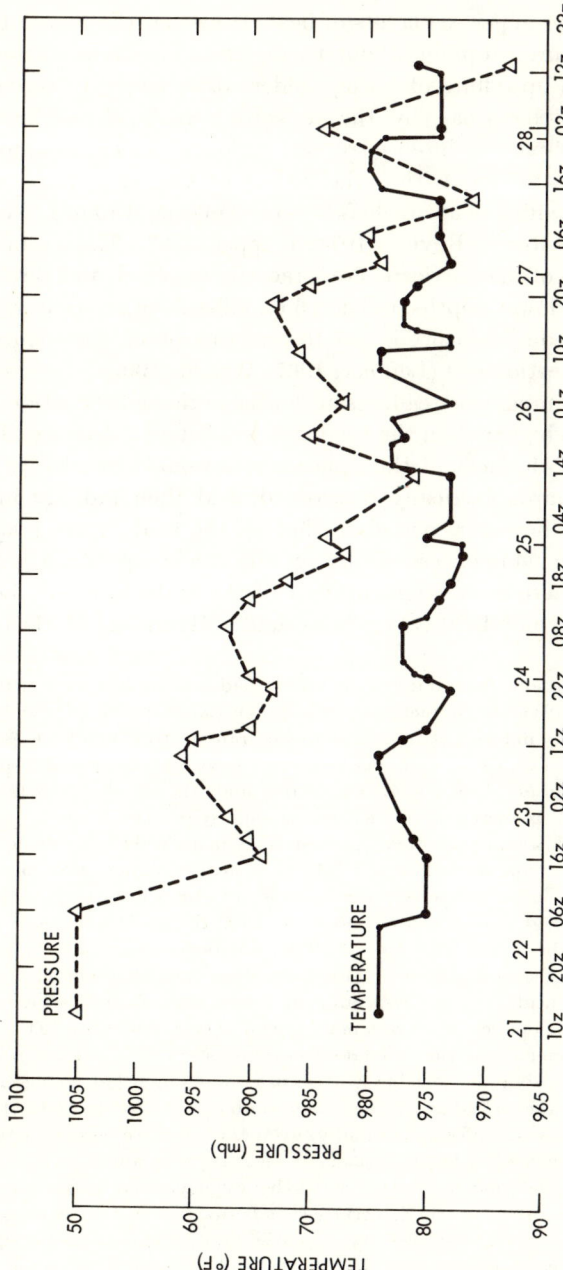

FIG. 18. The central sea level pressure is plotted as a function of the local sea surface temperature for Hurricane Ginny, 21–28 October 1963, which passed over the Gulf Stream on 24 October (from Perlroth, 1967, p. 266). Solid line indicates extrapolated temperature reading based on composite sea surface temperature chart, 11–19 October, 1963.

Finally, the fact that hurricanes leave cold wakes in the ocean might suggest that they do drain the upper sea layers of heat. However, the reason for the sea surface temperature dropping about two degrees Centigrade after hurricane passage is the upwelling of lower, colder water owing to convection currents induced in the ocean by the pressure deficits of the hurricane (Perlroth, 1967). Slow-moving intense tropical cyclones can cause upwelling, from depths of 40 to 65 m, within the radius of hurricane force winds; compensatory downwelling of as much as 80 to 100 m occurs 45 to 110 n miles from the path of the storm (Revesz, 1971; Leipper, 1967). The warm ocean water near the center of the hurricane flows radially outward, and is replaced by colder water from lower depths; outside 65 n miles from the center, below the outflow, there is compensatory inflow toward the center of the hurricane to complete the convection cell (Leipper, 1967; Wright, 1969).

Leipper finds the temperatue readings at depths in the outer portion of the hurricane path to be higher than the temperatures before the storm. Thus a significant portion of the heat of the upper layers would seem to be transported radially outward, vertically downward, and then radially inward. Leipper and Wright, however, postulate that all the heat of the relatively warm upper oceanic level in excess of that of the colder upwelling is transferred to the atmosphere as an augmentation of the ambient rate[8]; Leipper suggests a transfer rate of 4500 cal/cm^2-day under Hurricane Hilda (1964),

---

[8] In this regard deductions from data taken off Barbados in August 1968 may be of interest (Warsh, 1973). Warsh characterizes the convective activity of the tropical atmosphere on a given day in terms of graduated modes, ranging from mode one (severely depressed convection with almost no cumulus) to mode six (severely enhanced convection with mostly cumulonimbi and heavy showers). Warsh finds the sensible heat flux from the ocean to the atmosphere increases only extremely slightly from mode one to mode six. In contrast, the latent heat flux *decreases* to one-half in magnitude from mode one to mode four, then increases from mode four to mode six to nearly its mode one magnitude. Since the equivalent potential temperature in a mode six atmosphere is almost everywhere 335°K or higher, and since the latent heat flux from the sea in a mode six atmosphere is still less than the equivalent of one-half centimeter of evaporated water per day, the results would not seem to augur well for a sometimes postulated order of magnitude increase in latent heat transfer over ambient levels *under* even a severe storm. Ostapoff *et al.* (1973) also deduce, from oceanographic observations in the ITCZ during a broad light rainfall from cumuli, that the latent heat transfer was appreciably reduced owing to high humidities, but the sensible heat transfer was somewhat increased, with the result that the Bowen ratio had an average value of about 0.2 (over twice the tropical norm), and a maximum value of 0.247. Finally, even though the measurements were taken over free water surfaces in the laboratory or over lakes at low wind speeds, the results of Easterbrook (1969) are interesting as further evidence that intuitive expectations about latent heat transfer can err. Easterbrook finds that the vortical wind patterns created by waves on free water surfaces can suppress vertical turbulent transport of water vapor and reduce evaporation, such that larger height-to-period ratios for surface waves can lead to smaller mass transfer across the two-phase interface.

and Wright suggested 7200 cal/cm²-day under Typhoon Shirley (1965). The previously cited figure of Malkus and Riehl (1960) for a moderate hurricane was 2420 cal/cm²-day. Leipper acknowledges that the transfer so predicted is so much larger to the right of the path that even observed asymmetries in intensity are inadequate for his estimation. Further reasons for doubt concerning Leipper's calculation for Hurricane Hilda will be given in Section 3.9.1. [Most discussions emphasize the cold wake left in the upper oceanic layers by intense, slowly moving hurricanes; however, the reduction of surface layer temperature by upwelling and mixing may be occurring directly under the hurricane. Black and Mallinger (1972, p. 74) find evidence of radial outflow and downwelling of warm upper layer water for Hurricane Ginger (1971); the upwelling of colder water was found to reduce the sea surface temperature by 4°C on one day, and by 2.5°C on another day when the hurricane was translating at a higher speed. Reduced latent and sensible enthalpy transfer from sea to atmosphere ensued, with a reduction in intensity. Thus, the interaction of a hurricane with the ocean may be such that reduced transfer from the sea, not augmentation of transfer over the ambient level, may occur (cf. Section 3.9.1).] The hurricane-induced cooling in the ocean may be anticipated to be reduced to 1°C or less for particularly stable ocean stratification, as would be the case for very cold water below the well-mixed layer.

Leipper and Volgenau (1972) later introduced the concept of a hurricane heat potential, which they applied to oceanographic data for the Gulf of Mexico. The idea is that all the heat content of the upper oceanic layer (top 500 cm) implied by water temperature in excess of 26°C is available to sustain a hurricane. The authors acknowledge that their theory implies that in a hurricane, the air controls the sea temperature, although in almost every other physical situation, the sea is regarded as characterizing the air temperature. If all this heat were really available for transfer, why would a hurricane be required (rather than just the normally operative turbulent transfer mechanisms) to effect the transfer? One wonders how such a large potential could be developed, and why hurricanes are not always formed over waters with surface layer temperatures above 26°C.

3.9. Critique of the Riehl–Malkus Model

A critique of the Riehl–Malkus reasoning that an extra oceanic source of latent and sensible heat maintains the tropical cyclone might begin by noting that use of the adiabatic expansion relation $p \sim T^{\gamma/(\gamma-1)}$ is inappropriate for a (turbulent) frictional layer. Such a relation implies that the entropy is constant for each fluid element. Riehl (1954, p. 26) is correct when he assumes that neglect of diffusion gives almost a 6°C drop (with condensation) in passing from 1014 mb (autumnal sea level ambient) to 930 mb. In fact, the inflow layer is largely cloud-free, and the same adiabatic expansion taken as dry produces a 7.3°C drop. Such a large temperature drop does not occur (though whether the static temperature is really *constant* as Riehl believes is uncertain), and should never have been expected. But Riehl (1954, p. 286) was surprised at measurements he interpreted to imply constant static temperature during inflow in the frictional layer, and passed immediately from the (clearly inappropriate) isentropic expansion to the postulate of latent and sensible heat transfer from the ocean greatly in excess of ambient transfer levels.[9]

3.9.1. Evaluating the Arguments for Greatly Augmented Enthalpy Transfer. There are no indisputable semiempirical theories or direct measurements of the heat and mass transfer from the ocean to the atmosphere in a hurricane: "unfortunately there is little information on C_E ('the nondimensional coefficient for the air–sea energy interchange') under hurricane conditions, other than the semispeculative guess that the exchange coefficients of latent heat, sensible heat, and momentum are probably of the same magnitude" (Ooyama, 1969, p. 15). Hidy (1972, pp. 1086 and 1097) notes the lack of quantitative progress on air–sea transfer, cites the paucity of data on the drag coefficient at sea level for winds over 20 mph, and mentions the difficulties in extrapolating data accurately; in fact, one of few known results casts doubt on the validity of the Reynolds analogy relating turbulent transfer of heat and momentum at the air–sea interface. There have been attempts to extrapolate to hurricane conditions empircal laws relating ocean evaporation rates to wind speed. The actual measurements refer to low wind speed and involve relatively dry air, far from saturation; extrapolation to hurricane winds and nearly saturated low-level tropical air is unjustified and misleading. Garstang (1967) emphasizes that a critical assumption in the theory of Riehl and Malkus is that the sea-to-air transfer coefficients increase linearly with wind speed indefinitely into the hurricane regime. Zipser (1969, p. 813) found that using such linearly extrapolated bulk turbulent transfer relations, which imply an order of magnitude increase in latent heat transfer under moderately strong tropical disturbances over the ambient rate of transfer, would have produced clouds in an observed tropical disturbance with unsaturated convective downdrafts on a time scale an order of magnitude too short. Zipser also notes that Riehl and Malkus expected downdrafts around cumulonimbi

[9] Kraus emphasizes that the Riehl–Malkus theory depends on an isothermal inflow layer and discusses whether frictional effects are fully accounted for in that theory: "The importance of [an] additional energy supply to air in the hurricane, before it ascends in the warm core, was stressed particularly by Malkus and Riehl (1960). The argument would be weakened somewhat if a (verbal) suggestion by Carrier was found to be true. Carrier maintains that surface friction must cause a deceleration and compression of a surface air parcel along its trajectory; this compression compensates to some extent for the expansion that would be associated otherwise with the externally imposed pressure reduction" (Kraus, 1972, p. 208).

to be saturated, but the observed downdrafts were unsaturated and so details of the Riehl–Malkus predictions on sea-to-air transfers near cumulonimbi must be modified. In the absence of more quantitative evidence Riehl (1954, p. 287) asserts that much sea spray is tossed into the air in a hurricane. This does not in itself assure augmented net enthalpy transfer from sea to air, since heat must be drawn from the air to evaporate the drops for later condensation of the water vapor in the eyewall and inner rainbands. Evaporation of spray leads to a temperature decrease as well as a dew point increase, such that the total stagnation enthalpy remains unchanged. Heat loss from a spray particle (which acts like a wet-bulb thermometer) to surrounding air of relative humidity of 0.85 or higher will be small.

Possibly the closest approximation to a direct measurement of the Riehl–Malkus postulate of an augmented oceanic heat source in a hurricane are the tritium-tracing measurements by Östlund (1968, 1970) in Hurricane Faith (1966). The tritium deposited initially in the stratosphere following American and Russian fusion bomb tests in 1961 and 1962 temporarily created trace amounts in the troposphere that exceeded the amounts in the ocean; determining tritium concentrations could theoretically determine the source of water substance in a hurricane. While the experiment cannot be repeated without the resumption of atmospheric nuclear testing, and while the present author does not regard the experiment as highly reliable, Hidy (1972) reported the result without reservations. Östlund found that the ambient rate of latent heat transfer from sea to atmosphere in the tropics (which is equivalent to evaporation of about 0.5 cm/day of water) was increased by a factor of about three within the inner 100 km radius of the hurricane, as opposed to the order of magnitude increase predicted by the theory of Riehl and Malkus.

There seems to be no indisputable direct evidence for the internal ocean heat source postulate of the Riehl–Malkus theory; the question does remain whether or not Carrier and his co-workers have obviated the indirect justification, specifically, the assertion by Riehl and Malkus that the oceanic heat source is necessary to satisfy the laws describing conservation of energy and momentum in the storm.

In this respect the earlier quotation from Rosenthal (1971b, p. 772), citing Ooyama (1969, pp. 14–15), should be discussed. Rosenthal asserts that higher level air sinking down into the surface frictional layer will be drier and of lower equivalent potential temperature than the original boundary layer air. If the quasi-steady approximation of the Carrier model is justified, then in the mature stage the total stagnation enthalpy is a function mainly of position. According to Carrier and his co-workers, the plume and radiative transfer mechanisms that carry heat and moisture up into the ambient tropical atmosphere persist, with neither augmentation nor diminution to any appreciable degree, in the presence of the storm. These transfer mechanisms help compensate for rain-out in the outer spiral bands such that by the time a particle has *slowly* sunk into the boundary layer at 0.005 mph, it has been enriched in total stagnation enthalpy to the ambient value at its current height. In other words, much of the sea-to-air transfer of latent and sensible heat continues to pass across the boundary layer with little diminution, just as in the ambient, such that the enthalpy of air originally in the 700–900 mb strata increases to that of air originally in the boundary layer owing to enrichment of its total static enthalpy as it slowly descends. Only as the hurricane leaves the tropical oceans is this ambient transfer reduced; the air entering the boundary layer eventually is of lower total static enthalpy, since it comes from air which is originally higher in the tropical ambient (hence colder and drier), and since that air is not appreciably enriched as it descends. In this way traverse over ocean patches of varying temperature can help cause the well known nonmonotonic perturbations in hurricane intensity within the general level of strength computed from the spawning ambient as discussed earlier (see also Section 4.4).

In this critique of the Riehl–Malkus model, one should note a later modification of earlier statements. Malkus and Riehl (1960, p. 17) write: "The total added heat energy from the ocean is on the order (for the moderate storm) of 2.5 cal/gm, while the average normal heat content (latent plus sensible) of tropical air is about 80 cal/cm." Thus the enthalpy contribution added by the ocean to ambient air enthalpy is about three percent for a moderate storm. "The heat ... gained by the air is only a minute fraction of that carried inward through the cylinder at the distance of the 1000-mb isobar ... Nevertheless, it is these small increments that produce the strong inward warming. ... They are thus of utmost importance for generation and maintenance of tropical storms, even though they may be wholly neglected in a general energy balance. ..." (Palmén and Riehl, 1957, p. 156). That such a marginal amount of heating is critical to both generation and maintenance seems intuitively surprising. Later Malkus (1962, pp. 247–249) limited the role of greatly augmented enthalpy transfer strictly to the maintenance of the hurricane, since during development the high winds that supposedly permit the increased turbulent transfer do not exist.

Hawkins and Rubsam (1968) have computed the structure and budgets for Hurricane Hilda for October 1, 1964, when it was an intense storm over the Gulf of Mexico. An unprecedented five-level collection of data was available for exhaustive analysis, and the total static enthalpy budget was calculated very similarly to the approach used by Malkus and Riehl (1960). Hawkins and Rubsam found, however, that (upon inclusion of the postulated latent and sensible enthalpy flux from the sea) the amount of energy estimated to be radiated away from the top of the storm was twenty times what seems physically possible. The kinetic energy budget computed for Hilda was also at significant variance from the one computed by Riehl and Malkus (1961) for Hurricane Daisy (1958). In view of this, Leipper's (1967) suggestion that the oceanic heat transfer to Hilda would be *underestimated*, using Malkus's formulas, by 50–90 % seems incomprehensible.

3.9.2. Temperature Measurements in Hurricanes. If numerous accurate temperature measurements in rapidly swirling, heavily raining portions of the hurricane could be readily carried out, resolution of doubt about the existence of a large ocean heat source within hurricanes could be quickly accomplished. However, such measurements are still subject to controversy and inaccuracy. Aircraft measurements tend to be taken above the inflow layer for safety reasons: "... the preferred altitude for reconnaissance traverses of a hurricane is from 10,000 to 14,000 feet " (Meyer, 1971a, p. 58). Currently used instrumentation may suffer from slow response time, from wet-bulb effects, and from the need to deduce the static temperature from a measurement that includes the dynamic contribution (Meyer, 1971b, pp. 19–37). Riehl and Malkus (1961, p. 188) discuss possible unresolved errors in midtropospheric temperature measurements of a few degrees Centigrade in magnitude.

First, to a reasonable degree of accuracy a dry-bulb thermometer measures $[T + (q^2/2c_p)]$, and a wet-bulb thermometer (with knowledge of the pressure) yields $[T + (q^2/2c_p) + (LY/c_p)]$ (Hess, 1959, p. 61). At very low altitudes, the dry-bulb temperature is the total potential-like temperature (θ_t/c_p), and the wet-bulb temperature is then related to the total stagnation temperature (H_t/c_p), because the missing gravitational contribution (gz/c_p) is negligible. Here the recovery factor (usually denoted r) for the thermometers is taken as unity; for the relevant Mach and Reynolds numbers this seems satisfactory, since for a well-designed instrument $r > 0.9$ for turbulent flow in air. The belaboring of the dynamic contribution would normally be unjustified in meteorology, but a 200 mph wind is equivalent to a contribution of about 4°C; this amount is in excess of one percent of both the static temperature ($\sim 300°K$) and also the total stagnation temperature ($\sim 350°K$) for the sea level autumnal tropical ambient. Whether temperature measurements made several decades ago, when hurricanes generally were not believed to be capable of such speeds, are properly corrected for this dynamic contribution is not always clear. It should also be recalled that, while the theory needs some improvement, the still quite useful result of Carrier *et al.* (1971) implies that the wet-bulb temperature would remain *approximately* radially invariant in a traverse of a hurricane along a ray and at constant height up to about 700 mb altitude, from the outer eyewall to the outer edge.

Deppermann (1937, p. 5), whose work is frequently cited by Riehl, writes: "... remarkably uniform diurnal temperature oscillations are maintained both before and after a typhoon; but when the station is under typhoon influence, a reduction of the *maximum* temperature by about 3°C occurs, while the *minimum* temperature remains the same." *If* by maximum and minimum temperatures Deppermann means dry-bulb and wet-bulb temperatures, respectively, and if one assumes Deppermann correctly accounted for the dynamic contribution ($q^2/2c_p$), then Deppermann finds that the static temperature T decreases by about 3°K in a typical typhoon, while the total stagnation temperature increases by a few degrees Centigrade (5°–6°C at the very most). In any case, there is no evidence of large increases in (H/c_p), as an oceanic heat source would allegedly cause. Also, if (H/c_p) is roughly radially constant in the inflow layer, then a decrease in T of about 3.7°C from sea level autumnal tropical ambient is necessary for condensation, and this is reportedly not achieved in a typical typhoon, so a cloud-free frictional inflow layer is no surprise in Carrier's model.

Arakawa (1954, p. 119) reported that the air temperature dropped 1°C from 28°C to 27°C, and the wet-bulb temperature remained about constant radially at approximately 26.5°C, during the passage of an 898 mb typhoon over a Japanese weather ship in October 1944; only in the eye did the two temperatures rise sharply. Winds rose to about 120 mph in the eyewall. Again, this seems compatible with the Carrier–Hammond–George predictions, which include no oceanic heat transfer within hurricanes much above ambient level. Palmén and Newton (1969, p. 478) assert with no elaboration that the measurement implies an increase of the equivalent potential temperature from 360° to 385°K, and thus is a confirmation of the internal heat source postulate; the author does not understand how this result was achieved.

To avoid difficulties in boundary layer measurements, Riehl (1963a, p. 277) studied equivalent potential temperatures deduced from aircraft data taken at the 245–250 mb level at the inner edge of the eyewall. Little complication from moisture is expected at such heights, and the variation of equivalent potential temperature with height is not expected to be large. Riehl reports equivalent potential temperature of 370°K for a hurricane of central pressure of 960 mb, but provides no further details on the instrumentation or data reduction whatever. If the result is correct, very significant evidence exists for an oceanic heat source. However, Dr. R. H. Simpson of the National Hurricane Center in Miami, Florida has informed the author that a vortex thermometer was used (private correspondence). If one recalls that 250 mb corresponds very roughly to approximately 35,000 ft, then the remarks of Gentry (1964, p. 12) are illuminating: "The wind tunnel tests of the vortex probe under icing conditions (Ruskin and Schecter ...) indicated that the recorded temperature might be too high due to the change of state on impact at probe entry. ... No evidence is readily available as to whether much ice accumulated on the vortex probe of the aircraft flying at about 35,000 ft or higher. Ordinarily, little liquid water would be encountered at the higher elevations and the ice accumulation should be minor. Some data collected by the research aircraft in 1961 and 1962 indicated, however, that at times the ice accumulation might be significant. There is the possibility, therefore, that the temperatures recorded for the upper troposphere are too high by an unknown amount. There will probably always remain a question as to the accuracy of the temperature measurements until some absolute standard of comparison is developed."

3.9.3. *On the Magnitude of Possible Increases in the Wet-Bulb Temperature at Sea Level.* Some insight into possible increases in the total stagnation temperature $(H_t/c_p) = T + (gz/c_p) + (q^2/2c_p) + (LY/c_p)$ at sea level within a hurricane will now be sought. *Very near* $z = 0$, $q \doteq 0$. Also, the air temperature assumes the temperature of the sea; since the sea temperature is held spatially constant in all hurricane analyses known to the author [with one experimental exception (Anthes, 1972, pp. 473–474)], and since the sea temperature for an intense slowly translating hurricane decreases toward the center of the storm if it changes at all, the conservative approximation for current purposes is to hold T everywhere constant at the ambient sea level value for the tropical autumn. Then the only changes in (H/c_p) are due to changes in $(LY/c_p,)$ or more specifically, to changes in Y since L and c_p are taken constant for current purposes. Further, with only one percent error.

$$
\begin{aligned}
Y &\equiv \rho_v/\rho \\
&\equiv \sigma(\text{RH})P(T)/[p_a + \sigma(\text{RH})P(T)] \\
&\doteq \sigma(\text{RH})P(T)/p
\end{aligned}
\tag{43}
$$

so the vapor mass fraction Y increases as the relative humidity RH increases or as the total pressure p decreases, for temperature T fixed. In the autumnal tropical ambient, if L represents latent heat of condensation plus glaciation, at sea level $(LY/c_p) \doteq 50°$K. The ambient sea level relative humidity is 0.84 and the pressure is 1014 mb. [The relative humidity is unity at the air–sea interface but sea level measurement generally means readings at shipboard height, or more phenomenologically, it means measurement in the postulated, roughly thirty foot thick constant-stress layer near the air–sea interface (Hidy, 1972, p. 1084).] For a moderate hurricane with central pressure of 966 mb, the increase in (H/c_p) is 2.5°K, and for an intense hurricane with central pressure of 910 mb, the increase in (H/c_p) is 5.7°K—provided RH is held at 0.84. If RH increases from 0.84 to unity under the eyewall, then (H/c_p) increases by 12.5°K for the moderate hurricane and 16.3°K for an intense hurricane—although the justification for adopting such an increase in RH appears unestablished, and Anthes and Johnson (1968, p. 297) cite evidence

that the relative humidity of the boundary layer tends to remain between
80 and 90 %. Malkus and Riehl (1960, p. 16) claim the equivalent potential
temperature, which is clearly closely related to (H/c_p), increases by 12.5°K
from ambient to eyewall in the moderate hurricane [although Riehl (1963a,
p. 277) seems to augment this to 16°K], and by 35°K in the intense
hurricane. By the reasoning adopted here the physical basis of the 12.5°K
increase becomes evident, but the 35°K increase for an intense hurricane
represents a physically impossible relative humidity of over 100 %. Succinctly, while Carrier and his co-workers discuss increases in the first normal
derivative of H at sea level in hurricanes over ambient on the order of 10 %,
Riehl and Malkus discuss occasional increases in H itself on the order of 10 %.
Carrier and his co-workers neglected the full effect of decreased air density as
the eyewall is approached, and found H to be radially constant; the conjecture here is that the error in H so incurred is about one percent or so.

3.10. Numerical Simulation of Hurricanes on Digital Computers

In recent years there has been a proliferation of computer models of the
tropical cyclones (Yamasaki, 1968; Ooyama, 1969; Rosenthal, 1970; Sundqvist,
1970; Kurihara, 1971; Anthes, 1972). These models warrant a review in themselves, and only a limited discussion is undertaken here.

First, it must be understood that these models are by no means exact
solutions of the full conservation equations, subject to appropriate boundary
and initial conditions. In fact, to paraphrase Lorenz (1967), one might remark
that detailed reproduction of the hurricane lifespan by numerical methods
might not appreciably increase physical insight, because the total behavior
is so complex that the relative importance of various features might be no
more evident from examination of the computer solutions than from direct
observations of the real atmosphere. In any case, computer solutions are today
far from exact solutions; in fact, today no computer solution even attempts
to treat observed initial data (Ooyama, 1969, pp. 35–37; Rosenthal, 1971b,
p. 35), but rather describes how a general hurricane grows and dies—in fact,
works in computer simulation emphasize results for the *later* stages of intensification and the peak intensity portion of the lifetime. The adopted initial conditions almost invariably have pronounced axisymmetric circulations, and only a
major alteration of parameterizations prevents inevitable hurricane formation.

Parameterization also entails compromise with exact solution. Because
many phenomena on the cumulus scale and smaller are crucial to the
hurricane scale flow, computer solution today is feasible only if the roles of
turbulent diffusion, radiative transfer, and (especially) cumulus convection
are simulated by parameterization. No one knows with certainty how to
describe the role of these often shorter-time-scale phenomena in terms of

variables occurring in the hurricane scale solution, or even if adequate description of this type is possible. Reviewing the parameterizations of cumulus convection currently used in numerical simulations of hurricanes, Ooyama (1971, p. 744) acknowledges that "... it is generally agreed that these methods are essentially stopgap measures which are used for want of better alternatives." Gray (1972b, p. 57, 60) writes: "... the authors (Gray and Lopez) feel that it is very difficult (or impossible) at the present time for any numerical model of tropical motion to come to grips with the real problems of incorporating the cumulus convection in terms of the broader scale flow. ... In that the cumulus cloud is such a distinctive physical unit, it would appear that it should be independently treated. ... The physics of the cumulus–broadscale interaction may not necessarily be overcome by applying the primitive equations to even smaller grids and time steps without additional insight into the character of the individual convective elements." While there are some physical concepts that guide currently employed parameterizations (such as weak integral constraints on energy and water substance), the parameterizations remain rather arbitrary and experimental (for a summary of them, see Haltiner, 1971; Bates, 1972). Often the parameterizations are freely adjusted until the cyclone scale results are consistent with measurements and/or intuition. The computer results are very sensitive to the assumptions made concerning the formulation of cumulus convection and eddy transfer (Ramage, 1971, p. 40; Haltiner, 1971, p. 262; Garstang, 1972, p. 619). Without any disparagement. such adjustment of parameterizations may be termed curve-fitting. What is learned from current computer experiments is moot. Of course, no model can today entirely escape such curve-fitting. The question to be pondered is how refined a solution is currently warranted in view of the limited physical understanding of certain critical phenomena? The validity of a parameterization of cumulus convection might be established by fixing all constants and functional forms, and by showing that such a parameterization successfully describes a wide variety of atmospheric phenomena; such success would furnish reason to scrutinize the incorporated physics carefully.

A further complication in evaluating numerical models is that sometimes the finite differencing, intentionally or inadvertently, adds to the stated formulation in the sense that effects not in the differential equations are present in the difference equations. The most accurate differencing of the differential equations is not necessarily preferred, but rather numerical techniques are rated "... on intuitive meteorological inspection of results" (Anthes et al., 1971, p. 747). For instance, errors introduced by less accurate upstream-difference schemes were retained to simulate lateral mixing when more accurate central-difference schemes gave less realistic results (Rosenthal, 1970, pp. 657–658). Later the differential formulation was modified so that an accurate differencing of the modified expression for lateral diffusion gave the numerical contribution desired. However, the modified expression "... is not very satisfying from a physical point of view" (Anthes *et al.*, 1971, p. 747). What can be learned from such procedures again seems moot.

Even if the parameterizations for cumulus convection, turbulent mixing, and radiative transfer were adequately known, and the finite-differencing were adequately accurate, the question of feasibility in seeking a uniformly valid solution for the entire cyclone would remain. Large gradients occur over relatively small scales in important subregions of the storm (e.g., the frictional boundary layer and the eyewall), while small gradients occur over relatively large scales in the bulk of the storm (the rapidly swirling regions and the outflow layer aloft). On current computers usually a more or less fixed grid of 10 km or 20 km radial resolution and at most thirteen layers of vertical resolution is adopted by practical considerations; the overwhelming bulk of the grid points then lie outside the eye, the eyewall, and the frictional boundary layer—where important processes are occurring . The domain sizes are quite limited (typically 440 km) so relatively weak boundary conditions (requiring only that purely advective influx occur at the side boundaries) are employed and much of the storm lies outside the domain of computation. Garstang (1972, p. 619) states numerical results are often very sensitive to the treatment of the outer limits. An alternative to direct numerical solution of the uniformly valid equations is to subdivide the storm into natural portions where different gradients and phenomena are operative. This is the approach, with provision for interfacial compatibility, used by Carrier and his coworkers.

Some evidence for uncertainty concerning current numerical models for the hurricane is provided by results for the frictional inflow layer. First, with his model, Rosenthal (1971b) performs numerical experiments in which the latent heat transfer from the sea is greatly augmented over ambient but the sensible heat transfer is set to zero. The goal is to show that the latent heat transfer is the important portion of the total enthalpy transfer. Of course, Rosenthal is well aware that the sensible heat transfer is required to permit the latent heat transfer (Riehl, 1954, p. 336). The computation is cited as evidence of how incompletely the full conservation equations are represented in current models. Thus the current computer programs cannot prove the physical validity of the Riehl–Malkus oceanic heat source postulate; the solutions which the current computer programs generate merely reflect the incorporation of the Riehl–Malkus postulate in their formulation.[10] Next,

---

[10] Black and Mallinger (1972, p. 64) characterize most models as giving a 50 % reduction in maximum wind speed for a 2°C drop in sea surface temperature. Doubt concerning such sensitive dependence of hurricane intensity on sea surface temperature was discussed in Section 3.8. Recently the existence of a closed oceanic convection cell induced by hurricane winds, already discussed, has been further confirmed by additional

the study by Anthes (1971) of the incompressible axisymmetric flow in a frictional layer under an intense Rankine-like vortex furnishes an opportunity to test how well the finite-difference techniques used in computer simulations of hurricanes succeed in a problem of reasonable complication to which the answer is apparently known. Interestingly, Anthes introduces an initial value problem as a convenient device for achieving the desired steady solution, without claiming validity for the transient phase necessarily computed. There are three points about the solution that will be noted. (1) The multilayer solution reveals that the one-layer treatments of Ooyama, Rosenthal, and Yamasaki underestimate the tangential and (particularly) the radial velocity components of the frictional layer, and give a premature eruption from the Ekman layer at incorrectly large radial distances from the center. (2) While both diffusion coefficients are held constant, Anthes takes the radial diffusion coefficient four orders of magnitude larger than the axial. Either by this means, or by large truncation errors through forward-differencing of advective terms, for plausible solution radial diffusion plays a role in regions of the flow where boundary layer theory suggests radial diffusion should be negligible. (3) The Ekman layer divergence changes from a weak downdraft to a large updraft about 125 km from the axis of symmetry. The spuriousness of this result, and the actual two-part structure of the frictional layer at distances from the axis at which nonlinear terms are significant [as revealed by analytic and more sophisticated numerical techniques (Carrier *et al.*, 1971; Carrier, 1971a; Burggraf *et al.*, 1971; Dergarabedian and Fendell, 1972b)], remain obscured (Section 3.5).

Nevertheless, more elaborate computer simulations are continually developed. Anthes (1972) has attempted to account for departures from azimuthal symmetry. The axis of the simulated hurricane vortex undergoes a curious anticyclonic rotation at approximately 8 mph about the center of the storm.

---

measurements on Typhoons Trix (which was situated over the central South China Sea on October 23–24, 1952) and Wilma (which was similarly situated on October 28–29, 1952), and on other storms (Ramage 1972). Ramage (1972, p. 491) notes: "When they struck the Philippines *Trix* and *Wilma* were ... equally intense typhoons. *Trix* reintensified over the South China Sea and *Wilma* probably did. Paired aircraft reconnaissances on the 23rd and 28th and the 24th and 29th ... show little intensity difference, in contrast to Ooyama's numerical hurricane model ... in which a smaller surface temperature difference is associated with a significantly large intensity difference." Furthermore, the path for Wilma was much less influenced by oceanic cooling owing to the prior passage of Trix than the work of Brand (1971) (previously cited in Section 1.5) would suggest, even though Brand considered typhoons generally more separated in time. This is one more item of evidence suggesting that ocean temperatures are not solely determining influences on hurricane behavior. Ramage (1972, p. 493) concludes: "The history of typhoon *Wilma* suggests that surface temperature may not always control storm intensity nor significantly affect storm movement, for *Wilma* moved along the cold water wake of *Trix* and remained intense until reaching Indochina."

Anthes suggests the path of a cyclone is not influenced by horizontal sea surface thermal gradients, although Brand (1971) claims observational evidence to the contrary. Newer computer models are evolving that (unlike most of the models previously discussed) do not isolate the storm from its surroundings, which may largely determine the hurricane path. However, these newer models often must employ a grosser grid in order to describe a larger region.

3.11. Implications of Hurricane Models on Seeding

The current mode of attempting artificial modification of tropical cyclones is to introduce silver iodide crystals in the supercooled water believed to exist in appreciable, naturally unfrozen quantities high in the eyewall, and perhaps also in rainbands and cumulus clouds both close to and far from the eyewall. In a typical field experiment, over two hundred rocket cannisters, each containing about 150 gm of silver iodide, are released at about 33,000 ft over a 30 km distance (entirely outside the eye), along a ray 90° clockwise to the direction of storm translation; such an experiment takes two to three minutes to execute.[11] In laboratory studies, a gram of silver iodide produces 10^{12} to 10^{14} crystals, which serve as ice nuclei and which grow to $3\,\mu$ diameter. The goal is that there will be 10^3 nuclei/liter of space over a depth of several kilometers a few minutes later; this concentration is reduced to negligible levels in a couple of hours (Gentry, 1971a; Penner, 1972). The silver iodide will hopefully cause the supercooled water droplets to freeze, and thus to release the heat of fusion (Battan, 1969).

The proposed mechanism by which this heat release (and supposed attendant temperature rise of a couple of degrees Celsius, and density decrease) is efficacious has been altered on several occasions (Simpson and Malkus, 1964; Rosenthal, 1971a; Gentry and Hawkins, 1971), and at times several alternative possible mechanisms have been set forth (Gentry, 1971a,b). A concise history of the various suggestions has been given by Riehl (1972a, pp. 267–

[11] Typically five runs of the kind described, at two-hour intervals, starting from the radius of maximum winds and proceeding radially away from the center, characterize the so-called eyemod experiment. Most of the discussion is oriented toward this type of experiment. However, it may be noted that the seeding of Hurricane Ginger (1971) discussed below was a so-called rainsector experiment: all the water-containing clouds in a 45° sector from 50 to 100 n miles from the center are seeded at about 22,000 ft to divert low-level inflow to premature eruption and high-level outflow outside the eyewall, with resultant dispersal of latent heat energy over a wider region of the storm (Hawkins *et al.*, 1972). In concept, a rainsector experiment consists of four fifty-minute seedings, at fifty-minute intervals. The reason a rainsector, rather than eyemod, experiment was attempted on Ginger was that it was a large diffuse storm (moderate gale winds out to 250 n miles, circulation out to 400 n miles) with only an ill-defined eyewall (maximum winds of about 60–70 knots at 40 to 60 n miles from the center).

270). One accepted point is that seeding in the nascent eye of a developing tropical storm should be avoided since this procedure would probably abet intensification (Rosenthal, 1971a). The most recent proposal has been to seed cumuli *outside* the eyewall and above the 0°C isotherm (roughly, 600 mb or 15,000 ft) with silver iodide. (Of course, it is more difficult to find seedable clouds, and there is a greater expanse to cover, outside the eyewall.) The aim is not to capitalize directly upon the heat of fusion released by supercooled water droplet freezing, but rather to cause these cumuli outside the eyewall to grow several thousand feet in height from midtropospheric to upper tropospheric levels. This would hopefully initiate a chain of events from which an order of magnitude more heat than that associated with freezing will indirectly be realized. This added heat would come from some low-level inflowing warm moist air erupting not up the eyewall, but prematurely up the growing cumulus, with associated condensation and additional glaciation. If entirely successful, proponents have suggested that perhaps a substitute eyewall further from the center, and hence (from conservation of angular momentum) with reduced swirling speeds, could be realized from this premature eruption of boundary layer air (Gentry and Hawkins, 1971; Rosenthal, 1971c).

Fukuta (1972) suggests that the field experiments conducted to date have employed overseeding of supercooled water droplets to form small ice particles; the energy release would be small and the effect over in less than half an hour. Instead, Fukuta suggests reducing the seeding material to 1 % or even 0.1 % of that used in past field experiments, to form large precipitable ice particles. The latent heat of fusion these falling particles could extract from the inflow-updraft stream would supposedly raise the central pressure for hours, reduce swirling, and induce spreading out of the storm. A detailed analysis fully substantiating either the concepts of Gentry and Hawkins or the concepts of Fukuta has not appeared.

Clearly, like numerical simulation of hurricanes, the seeding of hurricanes warrants an entire review unto itself. In fact, with the proposed extension of the U.S. seeding effort [Project Stormfury, a joint effort of the Department of Commerce (NOAA) and the Department of Defense (Navy)], from the western North Atlantic to the western North Pacific Oceans (Mallinger, 1971), this might be an apt time for such a review.

Attention will be confined here to a few observations, especially to what the hurricane models suggest about the seeding procedure.

Even proponents of seeding anticipate only a 10 to 15 % decrease in maximum winds. In the one case in which larger decreases were noted, the anomaly is now attributed to synoptic peculiarities relating to upper level outflow (Hawkins, 1971). This is not to disparage such decreases; statistical treatment of a model suggests multimillion dollar annual savings in damage from hurricane seedings that would reduce peak winds by fifteen percent (Boyd et al., 1971). The problem is that the storm is naturally oscillating in intensity by the same order of magnitude, so it is very difficult to distinguish natural and artificially induced changes. In fact, the gradual increase of the maximum winds of Hurricane Ginger by 15 % during the day after initial seeding on September 26, 1971 was ascribed to natural forces by Project Stormfury personnel, as was the 11 % decrease after the second seeding on September 28 (Lieb, 1972). Fujita (1972) could find no evidence that seeding altered Ginger; since the central pressure of this weak hurricane was only 980 mb at the times of seeding, he suggests that perhaps the cloud-top heights were too low for the silver iodide to be effective. The threat of litigation has constrained the number of seeding experiments severely.

The National Hurricane Research Laboratory (Gentry, 1969a,b, 1970) has discussed a six to twelve-hour cycle of amelioration after seeding; a physical basis for this time scale has yet to come forth (see below). Further, if the central pressure deficit is reduced as reported, eyewall seeding must alter the eye in a still unidentified manner, although Black et al. (1972) suggest that one hour after seeding outside the eye, the eye may expand more than is consistent with natural variability. More complete post-seeding probing of the tropical cyclone would be helpful in evaluating these claims.

Rather similar computer models have produced different guidance with regard to current seeding practice. The reasons for the discrepancy are not fully available because some details remain unpublished. Rosenthal (1971c) notes that both the magnitude and the duration of the heating taken to simulate seeding in his computer experiments seem excessively large. Nevertheless, in those numerical experiments believed to most closely simulate the field experiments performed on Hurricane Debby, August 18 and 20, 1969, Rosenthal (1971a) predicts an increase after seeding in the maximum wind, and over wide radial extents an augmentation in wind level, at 700 mb; the comparable measurements at 12,000 ft indicated a reduction in winds (Hawkins, 1971). Rosenthal predicts a decrease in the maximum wind at sea level after seeding (though in substantial portions of the hurricane, surface winds are predicted to increase); no post-seeding measurements were made at sea level. Quantitative detail is omitted here because Rosenthal (1971a, p. 415) notes: "... *at best, the results should be considered qualitative guidance material.*" Sundqvist (1972) finds his numerical model predicts an increase in maximum winds from silver iodide seeding, which serves to release more effectively the latent heat in clouds below 0°C; the increase is about 10 % if seeding is carried out in the region of intense convection. Within 20 hr of cessation of seeding, the storm is predicted to return to its preseeding state; seeding does not initiate a cascade of effects that permanently alters the storm. Estoque (1971, p. 4), noting the tentative nature of his conclusions, remarks:

We have [studied] the properties of a hurricane model which is suitable for simulating the effects of artificial seeding. The main improvement of this model over previous ones is the explicit inclusion of cloud micro-physical processes, including the prediction of the liquid water distribution. Thus, the model is able to simulate more realistically the release of the latent heat of fusion due to artificial seeding. The model has been used to simulate artificial seeding over three different radial locations in the vicinity of the maximum in the surface wind. The expected reduction and outward displacement of the maximum radial surface pressure gradient did not materialize. Instead, the pressure gradient and also the wind increased in intensity. However, the seeded storm intensity decreased after the seeding stopped. The greatest effect occurs when seeding is done over a region just inside the surface wind maximum location.

An interesting feature of the unseeded hurricane is the occurrence of periodicity in time of the hurricane intensity. The magnitude and the period of this periodicity is about 10 m sec^{-1} and 8 hours, respectively. This magnitude is about the same order as that of changes induced artificially by seeding. If this periodicity is real, it should be taken into account in the interpretation of seeding experiments in the actual atmosphere.

The effectiveness of current seeding practices remains unresolved. If the Carrier model is valid, silver iodide seeding can only transiently upset the stable hurricane configuration. Under this model, warm-fog dispersal methods would have to be applied to the entire "throughput supply" layer of warm moist air to achieve the significant goal of premature rainout in outer spiral bands. The layer of warm moist air is so spatially extensive that such attempts seem somewhat impractical. Dergarabedian and Fendell (1971) have discussed the use of warm-fog dispersal methods for lower levels in a hurricane for premature rainout of the throughput flux; almost simultaneously, so did Matthews (1971, p. H-8), albeit more optimistically:

> The feasibility of warm cloud modification prior to cold cloud modification should be examined because warm cloud modification may permit growth of small warm clouds to temperatures at which cold cloud modification will be effective. The combined use of warm cloud and cold cloud modification techniques would permit selective seeding in all regions without cloud top temperature restrictions.

Seeding to divert a path seemingly holds little better promise than seeding to alleviate intensity, since there appears to be no way to discern what path alterations were due to human intervention under current understanding.[12]

---

[12] Dr. R. Cecil Gentry, director of the National Hurricane Research Laboratory and of Project Stormfury, stated recently that budgetary reductions had grounded the four aircraft used in cloud-seeding experiments, and probably no full-scale attempts at hurricane modification would be undertaken in the next three years (Anonymous, 1973). If and when seeding of hurricanes is resumed around 1976 perhaps activities will be extended to the North Pacific where more storms suitable for seeding occur annually.

Gray (1973) is also pessimistic about the effectiveness of silver iodide seeding to modify hurricanes artificially, especially since such seeding can be directly effective only as long as the lifetime of the cumulus cloud. Others have proposed the alternative of spreading evaporation retardants on tropical oceans to weaken incipient tropical cyclones (Gentry, 1969a), or even to preclude hurricanes altogether; casting aside questions about the survivability of such thin layers in the ocean and about pollution, one must wonder whether or not fewer, but more violent, hurricanes would result, since the storms may well play an essential (often, humanly beneficial) role in global energy and water substance transport. Since the damage from a hurricane in the United States tends to increase faster than the fourth power of the maximum sustained surface wind speed (Howard et al., 1972), and since (by extrapolation from data on typhoons in the northwestern Pacific) twenty percent of hurricanes have wind speeds in excess of 120 mph, a thirty percent reduction of maximum hurricane wind speed remains a very desirable goal. Gray (1973) believes a more safely executed and more effective (if more expensive) means of attaining such a goal than by silver iodide seeding is by dispersal of two to three million pounds of $0.1\ \mu$ sized carbon particles in the surface frictional layer at radial distances outside the outflow level cirrus shield (which extends out generally about 200 miles from the center for moderately severe storms) but inside the surrounding anticyclone circulations (which typically border the outer edge of the hurricane on the poleward side while in the trades). The carbon black dust (Downie, 1960; Frank, 1973) is envisioned to cause $0.5°–1°C$ heating per hour for ten hours over an area of 40,000–80,000 km^2 when dispersed to a 10 % areal coverage, because it absorbs 15 % of the solar radiation daily incident on the tropical surface and then rapidly transmits the heat by conduction to adjacent air flowing at low altitudes into the hurricane. Ideally, the resulting $5°–10°C$ rise in air temperature increases buoyancy in the frictional boundary layer, and the consequent convection leads to a dry downdraft which induces an increase in moisture flux from the ocean to the surface inflow layer. The result is hopefully that about ten percent of the mass inflow (which would reach the base of the eyewall from the edge of the cirrus shield in a day and a half or so) prematurely erupts, and the maximum wind speeds in the eyewall are significantly reduced on the time scale of about one day. Thus, the ultimate effect sought is that envisioned by Gentry and Hawkins (1971), but the means employed differs. Rosenthal's numerical model (Rosenthal, 1970, 1971a,b,c) predicts such a carbon dust seeding strategy will succeed in lowering the maximum wind speed, although rainfall and wind levels in the outer portion of the storm will increase. Three possible difficulties, besides expense of the operation, are dispersal problems owing to clumping of the carbon particles (which might be difficult to counter cheaply and without performance degradation); anticipating the storm path (so the particles remain outside the cirrus shield for ten hours but then flow into the storm, especially into the particularly severe right-hand semicircle with respect to an observor looking along the path); and being certain midtropospheric moisture levels at surface pressures of 995–1005 mb (where the cirrus shield often ends) can sustain tall cumulonimbi towers (otherwise, according to the Carrier model, the energy-enriched erupting air eventually returns to the surface layer and proceeds inward to the eyewall, perhaps to *intensify* the storm).

4. Theory of Tropical Cyclone Intensification

There exists no satisfactory theory explaining how, annually, a small number of tropical disturbances intensify to become hurricanes, and why in contrast most disturbances do not. Charney (1971) notes that, unlike in the midlatitudes, tropical temperature gradients are weak, and tropical cyclones most often form over the tropical oceans where the surface temperatures are particularly uniform. No theory of how pronounced baroclinicity develops in the barotropic tropics is likely to emerge until the detailed local balances and large-scale circulations of the tropical ambient are better understood quantitatively, because early stages of cyclogenesis will probably involve perturbation about that ambient.

Carrier (1971b) has presented an intensification process by which his quasi-steady mature model evolves in time from a tropical depression. Tracking back to even earlier evolution seems premature at the current state of understanding. Although Carrier's theory has been reproduced by Penner (1972) without objection, Carrier's theory of intensification is not currently entirely satisfactory. Thus, it is used here only as an interesting vehicle to raise some significant points about the briefer, later stage of intensification, as opposed to the longer, earlier stage. Ooyama (1969, pp. 35–37), among many others, has written about an initial very modest rate of pressure fall in intensifying tropical cyclones, followed by an appreciably accelerated rate of fall—it is this latter stage which is principally addressed here. [A particularly dramatic example of such rapid central sea level pressure fall after a slower decline in an earlier stage is furnished by Typhoon Irma, which moved northwestward in the central Philippine Sea in early November 1971 (Holliday, 1973). On November 8 Irma passed from a tropical depression to a storm; during the next two and one half days the central pressure fell about 20 mb. However, in the succeeding 24.5 hr the pressure fell from 981 to 884 mb, a minimum pressure exceeded by only two storms on record. During its peak intensity stage, the typhoon exhibited the occasionally observed double-eyewall structure, one ring of dense tall cloudiness occurring at a radius of only 3.5 n miles and the second concentric ring at a radius of 20 n miles.]

4.1. Carrier's Outline of Intensification

Carrier begins with a schematic view of an axisymmetric tropical depression in which there is a weak Rankine-vortex-like swirling, the maximum azimuthal speed lying relatively far from the axis of rotation. There is radial inflow in region I during the transient phase (Fig. 19). The swirling would quickly establish a shear layer beneath it; there is weak upflux out of (downflux into) the surface layer where the relative circulation increases (decreases) with radial distance. Since linear theory correctly predicts the *downflux* in the mature stage with rapid swirl, linear theory probably suffices to predict the boundary layer divergence during intensification. However, since *equilibration of the boundary layer requires times of* $O(\Omega^{-1})$, or roughly sixteen hours in the tropics, and since this is possibly an appreciable fraction of the later-stage, rapid-intensification time scale, a transient linear theory will be required. It must yield a radial influx through the boundary layer II in excess of the radial inflow speed in I. (Incidentally, the times characterizing signficant change in the thermodynamic state of the lower tropical ambient are probably on the order of 100 hr, a time span partly related to the typical magnitude of eddy viscosity appropriate for the tropical atmosphere. Hence, unless the intensification time from depression to cyclone greatly exceeds 100 hr— and this seems dubious—the thermodynamic state of the ambient may possibly be justifiably taken as fixed throughout the later stages of intensification. This statement by no means implies that the ambient can be taken as fixed throughout the entire intensification process. The very fact that descent speeds in the outer portions of even a mature hurricane are roughly comparable to ambient tropical descent speeds suggests that the tropical ambient probably cannot be held fixed throughout intensification.)

The air erupting from the boundary layer begins to displace the air initially in the central core of the developing storm. The air initially present in the core is of slightly lower pressure than the ambient air at the edge of the storm, but not vastly different in vertical stratification. The air erupting from the boundary layer under the Rankine-vortex-like swirl displaces the air initially present in the core vertically upward; since there is a "lid" on top of the storm, the vertically displaced initial air is, near the top of the core, squeezed radially outward.

The following *competition* develops. The new air rising out of the boundary layer is drawn entirely from relatively warm moist air near the bottom of the atmosphere. Thus, displacing the air initially present in the core is relatively light air. On the other hand, the *convective motion of new air* is small, especially at early times, and the *ambient processes* (*turbulent diffusion, radiational cooling, cumulus convection*) try to maintain the original, near-ambient stratification in the core. If the convective displacement wins out, then the core becomes lighter and lighter, relative to a column of air at the outer edge of the storm. This paragraph contains those points of Carrier's intensification model that seem most unsettling, and will be subject to close scrutiny in the critique below.

However, Carrier's basic proposition is that the swirling in I has led to a downflux into II, a spiraling inward in the boundary layer and an upflux into the core, and a lightening of the core by hydrostatic considerations. For dynamic consistency, the centrifugal force (anticipated to be the dominant inertial effect) must increase to balance the augmented radial pressure gradient. Since angular momentum is conserved in I, where friction is negligible, the fluid particles must necessarily move in closer to the axis of symmetry (axis of rotation). The result is that in time, in the Rankine-vortex-like swirl distribution, the maximum azimuthal speed increases in magnitude and the position of the maximum lies closer to the axis (Figs. 19 and 20). Hence, the more the pressure falls in the core, the more fluid sinks into the boundary layer to spiral inward, erupt upward, and cause further pressure reduction in the core. If the crucial early competition is resolved in favor of the organized convection, ultimately the particles erupting out of the boundary layer rise so quickly that they lie on a moist adiabat, and the greatly lightened core is entirely flushed of its original fluid.

While most descriptions (Palmén and Newton, 1969) tend to picture the eye as formed gradually as the pressure deficit develops, in the Carrier model the central core may well be competely flushed of ambient-like air, so that the air in the core lies on a moist adiabat, before much trace of an eye is to be found. At first, a Rankine-vortex-like swirl holds everywhere. From this fully developed one-cell structure with at most only the rudiments of an eye, a well-defined two-cell structure with a calm center region emerges rapidly, probably in much less than an hour, owing to inertial oscillation, in the following way.

As the pressure falls in the core relative to the ambient, the particles in I necessarily move in closer to the axis to permit a compensating centrifugal force to develop. Once the core is flushed and moist adiabatic ascent characterizes the full height of the core, no further pressure deficit can be generated. By inertia, the spinning particles continue to move in, a dynamic imbalance is created, and a radial acceleration develops to force the particles away from the axis of symmetry. This reverse motion creates a rarefaction at the center, and relatively dry warm motionless air sinks down the axial column to form an eye. This air may be air from above the storm or rained-out, slowly swirling air entrained out of the top of the moist adiabatic column (Fig. 21). Because there is no appreciable swirl (hence no associated pressure gradient) in the eye, there is no frictional boundary layer under the eye. The moist adiabatic column becomes an annulus displaced from the axis, i.e., the eyewall; the inertial oscillations of the eyewall eventually damp in time.

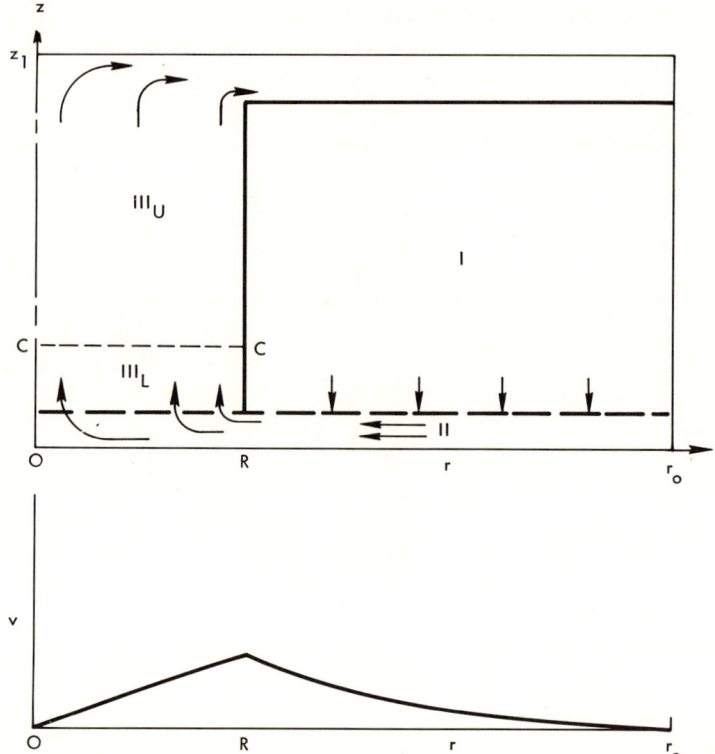

Fig. 19. In this schematic view by Carrier of the flow configuration and circumferential velocity distribution in an intensifying tropical depression at some early time $t = t_1$ (say), the interface C–C between the new and initial air in the core is idealized as horizontal for convenience. However, it is easy to show that for all but very modest circulations (i.e. except for angular velocities less than three times the ambient), a rigidly rotating core (which implies uniform updraft velocity at the top of the nonlinear frictional layer) could not accept all the fluid pumped through the Ekman boundary layer. Thus a uniform ascent in the core is dynamically impossible, and in fact ascent in the eyewall is concentrated in cumulonimbi. For the gross balances being discussed, many points can be made without accounting for such refinements in eyewall structure, though ultimately details of the structure are crucial. (From Carrier, 1971b, p. 146.)

4.2. Critique of the Carrier Model of Intensification

Carrier has offered more than the conventional linearized stability analysis to determine what scales of motion grow in what time interval (Yanai, 1964; Ogura, 1964; Yamasaki, 1969). [Bates (1970) did study the growth to finite amplitude in the ITCZ of nonzonally symmetric equatorial wave disturbances of 200 km longitudinal scale, which are observed to move westward at 13 knots. The time scale for early growth owing to barotropic

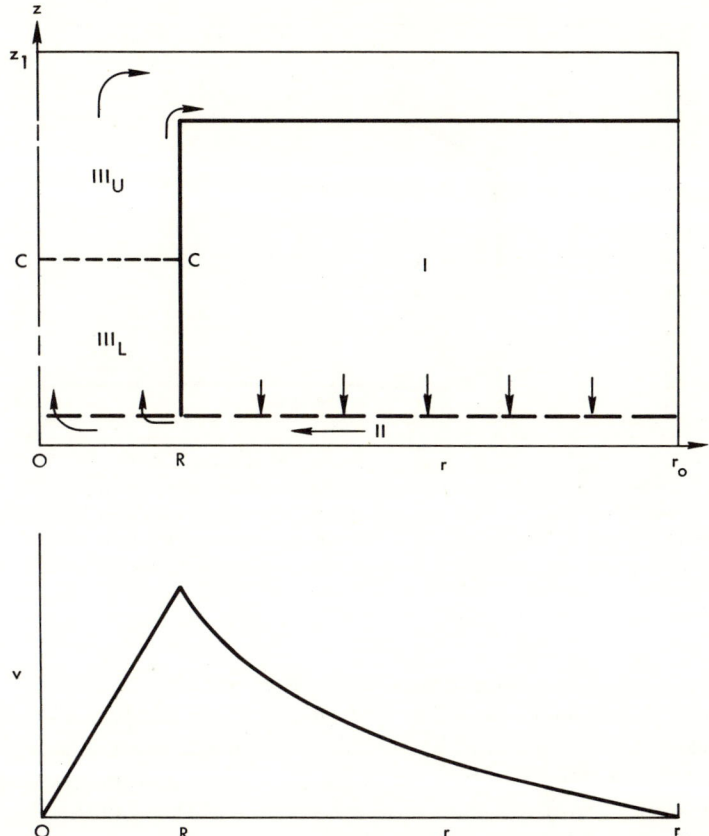

Fig. 20. Intensification from tropical depression to hurricane has progressed to a more advanced stage in this schematic diagram, holding at $t = t_2 > t_1$, according to Carrier's model. The magnitude of the maximum swirl is increased and its position lies closer to the axis of symmetry. (From Carrier, 1971b, p. 146.)

instability was about two days; frictional dissipation eventually curtails further growth effected by conversion of condensational heat to eddy available potential energy. Because Charney's concept that condensational heating in the tropics is dependent on convergence of moisture in the planetary boundary layer was incorporated, the disturbance had a warm core with maximum intensity in the lower atmosphere. Holton (1972) furnishes further lucid insight into equatorial wave disturbances, and in fact concisely discusses many other topics arising in tropical meteorology.] Carrier has also emphasized the role of lightening of the core rather than (say) large-scale horizontal turbulent diffusion. However, there are problems concerning the nature of the convective lifting in the core, as now discussed.

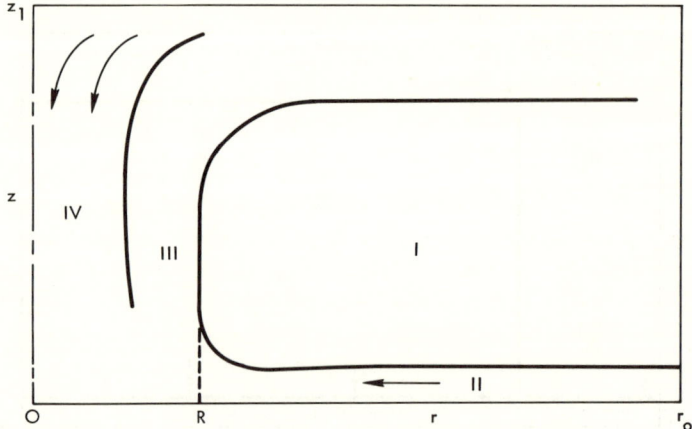

Fig. 21. Schematic picture by Carrier of the flow configuration which prevails when the radius of maximum swirl R is increasing and an incompletely formed eye is being filled with relatively dry and motionless air, which sinks down from the top of the storm under dry adiabatic compression. With the formation of an eye lighter in weight than the eyewall, the terminal stages of intensification and the beginnings of quasi-steady mature stage structure are realized. (From Carrier, 1971b, p. 150.)

The ambient distribution of total stagnation enthalpy is such that when a column stratified like the ambient is lifted as a whole adiabatically, warm air is squeezed radially outward at the top, even if warm air is added at the bottom (Fig. 7). The net effect for small convective motion, according to a method of characteristics numerical integration of the resulting hyperbolic boundary value problem, is *not* a lightening of the developing core (Dergarabedian and Fendell, 1972a, pp. 62–67).[13]

A closely related observation is to note that Carrier (1971b), in his discussion of intensification, seemed more concerned with how a hurricane is formed than whether it is formed. Hence, he adopted effectively the following proportionality:

$$\frac{\text{pressure fall from ambient}}{\text{total pressure fall for a moist adiabatic core}} = \frac{\text{new air into core}}{\text{total air required to flush core}}$$

This statement is true at the outset and, if a hurricane forms, at the end, of flushing the core. In fact, however, such a simplification lets a hurricane always be formed. In truth, the pressure fall for small times is less than that predicted by the fractional-volume proportionality. As the quasi-steady

---

[13] A calculation illustrating that undiluted lifting of the autumnal Caribbean ambient produces reduced temperatures at all tropospheric levels was also carried out by Riehl (1963b).

mature state approaches, the initially slow pressure fall probably accelerates to a rapid decrease, and the thermodynamic stratification of the emerging eyewall asymptotically approaches a near moist adiabat state.

The problem cannot be satisfactorily buried in parameterizations of turbulent transfer, radiative cooling, and cumulus convection in the core. What appears needed is explicit recognition that a broad uniform ascent in the core does not account for the actual mechanism of convection, which would seem to entail ascent in a concentrated number of intensified cumulonimbi. The somewhat tentative nature of this remark stems from statements like that by Palmén and Newton (1969, p. 490): "It seems clear from . . . observations and from descriptions of turbulent encounters that, in some cyclones, ascent in the inner region is dominated by cumulonimbus, . . . while in others a broad-scale and more uniform ascent takes place."

If cumulonimbi do play an essential role in the core ascent during intensification, some semblance of an eye would probably form at an earlier stage of hurricane development than Carrier suggests.[14] The eye would probably not be entirely created on a time scale of 20 min at the end of intensification, as Carrier suggests, although inertial oscillations at the end of intensification might rapidly delineate a previously ill-defined eye.

The tritium-tracing analyses by Östlund (1968) of Hurricane Hilda (1964) and Hurricane Betsy (1965) suggest that while the eye may have partially consisted of air entrained from the eyewall, there had been substantial subsidence of stratospheric air down into the eye. Thus, Carrier seems correct in citing two possible sources for the air in the eye.

4.3. The Distribution of Cumulonimbi during Intensification

Riehl (1972a, p. 249) notes: "Very fine weather, with hardly a cloud in the sky, often precedes the arrival of a hurricane by 1 day, when the storm is moving westward in the tropics. In the past this deceptively beautiful weather has been the cause of many disasters with heavy loss of human life in the Caribbean and elsewhere."[15] In a possibly related observation, Lopez (1972b) calls attention to the work of Oliver and Anderson (1969), in which satellite photographs are used to indicate that the majority of hurricanes form around a circulation center in the clear area ahead of (rather than directly under) a

---

[14] Consistent with his conception that it is the compressional heating of the compensatory downdraft that warms the air, because heat directly released by cumulus convection is almost entirely expended in raising the buoyant column, Gray (Shea, 1972, pp. 118–122) suggests that turbulent diffusion and mixing convey heat from the eye to warm the eyewall.

[15] From "Introduction to the Atmosphere" by Herbert Riehl. Copyright 1972 by McGraw-Hill Book Company. Used with permission.

convective region. The concentration of cloudy convective regions amid clear calm regions in tropical cyclogenesis may be related to characteristics of subsidence that, by continuity, must accompany convection.

However, perhaps also there is a concentrating of preexisting dispersed cumulonimbi that is a concomitant of large-scale tropical cyclogenesis. This admittedly speculative subject is included because it may be worth pursuing. Furthermore, Malkus *et al.* (1961) observed that for the inner rain area of radius 200 n miles in Hurricane Daisy (1958), in the prehurricane stage there were 60 cumulonimbi that covered one percent of the area; during intensification two and one-half percent of this area was covered by cumulonimbi. In the mature stage there was a persistence of cumulonimbus patterns relative to the center, and there were 200 cumulonimbi that covered four percent of the area. By studying measurements in Hurricanes Carrie (1957) and Cleo (1957), as well as Daisy, Gentry (1964) estimated five percent coverage of the inner eighty nautical mile radius area by cumulonimbi (total cloud coverage in this area was under 25%). Gentry adds that both aircraft penetration and radar returns suggest that the area of strong convective activity is larger in more intense hurricanes and is located closer to the center.

It might also be speculated [in contrast to conjectures by Gray (1972b, p. 64) on conditions for tropical cyclogenesis] that for a synoptic scale pressure difference to develop not all the air ascending in the cumulonimbi should locally subside, but some of the descent should be deferred to a larger horizontal scale.

4.4. *The Time-Dependent Flowfield during Intensification*

Although the energetics of the core may need revision, Carrier's modeling of the dynamics in the outer portions during the transient phase should remain substantially unchanged. Thus, this portion is developed here in some detail.

For axisymmetric intensification, a set of uniformly valid equations for the dynamics is now given in cylindrical polar coordinates. If subscripts r, z, and t denote partial differentiation, for a noninertial coordinate system rotating at the speed of the normal component of the rotation of the earth (i.e. normal to a plane locally tangent to the earth at about 15° latitude so $\Omega \doteq 0.06$ rad hr^{-1}):

(44) $$u_t + uu_r + wu_z - 2\Omega v - (v^2/r) + \rho^{-1}p_r = \nu u_{zz}$$

(45) $$(rv)_t + u(rv)_r + w(rv)_z + 2\Omega ru = \nu(rv)_{zz}$$

(46) $$p_z + \rho g = 0$$

and

(47) $$(ru)_r + rw_z = 0$$

These equations hold in $0 \leq r \leq r_0$, $0 \leq z \leq z_1$ where, again, z_1 is the top of the storm (a slippery lid) and r_0 is the outer edge. The boundary-initial conditions of relevance here are

(48) $\qquad r = r_0: u = v = 0; \qquad z = 0: u = v = w = 0$

(49) $\qquad t = t_0: u, v$ small

There is angular momentum in the initial flowfield in an inertial frame of reference, and this is conserved with the anticipated radial inflow toward the axis of symmetry; the picture of the flow is that given in Fig. 19.

For convenience the following notation is adopted. The radial, azimuthal, and axial velocity components will be denoted by capitals (U, V, W) in I and by lower case letters (u, v, w) in the frictional boundary layer II.

From results given earlier for the mature stage boundary layer dynamics, and from other considerations developed below for the transient boundary layer, it is anticipated that $\Psi = rV$ is a linear function of r^2, and independent of z; then there is a downflux into the frictional boundary layer II from the supply region I which is independent of r to lowest order. It follows from (45) and (47), together with (48) and (49):

(50) $\qquad \Psi = rV(r, t) = \Omega[R_0^2 - R^2(t)](r_0^2 - r^2)/(r_0^2 - R^2)$

(51) $\qquad \Phi = rU(r, t) = R\dot{R}(r_0^2 - r^2)/(r_0^2 - R^2)$

and

(52) $\qquad W(r, z, t) = \{2R\dot{R}z/[r_0^2 - R^2(t)]\}$

where $R(t_0) = R_0$. The associated pressure field from (44) and (46) is discussed later. This solution holds in I, i.e. in $R \leq r \leq r_0$. At $t = t_0$, in I, i.e. in $R_0 \leq r \leq r_0$, $\Psi = 0$. The particles at $r = R_0$ at $t = t_0$, at a later time $t > t_0$ lie at $r = R < R_0$. Typically, $r_0 = O(1000 \text{ miles})$, $R_0 = O(200 \text{ miles})$, and $(R_0/12) \leq R(t) \leq R_0$.

The next step is to develop the boundary layer in II that lies under the swirling flow in I given by (50)–(52). The equations for the viscous layer are taken to be the linear transient parabolic equations appropriate for low Rossby number flow. While the inequality $(\Phi^2 + \Psi^2) \ll \Omega^2 R^4$ may not formally hold everywhere in the flowfield, it is widely valid and, in addition, from experience with the mature quasi-steady storm, the key output, the boundary layer divergence, is adequately given by linearized theory. Hence,

(53) $\qquad \nu \phi_{zz} - \phi_t = -2\Omega\psi$

(54) $\qquad \nu \psi_{zz} - \psi_t = 2\Omega\phi$

where

(55)
$$\phi = r[u(r, z, t) - U(r, t)]$$
$$\psi = r[v(r, z, t) - V(r, t)]$$
$$w = W(r) + w'(r, z, t)$$

If $\chi = \phi + i\psi$,

(56) $$\nu\chi_{zz} - \chi_t = 2\Omega i\chi$$

(57) $$\Phi_r + \phi_r + rw_z' = 0$$

subject to

(58) $\chi(r, z = 0, t) \to -(\Phi + i\Psi), \qquad \chi(r, z \to \infty, t) \to 0, \qquad \chi(r, z, 0) = 0$

If $(r_0^2 - R^2) \to r_0^2$, an excellent approximation, then it is convenient to define

(59)
$$F(r, t) \equiv F_1(t)[1 - (r^2/r_0^2)]$$
$$\equiv [R_0^2 - R^2(t)][1 - (r^2/r_0^2)].$$

Carrier (1971a) studied the special form $F_1(t) = R_0^2 \exp(\alpha t)$. For the more general form (59), by Laplace transformation (Dergarabedian and Fendell, 1972a, pp. 56–57), the relevant boundary layer divergence is

(60) $w(r, z \to \infty, t)$
$$= (\nu^{1/2}/r_0^2)\mathrm{Re}(1/2\pi i) \int_{c-i\infty}^{c+i\infty} F_1(s)[(s - 2\Omega i)/(s + 2\Omega i)^{1/2}] [\exp(st)] \, ds$$
$$\doteq (\nu/\Omega)^{1/2}(1/r_0^2)[(6/7)\partial/\partial t - \Omega](R_0^2 - R^2)$$

for the $R(t)$ of practical interest. The divergence, a function of time only, is largest in magnitude in the steady state; if $R \to (R_0/12)$ typically,

$$w(r, z \to \infty) \doteq -(R_0/r_0)^2(\nu\Omega)^{1/2} = O(10^{-3} \text{ mph})$$

The net volumetric flux down into II, and hence (in an incompressible model) up into the core, $Q(t)$, is given by

(61) $$Q = Q_0[(6/7\Omega) \partial/\partial t - 1][1 - (R/R_0)^2]$$

where the quasi-steady discharge into the base of the core III is given by

(62) $$Q_0 = \pi(r_0^2 - R^2)(R_0/r_0)^2(\nu\Omega)^{1/2}$$

The previously discussed requirement for compatibility between (1) the nonlinear spin-up of the inviscid flow in I under conservation of angular momentum and (2) the pressure deficit generated between a column in the core and a column in the ambient owing to partial flushing of the core with boundary layer air driven radially inward by the outer inviscid flow, is now made quantitative. It is necessary to model the radial and azimuthal velocity components in the core; kinematic and continuity considerations suggest that, at least in the lower regions of III, for this purpose it is adequate to take

(63) $$ru = \phi = r^2 \dot{R}/R$$
(64) $$rv = \psi = \Omega(R_0^2 - R^2)r^2/R^2$$

Accordingly (44) is integrated from $r = 0$ to $r = r_0$ at some fixed z above the boundary layer; the profiles for u and v are given by (63) and (64) for $0 \leq r \leq R$ and by (50) and (51) for $R \leq r \leq r_0$. The term uu_r in (44) is a perfect differential and yields precisely zero; wu_z yields zero because u is independent of z and w is small anyway. It follows that the time-dependent, Coriolis, centrifugal, and pressure gradient terms yield, for $\rho \doteq$ const. and upon neglecting terms of $O(R^2/r_0^2)$ relative to those of order unity,

(65) $$p(r_0, z = 0, t) - p(r = 0, z = 0, t)$$
$$= [\rho \Omega^2 R_0^2 (R_0^2 - R^2)/R^2]\{1 + [2 \ln(r_0/R) - 1]R^2/R_0^2\}$$
$$+ \rho(d/dt)[R\dot{R} \ln(R/r_0)]$$

The last term on the right-hand side is characterized by the inverse square of the time for intensification; this time is probably much larger than the time for equilibration of the boundary layer. Thus normally the time-derivative term can be neglected.

However, when inertial oscillations associated with the continued radial influx and rebound of air in I occur, a shorter time scale enters and the time derivative term must be retained. Carrier (1971b, p. 157) suggests that the time scale of the oscillation is $O(4R_f^2/\Omega R_0^2)$ where R_f is the final equilibrium position of the eyewall $R(t)$. For typical values, with $R_0 \doteq 250$ miles, this time scale is 20 min. In fact, holding the convection-core pressure deficit from the ambient fixed at about 35 mb, holding ρ at its ambient sea level value, letting $R_0 \doteq 250$ miles and $r_0 \doteq 1000$ miles, one finds from (65)—by dropping the temporal derivative—that $R = R_f = O(35$ miles$)$. Restoring the temporal term, adopting as initial conditions $R = O(35$ miles$)$ and $\dot{R} \doteq -\Omega(R_0^2 - R^2)/2R$ (Carrier, 1971b, pp. 152 and 155), and holding the core pressure deficit relative to the ambient fixed at 35 mb (near quasi-steady conditions), one finds from (65) that R decreases to about 16 miles in about a half-hour before $\dot{R} = 0$. Thus, numerical values confirm the inertial oscillation period suggested by Carrier. Such oscillations probably reinforce the eye, rather than create it, as noted earlier (Section 4.2). The oscillation of the eyewall is dwelled upon because it seems to be the first proposal of a mechanism *internal* to the hurricane that might explain a crudely periodic variability in intensity of 10 to 20 % in peak winds, on the time scale of a few hours; such variability appears to be observed sometimes, and attribution to changes in the ambient atmospheric and underlying oceanographic states may not be always satisfactory.

The increased low-level radial inflow toward the center of the developing cyclone seems consistent with the previously discussed reorganization of the statistically steady cumulonimbus activity which helps maintain the ambient enthalpy transport process. Such radial inflow implies a stronger circulation (or vorticity), and this implies a decreased mixing of updraft air with surrounding air; the result would be a decreased density in the updraft region and hence a decreased pressure. As cumulonimbus activity increases in the central region of the incipient cyclone, there is probably diminution of such activity in the surrounding region. However, the downdraft in the surrounding region should differ little from the ambient since the total low-level inflow into the whole region will not have changed appreciably. Thus, the anticipation is that the flowfield presented in this section probably remains substantially correct, within the axisymmetric approximation.

5. Concluding Remarks

Several viewpoints not universally accepted have been adopted in this review. The existence of, and necessity for, a greatly augmented oceanic heat source in the inner regions of a hurricane as the basic sustaining mechanism has been questioned. The validity of the hot-tower thesis of heating from the tropical cumulonimbus has been scrutinized. The incomplete understanding of the processes and time scales of tropical cyclogenesis and intensification has been emphasized. The conviction that improved quantitative understanding of the tropical ambient is prerequisite to further progress on the deepening of tropical disturbances has been stated. Reasons for reservations concerning the probable success of current hurricane seeding techniques have been offered.

This unconventionality hopefully conveys two major points to the reader. First, any serious worker in hurricanes would be very remiss not to consult the referenced works directly and decide for himself whether or not the viewpoints adopted here are in fact valid. Second, this review on tropical cyclones describes a vital subject in rapid flux to which many basic physical and mathematical contributions are still to be made. This is not a field in which only exquisite refinements to elaborate computer programs remain. In fact, many valuable theoretical contributions in recent years have been achieved by Carrier and Gray via an approximate solution of simplified models with only modest computing, as opposed to attempting direct numerical simulation of the full conservation laws. The meteorologist with insight will hopefully agree that the tropical cyclone is neither a closed subject, nor a subject to which only those with access to very sophisticated computing facilities can aspire to contribute.

APPENDIX A. ESTIMATING THE KINETIC ENERGY AND WATER CONTENT OF HURRICANES

To estimate very quickly and roughly the total kinetic energy content of a mature hurricane of moderate intensity, one must evaluate I where

(A.1) $$I = \frac{1}{2} \int_V \rho v^2 \, dV$$
$$= \pi \int_{z=0}^{z=z_{1/2}} dz \int_{r=r_e}^{r=r_0} \rho [v(r)]^2 r \, dr$$

The speed is virtually all swirl, and about half the nine mile high, axisymmetric storm is rotating like (say) a potential vortex. The swirl is (suppose) 100 mph at the edge of the (nonrotating) eye $r_e = 30$ miles, and the storm extends out to $r_0 = 500$ miles. The density may be set to a constant average value of 5×10^{-4} gm cm^{-3}. Then

(A.2) $$I \doteq \tfrac{1}{2}\pi \rho z_1 (v_{\max})^2 r_e^2 \int_{r_e}^{r_0} dr/r \doteq 10^{24} \text{ ergs}$$

Battan (1961, p. 21) characterizes the kinetic energy of a hurricane as 10^{10} kW-hr, or 3.6×10^{23} erg.

An *upper bound* on the rainfall per day for a hurricane hovering over a spot is now calculated according to the Carrier model. The vertical velocity down into the boundary layer in the mature stage is

(A.3) $$w(r, z \to \infty, t \to \infty) \doteq -(R_0^2/r_0^2)(\nu \Omega)^{1/2}$$

The volumetric flux down into the boundary layer is, for $r_0^2 \gg R^2$,

(A.4) $$Q_0 = \pi r_0^2 w(r, z \to \infty, t \to \infty) \doteq -\pi R_0^2 (\nu \Omega)^{1/2}$$

this is also the flux up into the eyewall. The mass of air per unit time is given by ρQ_0 where ρ is fairly constant at a fixed height. The water content of this mass of air, which falls out as precipitation, is given in mass per time by

(A.5) $$J = \pi \rho R_0^2 (\nu \Omega)^{1/2} Y(z = z_t)$$

where z_t is a typical ambient-profile height for characterizing the mass fraction of water vapor for the flux into the eyewall. For $\rho \doteq 1.1 \times 10^{-3}$ gm cm^{-3}, $\nu \doteq 2 \times 10^5$ cm^2 sec^{-1}, $\Omega \doteq 6.25 \times 10^{-2}$ hr^{-1}, and [from the data of Jordan (1957)] $Y(z_t = 1000$ ft$) \doteq 1.65 \times 10^{-2}$, $J \doteq 1.8 \times 10^{11}$ gm sec^{-1}. If this falls entirely in an annulus from 20 to 30 miles, one gets 4.3×10^{-3} gm cm^{-2} sec^{-1}, or since 1 gm of liquid water occupies about 1 cm^3, about 370 cm day^{-1}. The record rainfall over a spot in one day is associated with a hurricane: 117 cm (Baguio, Philippines in July 1911) (Paulhaus, 1965). Of course, for a hurricane translating at 20 mph, no point is likely to lie under the eyewall for more than an hour, and $(370/24) \doteq 15$ cm hr^{-1}. (Typically, a hurricane is observed to drop a total of 15–30 cm of rain on coastal areas which pass under the central portions of the storm.)

Because of the flood-spawning torrential rains over Virginia associated with Hurricane Camille, it is interesting to consider whether the water vapor content of this hurricane at landfall is consistent with the large total rainfall over the Southeast. From isohyetal maps published by Schwarz (1970) it may be crudely estimated that in eastern Kentucky, Virginia, and West Virginia an area of 150 × 180 miles received an average of five inches of rain and an area of 100 × 360 miles received three inches. In addition 5×10^4 miles^2 of Mississippi, Alabama, Louisiana, and Tennessee received an average of three inches. This totals 2.58×10^{15} gm, which is probably a conservative estimate. From Jordan's data one can estimate that the Caribbean hurricane contained

(A.6) $$Q = \pi r_0{}^2 \int_0^{z_1} \rho Y\, dz = \pi r_0{}^2 I_1$$

where $\rho Y = \rho_v$ = density of water vapor. For an August–September ambient, $I_1 = 4.71$ gm cm^{-2}. Equating Q to the total rainfall, one finds $r_0 \doteq 260$ miles. This is an interesting value in that Camille was a small intense hurricane (De Angelis, 1969) with hurricane winds extending outward from the center about 60 miles and gale winds, 110 miles (Meyer, 1970, p.5). However, the results here indicate merely that $r_0 = O(260$ miles$)$ satisfies a conservative estimate of the total precipitation requirement; it is *not* intended to imply r_0 could necessarily be taken so small, though if one fits Carrier's proposed swirl profile [given below (35)] to the data, $r_0 = O(225$ miles$)$.

Appendix B. The Moist Adiabat

Because of the significant role it plays in understanding tropical cyclones, a derivation of the relation describing the moist adiabatic process seems worth including. The following is based on unpublished notes of G. F. Carrier of Harvard University.

When moist air expands adiabatically, the temperature eventually decreases to a value at which the water vapor density reaches its saturation level. For all further decreases in temperature, condensation occurs. Here it is assumed that the condensed water has zero volume (a good approximation), and that the liquid precipitates out as it is formed.

An air–vapor mixture is at temperature T, the mass of air in volume V being m_a, and the saturated vapor mass in V being m_v. The corresponding densities are $\rho_a \equiv (m_a/V)$ and $\rho_v \equiv (m_v/V)$; the pressure $p = p_a + p_v$. A relationship between T and p (say) as T changes is obtained by studying the moist adiabatic process in several steps (Fig. 22). The mixture is separated into two fluids (dry air and vapor), to be later recombined; this step is taken to require no expenditure of work.

Initially, the mass m_v of vapor occupies the volume V as in the top box of Fig. 22. That volume is decreased by a process in which dm_v of the vapor is condensed while it and the remaining vapor are at temperature T and pressure p_v. Thus, the new volume (that of the second box) is $V[1 - (dm_v/m_v)]$. The work done by external forces during this process is $(p_v V dm_v/m_v)$, and is indicated by the arrow to the left of the boxes. The heat gained $L\, dm_v$ is shown at the right. The vapor now expands adiabatically and isentropically to volume $(V + dV)$ as shown in box 3; the new temperature T' is given by

(B.1) $$T'/T = \{V[1 - (dm_v/m_v)]/(V + dV)\}^{\gamma_v - 1}$$
$$= 1 - (\gamma_v - 1)(dm_v/m_v + dV/V) + \cdots$$

In the third step, the gas in box 3 is heated to temperature $(T + dT)$, as indicated in box 4. The heat that must be added during this process is

(B.2) $$(m_v - dm_v)(c_v)_v[dT + (\gamma_v - 1)(dm_v/m_v + dV/V)T]$$
$$= m_v[(c_v)_v\, dT + R_v T\, dV/V] + R_v T\, dm_v + \cdots$$

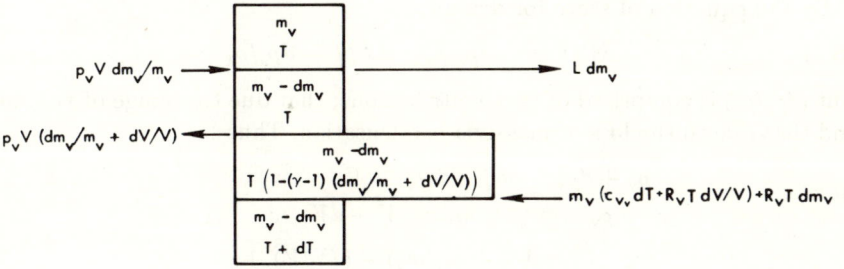

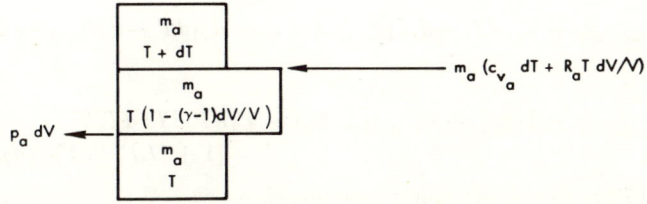

FIG. 22. Schematic diagrams of the sequence of thermodynamic steps by which the equation describing the moist adiabatic process is derived. The upper diagram sketches operations on the water vapor component, and the lower diagram sketches operations on the dry air component; the diagrams read from top to bottom. Arrows on the left of the diagrams denote work done on the system (arrow in) and work done by the system (arrow out). Arrows on the right of the diagrams indicate heat removed from the system (arrow out) and heat added to the system (arrow in).

where $(c_v)_v$ is the heat capacity at constant volume of the vapor, $(c_p)_v$ the heat capacity at constant pressure of the vapor, $R_v = (c_p)_v - (c_v)_v$, and $\gamma_v = (c_p)_v/(c_v)_v$.

The dry air, on the other hand, starts off in the bottom box of volume V with mass m_a at temperature T. It expands isentropically to volume $(V + dV)$ so that its temperature becomes $T[1 - (\gamma_a - 1)(dV/V)]$. Henceforth $\gamma_v = \gamma_a = \gamma$. The dry air is then heated at constant volume to temperature $(T + dT)$ by the addition of heat in the amount $m_a[(c_v)_a\, dT + R_a T(dV/V)]$.

Thus at the end of the process, a vapor mass $(m_v - dm_v)$ and a dry air mass m_a occupy a volume $(V + dV)$ at temperature $(T + dT)$. By definition, the process undergone will have been the moist adiabatic one if the net heat added is zero, and if the final $(\rho_v + d\rho_v)$ is the saturation value for the final temperature $(T + dT)$:

(B.3) $m_a[(c_v)_a\, dT + R_a T\, dV/V] +$
$$m_v[(c_v)_v\, dT + R_v T\, dV/V] + R_v T\, dm_v = -L\, dm_v$$

(B.4) $$\rho_v = m_v/V = \rho_v(T)$$

where $\rho_v(T)$ is the function that describes saturation density.

By the equation of state for dry air,

(B.5) $$dV/V = -d\rho_a/\rho_a = dT/T - dp_a/p_a$$

But $(d\rho_v/\rho_v)$ is comprised of two contributions: that due to change of volume and that due to the loss of mass via condensation. Thus

(B.6) $$\frac{\rho_v + d\rho_v}{\rho_v} = \frac{m_v - dm_v}{m_v} \frac{V}{V + dV}$$
$$= 1 - (dm_v/m_v) - (dV/V) + \cdots$$

Hence, after division by V and using the equations of state,

(B.7) $\{\rho_a[(c_v)_a + R_a]dT - dp_a\} + \{\rho_v[(c_v)_v + R_v]dT - dP\} = -(L/V)d(\rho_v V)$

or

(B.8) $\{[\gamma/(\gamma - 1)](p_a + P) + [(L/R_v) - T]\, dP/dT\}\, dT/T$
$= [1 + (LP/R_v T p_a)]\, dp_a$

where $P(T)$ is the saturation vapor pressure.

Alternatively, the moist adiabat may be derived as a special case of the equation for conservation of energy, written as the sum of internal plus kinetic energy for a *multicomponent* mixture. Here Cartesian tensor notation will be used instead of vector notation, and a comma denotes partial differentiation. Furthermore, the symbolism previously adopted in the main text for describing the dynamics of an effectively one-component gas will *not* be used here.

The air velocity is denoted u_i, with $u_i u_i \equiv q^2$ and $u_i \delta_{i3} = w$, where the gravitational acceleration $g_i = -g\, \delta_{i3}$. The water vapor velocity is $(u_i + v_i)$. The vapor quantities ρ_v, $(c_v)_v$, $(c_p)_v$, R_v and the air quantities ρ_a, $(c_v)_a$, $(c_p)_a$, R_a have the same designations as above. The specific internal energy of the air is e_a, and of the vapor e_v. The specific heat of the liquid water substance is c_l. The heat lost by radiation and all other processes is Q, with units (mass/length-time^3). The conservation of energy for current purposes may be written (upon neglect of the kinetic energy contribution relative to internal energy contributions)

(B.9) $$[\rho_a e_a + \rho_v e_v]_{,t} + [\rho_a e_a u_i + \rho_v e_v (u_i + v_i)]_{,i}$$
$$= -(\rho_a + \rho_v) g w + (\sigma_{ij} u_i)_{,j} + (k T_{,i})_{,i} - c_l T \omega - Q$$

where σ_{ij} is the total stress tensor, k is the thermal conductivity, and

(B.10) $$\rho_{v,t} + \rho_v (u_i + v_i)_{,i} = -\omega$$

In general, ω (the mass of water vapor condensed per unit time) must be specified explicitly or through a prescribed mechanism; when the fluid is saturated and $(dT/dt) < 0$, ω is known.

For steady flow with densities, pressures, and temperature dependent on the vertical coordinate z only, with negligible diffusion so $v_i = 0$ and $\sigma_{ij} = -p \delta_{ij}$, in the hydrostatic approximation $p_{,z} + \rho g = 0$, it follows from continuity that

(B.11) $$(\rho_a w)_{,z} = 0, \quad (\rho_v w)_{,z} = -\omega$$

hence,

(B.12) $$\rho_a w e_{a,z} + \rho_v w e_{v,z} - \omega e_v + p w_{,z} + c_l T \omega = 0$$

With the aid of the continuity equations and the equations of state,

(B.13) $$\rho_a w [(c_p)_a T]_{,z} + \rho_v w [(c_p)_v T]_{,z} - w p_{,z} = \omega (e_v + R_v T - c_l T)$$

Since

(B.14) $$e_v = c_l T + L(T) - R_v T$$

and

(B.15) $$\omega = -w \rho_a (\rho_v / \rho_a)_{,z} = -(1/V)\, d(\rho_v V)$$

upon substitution and cancellation of the common factor w, one recovers (B.7).

Acknowledgments

The author is very deeply indebted to Prof. George F. Carrier of Harvard University for generous and indispensable guidance and assistance on every aspect of the work described here. He also wishes to thank Dr. Paul Dergarabedian of The Aerospace Corporation for many stimulating discussions. However, full responsibility for all errors lies entirely with the author. All computations were programmed by Mr. Phillip S. Feldman, of TRW Systems. This study was initiated owing to the interest of Capt. Hugh Albers of the Interdepartmental Committee on Atmospheric Sciences (Department of Commerce); the encouragement and helpful suggestions of Drs. John Perry and Lawrence G. Roberts of ARPA Information Processing Techniques (Department of Defense), Drs. Robert Rapp and W. L. Gates of Rand Corporation, and Mr. James Murray of the Army Research Office (Durham) are gratefully acknowledged. Finally, the author wishes to thank Dr. H. E. Landsberg for the opportunity to present this paper. This work was carried out under contracts DAHC04-71-C-0025 and DAHC04-67-C-0015 with the U.S. Army Research Office (Durham, North Carolina). Although this work was sponsored in part by the Advanced Research Projects Agency of the Department of Defense under ARPA Order 1786, the views and conclusions contained in this paper are those of the author and should not be interpreted as necessarily representing the official policies, either expressed or implied, of ARPA or the U.S. Government.

Partial List of Symbols

c_l Heat capacity of liquid water
c_p Heat capacity at constant pressure
c_v Heat capacity at constant volume
e Internal energy
E Ekman number, ν/Ψ_0
g Magnitude of gravitational acceleration, $|g_i|$
g_i Acceleration of gravity
H Total static enthalpy, $c_p T + gz + LY$
H_t Total stagnation enthalpy, $H + q^2/2$
L Specific latent heat of phase transition
m_i Molecular weight of species i
n Power of algebraic decay of swirl with radial distance
p Pressure
P Saturation vapor pressure
q Wind speed, $|q_i|$
q_i Velocity vector for a fluid in noninertial frame rotating with earth
Q Net volumetric flux per time; heat loss per voume-time
Q_0 Steady-state net volumetric flux per time
r Radial coordinate in cylindrical polar coordinates
R Radial position of maximum azimuthal velocity v
R_i Gas constant for species i, $\bar{R}/m_i$
$\bar{R}$ Universal gas constant
RH   Relative humidity
t Time
T Temperature
u Radial velocity component of q_i
u_i Velocity vector for dry air
U Radial velocity component of q_i in the inviscid flow exterior to the frictional boundary layer
v Azimuthal velocity component of q_i
v_i Velocity vector for water vapor relative to dry air; velocity vector for a fluid in a noninertial system
V Azimuthal velocity component of q_i in the inviscid flow exterior to the frictional boundary layer; volume
w Vertical velocity component of q_i
W Vertical velocity component of q_i in the inviscid flow exterior to the frictional boundary layer
x $\Omega r^2/\psi_0$; a factor in the moist adiabat based on $dH = 0$
Y Water vapor mass fraction, ρ_v/ρ
z Altitude above sea level

Greek Symbols

γ c_p/c_v
ρ Density
ϕ ru [sometimes, $r(u - U)$; sometimes, nondimensional]
Φ rU
ψ rv [sometimes, $r(v - V)$; sometimes, nondimensional]
Ψ rV
σ Ratio of molecular weight of water vapor to dry air
σ_{ij} Total stress tensor
ν Kinematic viscosity (usually given turbulent transfer values)
ω Mass of water vapor condensed per volume-time
Ω Coriolis parameter (component of the angular velocity of the earth perpendicular to a plane locally tangential to the sea surface in the tropics)

θ azimuthal coordinate in cylindrical polar coordinates; potential-like enthalpy, $c_p T + gz$
θ_t Stagnation potential-like enthalpy, $\theta + q^2/2$
ζ $z/E^{1/2}$

Superscript
˙ Total derivative with respect to time

Subscripts
a Dry air; ambient
c At the axis of symmetry at sea level
e Eyewall at sea level
i Vector (Cartesian tensor notation)
o Outer edge of hurricane; initial value; typical value
s Ambient at sea level
t Total
v Vapor

References*

Agee, E. M. (1972). Note on ITCZ wave disturbances and formation of Tropical Storm Anna. *Mon. Weather Rev.* **100**, 733–737.

Alaka, M. A. (1968). "Climatology of Atlantic Tropical Storms and Hurricanes," ESSA Tech. Rep. WB-6. US Govt. Printing Office, Washington, D.C.

Anderson, L. G., and Burnham, J. M. (1973). Application of economic analyses to hurricane warnings to residential and retail activities in the U.S. Gulf of Mexico coastal region. *Mon. Weather Rev.* **101**, 126–131.

Anonymous (1972). Hurricane Agnes: The most costly storm. *Weatherwise* **25**, 174–184.

Anonymous. (1973). U.S. hurricane control test program cut. *Los Angeles Times* **92**, 4 February, Part 8, p. 5.

Anthes, R. A. (1971). Iterative solutions to the steady-state axisymmetric boundary-layer equations under an intense pressure gradient. *Mon. Weather Rev.* **100**, 261–268.

Anthes, R. A. (1972). Development of asymmetries in a three-dimensional numerical model of a typical cyclone. *Mon. Weather Rev.* **100**, 461–476.

Anthes, R. A., and Johnson, D. R. (1968). Generation of available potential energy in Hurricane Hilda (1964). *Mon. Weather Rev.* **96**, 291–302.

Anthes, R. A., Rosenthal, S. L., and Trout, J. W. (1971). Preliminary results from an asymmetric model of the tropical cyclone. *Mon. Weather Rev.* **99**, 744–758.

Arakawa, H. (1954). On the pyramidal, mountainous, and confused sea in the right or dangerous semi-circle of typhoons. *Pap. Meteorol. Geophys.* **5**, 114–123.

Aspliden, C. I. (1971). "On Energy Distribution in the Tropical Troposphere," Rep., Dept. Meteorol., Florida State University, Tallahassee.

Atkinson, G. D. (1971). "Forecasters Guide to Tropical Meteorology," Tech. Rep. No. 240. USAF Air Weather Serv., Scott AFB, Illinois.

Bates, J. R. (1970). Dynamics of disturbances on the Intertropical Convergence Zone. *Quart. J. Roy. Meteorol. Soc.* **96**, 677–701.

Bates, J. R. (1972). Tropical disturbances and the general circulation. *Quart. J. Roy. Meteorol. Soc.* **98**, 1–16.

Battan, L. J. (1961). "The Nature of Violent Storms." Doubleday, Garden City, New York.

Battan, L. J. (1969). "Harvesting the Clouds—Advances in Weather Modification." Doubleday, Garden City, New York.

Black, P. G., and Anthes, R. A. (1971). On the asymmetric structure of the tropical cyclone outflow layer. *J. Atmos. Sci.* **28**, 1348–1366.

Black, P. G., and Mallinger, W. D. (1972). "The Mutual Interaction of Hurricane Ginger

* This article was completed in February 1973.

and the Upper-Mixed Layer of the Ocean," Appendix D, Project Stormfury Annual Report 1971. Nat. Hurricane Res. Lab., Coral Gables, Florida.

Black, P. G., Senn, H. V., and Courtright, C. L. (1972). Airbourne radar observations of eye configuration changes, bright band distribution, and precipitation tilt during the 1969 multiple seeding experiments in Hurricane Debbie. *Mon. Weather Rev.* **100**, 208–217.

Boyd, D. W., Howard, R. A., Matheson, J. E., and North, D. W. (1971). "Decision Analysis of Hurricane Modification," Proj. 8503 Final Rept. Stanford Res. Inst., Menlo Park, California.

Bradbury, D. L. (1971). "The Filling Over Land of Hurricane Camille, August 17–18, 1969," Satellite and Mesometeorol. Res. Proj. Res. Pap. 96. Dept. Geophys. Sci., University of Chicago, Chicago, Illinois.

Brand, S. (1970a). "Geographic and Monthly Variation of Very Large and Very Small Typhoons of the Western North Pacific Ocean," Tech. Pap. No. 13–70. Nav. Weather Res. Facility, Norfolk, Virginia.

Brand, S. (1970b). Interaction of binary tropical cyclones of the western North Pacific. *J. Appl. Meteorol.* **9**, 433–441.

Brand, S. (1971). The effects on a tropical cyclone of cooler surface waters due to upwelling and mixing produced by a prior tropical cyclone, *J. Appl. Meteorol.* **10**, 865–874.

Brand, S., and Bellock, J. W. (1972). "Changes in the Characteristics of Typhoons Crossing the Philippines" Tech. Pap. No. 6–72. Environ. Prediction Res. Facility, Nav. Postgrad. School, Monterey, California.

Burggraf, O. R., Stewartson, K., and Belcher, R. (1971). Boundary layer induced by a potential vortex. *Phys. Fluids* **14**, 1821–1833.

Burroughs, L. D., and Brand, S. (1973). Speed of tropical storms and typhoons after recurvature in the western North Pacific Ocean. *J. Appl. Meteorol.* **12**, 452–458.

Byers, H. R. (1944). "General Meteorology," 2nd ed. McGraw-Hill, New York.

Carlson, T. N. (1969). Some remarks on African disturbances and their progress over the tropical Atlantic. *Mon. Weather Rev.* **97**, 716–726.

Carpenter, T. H., Holle, R. L., and Fernandez-Partagas, J. J. (1972). Observed relationships between lunar tidal cycles and formation of hurricanes and tropical storms. *Mon. Weather Rev.* **100**, 451–460.

Carrier, G. F. (1970). Singular perturbation theory and geophysics. *SIAM (Soc. Ind. App . Math.) Rev.* **12**, 175–193.

Carrier, G. F. (1971a). Swirling flow boundary layers. *J. Fluid Mech.* **49**, 133–144.

Carrier, G. F. (1971b). The intensification of hurricanes. *J. Fluid Mech.* **49**, 145–158.

Carrier, G. F., Hammond, A. L., and George, O. D. (1971). A model of the mature hurricane. *J. Fluid Mech.* **47**, 145–170.

Charney, J. (1971). Tropical cyclogenesis and the formation of the Intertropical Convergence Zone. *In* "Mathematical Problems in the Geophysical Sciences. I. Geophysical Fluid Dynamics (Lectures in Applied Mathematics, Vol. 13)" (W. H. Reid, ed.), Vol. 1, pp. 335–368. Amer. Math. Soc., Providence, Rhode Island

Charney, J. G., and Eliassen, A. (1964). On the growth of the hurricane depression. *J. Atmos. Sci.* **21**, 68–75.

Cox, J. L., and Jager, G. (1969). "A Satellite Analysis of Twin Tropical Cyclones in the Western Pacific," Tech. Memo. WBTM SOS 5. ESSA Space Operations Support Div., Silver Spring, Maryland.

Day, J. A. (1966). "The Science of Weather." Addison-Wesley, Reading, Massachusetts.

DeAngelis, R. M. (1969). Enter Camille. *Weatherwise* **22**, 173–179.

Denny, W. J. (1972). Eastern Pacific hurricane season of 1971. *Mon. Weather Rev.* **100**, 276–293.

Deppermann, C. E. (1937). "Wind and Rainfall in Selected Philippine Typhoons."

Manila Weather Bureau Central Observatory Report, Bureau of Printing, Manila, Philippines.

Dergarabedian, P., and Fendell, F. (1970). On estimation of maximum wind speeds in tornadoes and hurricanes. *J. Astronaut. Sci.* **17**, 218–236.

Dergarabedian, P., and Fendell, F. (1971). "Tornado and Hurricane Thermodydrodynamics," Tech. Rep. TRW Systems, Redondo Beach, California.

Dergarabedian, P., and Fendell, F. (1972a). "Tropical Cyclones," Rep. 18524-6000-R0-01. TRW Systems, Redondo Beach, California.

Dergarabedian, P., and Fendell, F. (1972b). The surface frictional layer under a hurricane vortex. *J. Astronaut. Sci.* **20**, 9–34.

Downie, C. S. (1960). Cloud modification with carbon black. *In* "Cumulus Dynamics" (C. E. Anderson, ed.), pp. 191–209. Pergamon, Oxford.

Easterbrook, C. C. (1969). "A Study of the Effects of Waves on Evaporation from Free Water Surfaces," Res. Rep. No. 18. Bureau of Reclamation, U.S. Department of the Interior, US Govt. Printing Office, Washington, D.C.

Eliassen, A., and Kleinschmidt, E. (1957). Dynamic meteorology. *In* "Handbuch der Physik" (S. Flügge, ed.), Vol. 48, pp. 1–54. Springer-Verlag, Berlin and New York.

Emmons, H. W., and Ying, S. J. (1967). The fire whirl. *In* "Eleventh Symposium (International) on Combustion." pp. 475–486. Combustion Institute, Pittsburgh, Pennsylvania.

Erickson, C. O., and Winston, J. S. (1972). Tropical storm, midlatitude cloud-band connections and the autumnal buildup of the planetary circulation. *J. Appl. Meteorol.* **11**, 23–36.

Estoque, M. A. (1971). "Hurricane Modification by Cloud Seeding," Rep., Rosenstiel School of Marine and Atmospheric Science, University of Miami, Coral Gables, Florida.

Fletcher, R. D. (1955). Computation of maximum surface winds in hurricanes. *Bull. Amer. Meteorol. Soc.* **36**, 247–250.

Frank, W. M. (1973). "Characteristics of Carbon Black Dust as a Large-scale Tropospheric Heat Source," Pap. No. 195. Dept. Atmos. Sci., Colorado State University, Fort Collins.

Fujita, T. T. (1972). "Use of ATS Pictures in Hurricane Modification," Satellite and Mesometeorol. Res. Proj. Res. Pap. No. 106. Dept. Geophys. Sci., University of Chicago, Chicago, Illinois.

Fukuta, N. (1972). "Modification of Hurricanes by Cloud Seeding," Rep., Denver Res. Inst., Denver, Colorado.

Garstang, M. (1967). Sensible and latent heat exchange in low latitude synoptic scale systems. *Tellus* **19**, 492–508.

Garstang, M. (1972). A review of hurricane and tropical meteorology. *Bull. Amer. Meteorol. Soc.* **53**, 612–630.

Garstang, M., La Seur, N. E., Warsh, K. L., Hadlock, R., and Peterson, J. R. (1970). Atmospheric-oceanic observations in the tropics. *Amer. Sci.* **8**, 482–495.

Gates, W. L., Batten, E. S., Kahle, A. B., and Nelson, A. B. (1971). "A Documentation of the Mintz-Arakawa Two-level Atmospheric General Circulation Model," Rep. R-877-ARPA. Rand Corp., Santa Monica, California.

Geissler, J. E. (1972). On the vertical distribution of latent heat release and the mechanics of CISK. *J. Atmos. Sci.* **29**, 240–243.

Gentry, R. C. (1964). "A Study of Hurricane Rainbands," Nat. Hurricane Res. Proj. Rep. No. 69. US Weather Bureau, Washington, D.C.

Gentry, R. C. (1969a). Project Stormfury. *Bull. Amer. Meteorol. Soc.* **50**, 404–409.

Gentry, R. C. (1969b). Project Stormfury, 1969. Pap., *Tech. Conf. Hurricanes. 6th, 1969.*

Gentry, R. C. (1970). Hurricane Debbie modification experiments. *Science* **168**, 473–475.

Gentry, R. C. (1971a). To tame a hurricane. *Sci. J.* **7**, 49–55.
Gentry, R. C. (1971b). Project Stormfury, 1971. *Bull. Amer. Meteorol. Soc.* **52**, 775.
Gentry, R. C., and Hawkins, H. F. (1971). "A Hypothesis for Modification of Hurricanes," Appendix B, Project Stormfury Annual Report 1970. Nat. Hurricane Res. Lab., Coral Gables, Florida.
Godshall, F. A. (1968). Intertropical Convergence Zone and mean cloud amount in the tropical Pacific Ocean. *Mon. Weather Rev.* **96**, 172–175.
Gray, W. M. (1968). Global view of the origin of tropical disturbances and storms. *Mon. Weather Rev.* **96**, 669–700.
Gray, W. M. (1972a). "A Diagnostic Study of the Planetary Boundary Layer Over the Oceans," Pap. No. 179. Dept. Atmos. Sci., Colorado State University, Fort Collins.
Gray, W. M. (1972b). "Cumulus Convection and Larger-scale Circulations. Part 3. Broadscale and Mesoscale Circulations," Pap. No. 190. Dept. Atmos. Sci., Colorado State University, Fort Collins.
Gray, W. M. (1973). "Feasibility of Beneficial Hurricane Modification by Carbon Dust Seeding," Pap. No. 196. Dept. Atmos. Sci., Colorado State University, Fort Collins.
Greenspan, H. P. (1968). "The Theory of Rotating Fluids." Cambridge Univ. Press, London and New York.
Grossman, G., and Rodenhuis, D. (1972). "The Effect of Release of Latent Heat on the Vorticity of a Tropical Storm Over Land," Tech. Note BN-722. Inst. Fluid Dyn. Appl. Math., University of Maryland, College Park.
Haltiner, G. J. (1971). "Numerical Weather Prediction." Wiley, New York.
Haurwitz, B. (1935). The height of tropical cyclones and the "eye" of the storm. *Mon. Weather Rev.* **63**, 45–49.
Hawkins, H. F. (1971). Comparison of results of the Hurricane Debbie (1969) modification experiments with those from Rosenthal's numerical model simulation experiments. *Mon. Weather Rev.* **99**, 427–434.
Hawkins, H. F., and Rubsam, D. T. (1968). Hurricane Hilda, 1964. II. Structure and budgets of the hurricane on October 1, 1964. *Mon. Weather Rev.* **96**, 617–636.
Hawkins, H. F., Bergman, K. H., and Gentry, R. C. (1972). "Report on Seeding of Hurricane Ginger," Appendix B, Project Stormfury Annual Report 1971. Nat. Hurricane Res. Lab., Coral Gables, Florida.
Hess, S. L. (1959). "Introduction to Theoretical Meteorology." Holt, New York.
Hidy, G. M. (1972). A view of recent air-sea interaction research. *Bull. Amer. Meteorol. Soc.* **53**, 1083–1102.
Holliday, C. R. (1973). Record 12- and 24-hour deepening rates in a tropical cyclone. *Mon. Weather Rev.* **101**, 112–114.
Holton, J. R. (1972). "An Introduction to Dynamic Meteorology." Academic Press, New York.
Holton, J. R., Wallace, J. M., and Young, J. A. (1971). On boundary layer dynamics and the ITCZ. *J. Atmos. Sci.* **28**, 275–280.
Howard, R. A., Matheson, J. E., and North, D. W. (1972). The decision to seed hurricanes. *Science* **176**, 1191–1202.
Hubert, L., and Timchalk, A. (1969). Estimating hurricane wind speeds from satellite pictures. *Mon. Weather Rev.* **97**, 382–383.
Hughes, L. A. (1952). On the low-level wind structure of tropical storms. *J. Meteorol.* **9**, 422–428.
Jennings, G. (1970). "The Killer Storms—Hurricanes, Typhoons, and Tornadoes." Lippincott, Philadelphia, Pennsylvania.
Johnson, D. H. (1969). The role of the tropics in the global circulation. *In* "The Global Circulation of the Atmosphere" (G. A. Corby, ed.), pp. 113–136. Roy. Meteorol. Soc., London.

Jordan, C. L. (1957). "A Mean Atmosphere for the West Indies Area," Nat. Hurricane Res. Proj. Rep. No. 6. US. Dept. of Commerce, Washington, D.C.
Keenan, J. H., and Keyes, F. G. (1936). "Thermodynamic Properties of Steam Including Data for Liquid and Solid Phases." Wiley, New York.
Kraus, E. B. (1972). "Atmosphere-Ocean Interaction." Oxford Univ. Press (Clarendon), London and New York.
Kuettner, J. P. (1971). Cloud bands in the earth's atmosphere. Observations and theory. *Tellus* **23**, 404–426.
Kurihara, Y. (1971). An eleven-layer, axisymmetric, primitive equation model of a tropical cyclone. *Bull. Amer. Meteorol. Soc.* **52**, 769.
Leipper, D. F. (1967). Observed ocean conditions and Hurricane Hilda, 1964. *J. Atmos. Sci.* **24**, 182–196.
Leipper, D. F., and Volgenau, D. (1972). Hurricane heat potential of the Gulf of Mexico. *J. Phys. Oceanogr.* **2**, 218–224.
Lieb, H. S. (1972). Project Stormfury Director Reveals Results of Hurricane Ginger Seeding Experiments," News Release NOAA 72-4. U.S. Dept. of Commerce, Washington, D.C.
Lopez, R. E. (1972a). Cumulus Convection and Larger-scale Circulations. Part 1. A Parametric Model of Cumulus Convection," Pap. No. 188. Dept. Atmos. Sci., Colorado State University, Fort Collins.
Lopez, R. E. (1972b). "Cumulus Convection and Larger-scale Circulations. Part 2. Cumulus and Meso-scale Circulations," Pap. 189. Colorado State Univ. Fort Collins.
Lorenz, E. N. (1966). The circulation of the atmosphere. *Amer. Sci.* **54**, 402–420.
Lorenz, E. N. (1967). "The Nature and Theory of the General Circulation of the Atmosphere." World Meteorol. Organ., Geneva.
McWilliams, J. C. (1971). The boundary layer dynamics of symmetric vortices. Ph.D. Thesis, Harvard University, Cambridge, Massachusetts.
Mahrt, L. J. (1972a). A numerical study of the influence of advective accelerations in an idealized, low-latitude, planetary boundary layer. *J. Atmos. Sci.* **29**, 1477–1484.
Mahrt, L. J. (1972b). A numerical study of coupling between the boundary layer and free atmosphere in an accelerated low-latitude flow. *J. Atmos. Sci.* **29**, 1485–1495.
Malkus, J. S. (1958). On the structure and maintenance of the mature hurricane eye. *J. Meteorol.* **15**, 337–349.
Malkus, J. S. (1960). Recent developments in studies of penetrative convection and an application to hurricane cumulonimbus towers. *In* "Cumulus Dynamics" (C. E. Anderson, ed.), pp. 65–84. Pergamon, Oxford.
Malkus, J. S. (1962). Large-scale interactions. *In* "The Sea" (M. N. Hill, ed.), Vol. 1, pp. 88–294. Wiley (Interscience), New York.
Malkus, J. S., and Riehl, H. (1960). On the dynamics and energy transformation in steady-state hurricanes. *Tellus* **12**, 1–20.
Malkus, J. S., Ronne, C., and Chafee, M. (1961). Cloud patterns in Hurricane Daisy, 1968. *Tellus* **13**, 8–30.
Mallinger, W. D. (1971). "Project Stormfury Experimental Eligibility in the Western North Pacific," Appendix L, Project Stormfury Annual Report 1970. Nat. Hurricane Res. Lab., Coral Gables, Florida.
Malone, M. J. and Leimer, D. R. (1971). "Estimation of the Economic Benefits to DOD from Improved Tropical Cyclone Forecasting," Rep. Headquarters Military Airlift Command, Scott AFB, Illinois.
Manabe, S., Holloway, J. L., Jr., and Stone, H. M. (1970). Tropical circulation in a time-integration of a global model of the atmosphere. *J. Atmos. Sci.* **27**, 580–613.
Matano, H., and Sekioka, M. (1971). Some aspects of extratropical transformation of a tropical cyclone. *J. Meteorol. Soc. Jap.* **49**, 736–743.

Matthews, D. A. (1971). "Ice-phase Modification Potential of Cumulus Clouds in Hurricanes," Appendix H, Project Stormfury Annual Report 1970. Nat. Hurricane Res. Lab., Coral Gables, Florida.
Meyer, J. W., ed. (1970). "Airborne Severe Storm Surveillance," Vol. 1. Tech. Note 1970-43. Lincoln Lab., Lexington, Massachusetts.
Meyer, J. W. (1971a). Toward hurricane surveillance and control. *Technol. Rev.* **74**, 53–60.
Meyer, J. W. (1971b). "Airborne Severe Storm Surveillance," Vol. 2, Tech. Note 1970-43. Lincoln Lab., Lexington, Massachusetts.
Miller, B. I. (1958). On the maximum intensity of hurricanes. *J. Meteorol.* **15**, 184–195.
Miller, B. I. (1967). Characteristics of hurricanes. *Science.* **157**, 1389–1399.
Munk, W., Snodgrass, F., and Carrier, G. (1956). Edge waves on a continental shelf. *Science* **123**, 127–132.
Murray, F. W. (1967). On the computation of saturation vapor pressure. *J. Appl. Meterol.* **6**, 203–204.
Ogura, Y. (1964). Frictionally controlled, thermally driven circulations in a circular vortex with application to tropical cyclones. *J. Atmos. Sci.* **21**, 610–621.
Oliver, V. J., and Anderson, R. K. (1969). Circulation in the tropics as revealed by satellite data. *Bull. Amer. Meteorol. Soc.* **50**, 702–707.
Ooyama, K. (1969). Numerical simulation of the life-cycle of tropical cyclones. *J. Atmos. Sci.* **26**, 3–40.
Ooyama, K. (1971). A theory on parameterization of cumulus convection. *J. Meteorol. Soc. Jap.* **49**, 744–756.
Orton, R. (1970). Tornadoes associated with Hurricane Beulah on September 19–23, 1967. *Mon. Weather Rev.* **98**, 541–547.
Ostapoff, F., Tarbeyev, Y., and Worthem, S. (1973). Heat flux and precipitation estimates from oceanographic observations. *Science* **100**, 960–962.
Östlund, H. G. (1968). Hurricane tritium. II. Air-sea exchange of water in Betsy 1965. *Tellus* **20**, 577–594.
Östlund, H. G. (1970). Hurricane tritium. III. Evaporation of sea water in Hurricane Faith 1966. *J. Geophys. Res.* **75**, 2303–2309.
Palmén, E. (1948). On the formation and structure of tropical cyclones. *Geophysica* **3**, 26–38.
Palmén, E., and Newton, C. W. (1969). "Atmospheric Circulation Systems." Academic Press, New York
Palmén, E , and Riehl, H. (1957). Budget of angular momentum and energy in tropical storms. *J. Meteorol.* **14**, 150–159.
Paulhaus, J. L. H. (1965). Indian Ocean and Taiwan rainfalls set new records. *Mon. Weather Rev.* **93**, 331–335.
Penner, S. S. (1972). Elementary considerations of the fluid mechanics of tornadoes and hurricanes. *Astronaut. Acta* **17**, 351–362.
Perlroth, I. (1967). Hurricane behavior as related to oceanic environmental conditions. *Tellus* **19**, 258–268.
Perlroth, I. (1969). Effects of oceanographic media on equatorial Atlantic hurricanes. *Tellus* **21**, 230–244.
Petrosyants, M. A. (1972). The national tropical experiment TROPEX-72. *Izv. Acad. Sci., USSR, Atmos. Oceanic Phys.* **8**, 519–522.
Phillips, N. A. (1970). Models for weather prediction. *Annu. Rev. Fluid Mech.* **2**, 251–292.
Ramage, C. S. (1971). "Monsoon Meteorology." Academic Press, New York.
Ramage, C. S. (1972). Interaction between tropical cyclones and the China Seas. *Weather* **27**, 484–494.
Rapp, R. (1970). Climate modification and national security. Pap., *Metorol. Tech. Exchange Conf., 1970*.

Revesz, W., Jr. (1971). Comparison of effects of various tropical storms on the vertical temperature of the ocean using pictorial representation. Master's Thesis, Naval Postgraduate School, Monterey, California.

Riehl, H. (1954). "Tropical Meteorology." McGraw-Hill, New York.

Riehl, H. (1963a). Some relations between wind and thermal structure of steady state hurricanes. *J. Atmos. Sci.* **20**, 276–287.

Riehl, H. (1963b). On the origin and possible modification of hurricanes. *Science* **141**, 1001–1010.

Riehl, H. (1969a). On the role of the tropics in the general circulation of the atmosphere. *Weather* **24**, 288–303.

Riehl, H. (1969b). Some aspects of cumulonimbus convection in relation to tropical weather disturbances. *Bull. Amer. Meteorol. Soc.* **50**, 587–595.

Riehl, H. (1972a). "Introduction to the Atmosphere," 2nd ed. McGraw-Hill, New York.

Riehl, H. (1972b). Intensity of recurved typhoons. *J. Appl. Meteorol.* **11**, 613–615.

Riehl, H., and Malkus, J. (1961). Some aspects of Hurricane Daisy, 1958. *Tellus* **13**, 181–213.

Rosenthal, S. L. (1970). A circularly symmetric primitive equation model of tropical cyclone development containing an explicit water vapor cycle. *Mon. Weather Rev.* **98**, 643–663.

Rosenthal, S. L. (1971a). A circularly symmetric primitive equation model of tropical cyclones and its reponse to artificial enhancement of the convective heating functions. *Mon. Weather Rev.* **99**, 414–426.

Rosenthal, S. L. (1971b). The response of a tropical cyclone model to variations in boundary layer parameters, initial conditions, lateral boundary conditions, and domain size. *Mon. Weather Rev.* **99**, 767–777.

Rosenthal, S. L. (1971c). "Hurricane Modeling at the National Hurricane Research Laboratory (1970)," Appendix C, Project Stormfury Annual Report 1970. Nat. Hurricane Res. Lab., Coral Gables, Florida.

Saffir, H. S. (1973). Hurricane wind and storm surge. *Mil. Eng.* **65**, 4–5.

Sanders, F., and Burpee, R. W. (1968). Experiments in barotropic hurricane track forecasting. *J. Appl. Meteorol.* **7**, 313–323.

Schwarz, K. (1970). The unprecedented rains in Virginia associated with the remnants of Hurricane Camille. *Mon. Weather Rev.* **98**, 851–859.

Shea, D. J. (1972). "The Structure and Dynamics of the Hurricane's Inner Core Region," Pap. No. 182. Dept. Atmos. Sci., Colorado State university, Fort Collins.

Sheets, R. C. (1972). "Some Statistical Characteristics of the Hurricane Eye and Minimum Sea-level Pressure," Appendix I, Project Stormfury Annual Report 1971. Nat. Hurricane Res. Lab., Coral Gables, Florida.

Shuleykin, V. V. (1970). The power of a tropical hurricane as a function of the underlying sea surface temperature. *Izv., Acad. Sci., USSR, Atmos. Oceanic Phys.* **6**, 729–739.

Shuleykin, V. V. (1972). Development and decay of a tropical hurricane under various thermal conditions. *Izv., Acad. Sci., USSR, Atmos. Oceanic Phys.* **8**, 1–17.

Simpson, R. H. (1971). "The Decision Process in Hurricane Forecasting," NOAA Tech. Memo. NWS SR-53. Nat. Weather Serv. Southern Region Headquarters, Fort Worth, Texas.

Simpson, R. H., and Frank, J. R. (1972). Atlantic hurricane season of 1971. *Mon. Weather Rev.* **100**, 256–267.

Simpson, R. H., and Malkus, J. S. (1964). Experiments in hurricane modification. *Sci. Amer.* **211**, 27–37.

Simpson, R. H., and Pelisser, J. M. (1971). Atlantic hurricane season of 1970. *Mon. Weather Rev.* **99**, 269–277.

Spark, E. H. (1971). The willy-willy. *Bull. Amer. Meteorol. Soc.* **52**, 574–575.

Spiegler, D. B. (1972). Cyclone categories and definitions: Some proposed revisions. *Bull. Amer. Meteorol. Soc.* **53**, 1174–1178.

Sundqvist, H. (1970). Numerical simulation of the development of tropical cyclones with a ten-layer model. Part I. *Tellus* **22**, 359–390.

Sundqvist, H. (1972). Model tropical cyclone behavior in experiments related to modification attempts. *Tellus* **24**, 6–12.

Varga, G. (1971). Some observations on the structure of shallow typhoons. Master's Thesis, Naval Postgraduate School, Monterey, California.

Vollsprecht, R. (1972). Willy-willy, hurricane or cockeyed bob? *Bull. Amer. Meteorol. Soc.* **53**, 888.

Walker, J. M. (1972). The monsoon of southern Asia: A review. *Weather* **27**, 178–189.

Warsh, K. L. (1973). Relation of sea-air interface energy fluxes to convective activity in the tropical Atlantic Ocean. *J. Geophys. Res.* **78**, 504–519.

Warsh, K. L., Echternacht, K. L., and Garstang, M. (1971). Structure of near-surface currents east of Barbados. *J. Phys. Oceanogr.* **1**, 123–129.

White, R. M. (1972). The national hurricane warning program. *Bull. Amer. Meteorol. Soc.* **53**, 631–633.

Wright, R. (1969). Temperature structure across the Kuroshio before and after Typhoon Shirley. *Tellus* **21**, 409–413.

Yamasaki, M. (1968). Detailed analysis of a tropical cyclone simulated with a 13-layer model. *Pap. Meteorol. Geophys.* **19**, 559–585.

Yamasaki, M. (1969). Large-scale disturbances in the conditionally unstable atmosphere in the low latitudes. *Pap. Meteorol. Geophys.* **20**, 289–336.

Yanai, M. (1964). Formation of tropical cyclones. *Rev. Geophys.* **2**, 367–414.

Zipper, E. J. (1969). The role of organized unsaturated convective downdrafts in the structure and rapid decay in an equatorial disturbance. *J. Appl. Meteorol.* **8**, 799–814.

Note Added in Proof. Several recent measurements in hurricanes seem to cast doubt on the Riehl–Malkus oceanic heat source postulate. First, properly adjusted measurements in the frictional boundary layer of Hurricane Ginger (September, 1971), a minimal hurricane in which winds did not much exceed 90 mph, indicate that at the height of 500 ft above sea level the static temperature decreased 1°–1.5°C from edge to eyewall. The need for oceanic heat transfer to maintain a supposedly isothermal boundary layer appears nonexistent, if this measurement proved typical. Also, both radar measurements *and* satellite photography of the eyewalls of Eastern Pacific Hurricane Irah on 23 September 1973 (with maximum wind speed of 130 kt) and of North Atlantic Hurricane Ellen on 21 September 1973 (with maximum wind speed of 120 kt and minimum central pressure of 962 mb) indicated significant outward sloping of the eyewall [see, e.g., NASA Johnson Space Center Skylab Program photograph SL3-118-2189]. In fact, radar suggests that the eyewall of the moderately intense cyclone Ellen sloped outward at 45° above 15,000 ft; past dismissal of radar data reporting significant outward eyewall sloping may have been unjustified. As a hurricane matures, the data suggest the intense convective updrafts in the core tilt more from the vertical (perhaps permitting precipitation, re-evaporated during descent, to enrich the low-level inflow), and persist longer (50 min, as opposed to 10-min lifespan further from the core) [Black, P. G. (1974). Preliminary assessment of handheld photographic observation of tropical storms from space. *In* "Skylab Visual Observations," Chap. 16. NASA Johnson Space Center, Houston, Texas]. The fact that compressional heating in the eye can directly contribute to the lightening of a column of air whose base lies in the highly swirling eyewall, obviates the need for an oceanic heat source to achieve that pressure deficit from ambient compatible with observed hurricane wind speeds. On the other hand, if the low-level inflow layer were to prove saturated near the eyewall, augmentation within the hurricane of ambient-level heat transfer from the ocean to the atmosphere would still be required.

A NUMERICAL STUDY OF VACILLATION

A. Quinet
Institut Royal Météorologique de Belgique, Bruxelles, Belgium

1. Introduction 101
2. The Laboratory Simulation of Large-Scale Atmospheric Flow 102
   2.1. The Various Simulated Geophysical Flows 102
   2.2. The Essential Distinction between Hadley and Rossby Flows 106
   2.3. The Characteristic Features of Vacillation 108
3. A Model Atmosphere for the Study of Vacillation 113
   3.1. General Comments on the Theoretical Approach to the Study of Vacillation 113
   3.2. The Tensor Formulation of Quasi-Hydrostatic Flows 114
   3.3. The Equations of the Two-Layer Balanced Model 120
4. The Spectral Dynamics and Energetics of the Model....................... 131
   4.1. The Spectral Representation 131
   4.2. The Spectral Equations 135
   4.3. Some Dynamical Properties of the Fields of Motion 138
   4.4. The Spectral Energetics of the Model 141
   4.5. Comments on the Various Physical Processes 148
5. The Numerical Study of Vacillation 149
   5.1. The Numerical Procedure 149
   5.2. The Potential Energy Vacillation 150
   5.3. The Kinetic Energy Vacillation 162
   5.4. The Various Vacillation Cycles 175
6. Vacillation in the Atmosphere 178
   6.1. Potential Energy Vacillation in the Atmosphere 178
   6.2. Kinetic Energy Vacillation in the Atmosphere 180
   6.3. Conclusion... 184
List of Symbols 182
References 182

1. Introduction

It is common knowledge that the motion of the atmosphere is extremely complex. However, the upper level weather maps exhibit flow patterns with more or less permanent and characteristic properties. Perhaps the most surprising observation is the extremely restricted number of scales of motion as compared with the *a priori* infinite number of degrees of freedom involved in the general thermohydrodynamical equations. The existence of moving waves whose progression is associated with a change in the tilt of their through and ridge lines, as well as the alternation between quasi-zonal and essentially meridional (lobed) circulations, are also typical features of the large-scale atmospheric flow.

It may therefore be believed that the apparent complexity of the weather maps should result from a few, highly selective processes linked in a more or less simple way. To identify these basic mechanisms and to state clearly how they interplay is the fundamental problem of large-scale atmospheric dynamics. To achieve this aim it is extremely useful to study simple geophysical flows that can be generated in the laboratory. In this respect, vacillation constitutes one of the most interesting phenomena.

The following text is devoted to the numerical study of vacillation. Section 2 gives a description of this kind of flow as it is known from laboratory experiments. Section 3 considers the minimum set of conditions to be fulfilled by the simplest possible numerical model to be able to simulate the phenomenon. Since vacillation appears in flows with various geometrical constraints, the tensor formulation of the equations has been adopted. The reader who is not, and does not want to become familiar with this technique can go directly to Section 3.3.2 where he will find the classical two-layer quasi-geostrophic model equations. Section 4 gives the spectral form of these equations when the fluid moves in a channel. Section 4 describes also the various processes capable of taking place in the model and interprets them in terms of the flow pattern. In order to allow an interpretation based on energy processes too, the spectral energetics of the model is formulated. Section 5 is a detailed numerical study, from both the dynamics and the energetics points of view, of two types of vacillation of the kind emphasized from laboratory experiments. Finally, some cases of atmospheric situations presenting striking similarities with vacillation are pointed out in Section 6.

2. The Laboratory Simulation of Large-Scale Atmospheric Flow

2.1. The Various Simulated Geophysical Flows

Laboratory experiments are one of the primary sources of knowledge in physics. Having carefully collected sufficient experimental observations, a theoretical explanation may be elaborated, leading to new results which in turn must be checked against new experiments. In this way, theory and experiment usually interplay in the study of physical phenomena.

Direct experimentation however is not possible in the case of some geophysical phenomena such as the dynamical processes associated with the large-scale atmospheric motion systems having linear horizontal dimensions of several thousands of kilometers. Nevertheless, making use of the concept of dynamic similarity (Rossby, 1926), one may try to build up laboratory devices which are capable of simulating some features of the actual atmospheric flow. There have been many attempts in this direction in the past (for an historical review, see Fultz et al., 1959).

Almost twenty-five years ago, such experiments, first suggested by Rossby (1947) and Starr were carried out by D. Fultz at the University of Chicago. In his initial experiments, Fultz (1949) generated thermally driven circulation in a rotating hemispherical shell of liquid. For suitable conditions of rotation rate and thermal forcing, he observed flow patterns for which the speeds of flow and changes with latitude compare reasonably well with the wind speeds observed in the atmosphere, provided the speeds are expressed in an appropriate unit, i.e. the equatorial rim speed.

This quantitative agreement stimulated further work in the laboratory modeling of the atmosphere. It should be recalled that, in analogy with the earth, where the resultant gravity field is nearly normal to the earth's mean sea level surface, Rossby (1926) suggested using as a bottom surface for the tank the paraboloid whose shape is the equipotential surface of the resultant of the local gravity field and of the centrifugal force due to the rotation of the vessel. For the rotation rate imposed in laboratory arrangements to simulate the atmospheric conditions, the slope of this equipotential surface is extremely small.

If one now considers a nearly radial gravity field operating on a fluid in motion on a sphere, it is clear that this configuration of the force field is impossible to reproduce with laboratory devices. It should however be noted that the effect of the curvature of the earth can be taken into account by other factors. For instance, in the case of a *barotropic* flow in a rotating tank, the theorem of conservation of absolute vorticity shows that an inward radial depth *decrease* of the fluid is dynamically analogous to an inward radial *increase* of entrainment vorticity (usually denoted by f in meteorology). It should be noted that the use of a bottom surface normal to the resultant force field operating on the fluid in the tank suppresses the radial depth gradient. On the other hand, the radial increase of depth in an hemispherical shell operates in an opposite direction to the one imposed to simulate the poleward increase of the earth's vorticity in barotropic flows.

This is perhaps one of the reasons why Fultz quickly dispensed with the hemispherical arrangement and used a cylindrical vessel, the so-called dishpan, which is, for a given experiment, rotating at constant angular velocity Ω (simulating the west to east rotation of the earth) around the vertically oriented axis of the cylinder. Differential heating is provided, at the bottom disk, by steady symmetrically distributed heating at the periphery (simulating equatorial conditions) and cooling at the center of the disk (simulating polar conditions).

Experiments with a cylindrical annulus rotating around its vertical axis were initiated in 1950 by R. Hide at Cambridge University. In these experiments, the heat source is at the outside wall which is at constant temperature T_b (simulating low latitude heating) and the cold source is at the inside wall at constant temperature T_a (simulating high latitude cooling, $T_a < T_b$).

When experiments are performed at different rotation rates and different intensities of the thermal forcing (i.e. for different south–north temperature contrasts) different types of flow patterns are generated. Roughly speaking, the dishpan experiments and those with a cylindrical annulus give the same qualitative observations. However, the reproducibility of the results is much better in the annulus experiments, presumably because the flow is much more constrained by the geometry of the system and consequently less sensitive to random perturbation in the initial conditions of operation.

For some intensities of thermal forcing and some rotation rates, the flow is symmetric with respect to the rotation axis and proceeds as in a trade wind cell, with upward motion at the warm rim and downward motion at the cold source. While moving northward in the upper layer (from the heat source to the cold source) the fluid is deflected by the rotation so that southerly winds are gradually changed into winds with a westerly component. At the "pole," the fluid particles subside and give rise, when moving to the south (from the cold source to the warm source), to northeasterly trade winds in the bottom layer. From these observations presenting obvious similarities with what is known about the tropical atmosphere, Fultz et al. (1959) have called this flow a *Hadley flow*.

For other operating conditions, unsymmetric flows with traveling waves arise. Some of these have a striking resemblance to the Rossby waves which are observed on upper level synoptic maps at middle and high latitudes. This similarity has led to the term: *Rossby flows*.

Careful experiments have shown that it was possible to distinguish several types of Rossby flows, each of them constituting a *Rossby regime*. Thus, Hide (1958) was the first (1950) to observe a *steady Rossby regime* in which the waves progress at constant speed and without changing their shape. The existence of another Rossby regime was also discovered by Hide (1953, 1958), namely a periodic regime in which the Rossby waves undergo periodic changes in their amplitude, shape, and progression (wave speed). Hide called this phenomenon *vacillation*. Finally, nonperiodic flows have also been observed, first (1950) by Long and Owens working with the dishpan arrangement at Chicago (Fultz et al., 1964) and are classified into the *irregular Rossby regime*. To sum up, the Hadley flow involves a single type of regime, called the Hadley regime, while Rossby flows can fall into one of the three types: steady, vacillating, or irregular Rossby regimes.

The critical conditions of thermal forcing and rotation rate for the transition between the Hadley and the steady Rossby regimes as well as between steady Rossby regimes with different wavenumbers have been experimentally

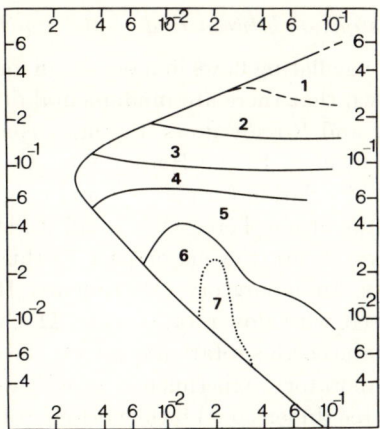

Fig. 1. Spectrum diagram of transition curves obtained by Fultz et al. (1959) in an annulus of water; abscissa is proportional to the square of the rotation rate Ω and the ordinate is proportional, for a given rotation rate, to the imposed thermal contrast ΔT. The "knee" curve separates the symmetric Hadley flow to the left of the curve and the Rossby flow. Figures indicate the wavenumber prevailing in the Rossby flow. Each experiment was performed in such a way that the radial temperature difference was raised slowly from zero after reaching the desired constant rotation rate.

determined by Fultz et al. (1959). Figure 1 gives the curves obtained by these authors working with a cylindrical annulus. Moreover, Fowlis and Hide (1965) have also given the critical transition conditions between the steady wave regime and the irregular regime. Vacillation arising between these two types of regimes is considered by these last authors as a transitional phenomenon whose domain of appearance in the plane of external parameters (Ω and $\Delta T = T_b - T_a$) is rather restricted as compared with the domain of existence of the other regimes.

It is interesting to recall that the experiments of Fultz et al. were performed in such a way that the fluid was first brought into solid rotation with the cylindrical annulus whose walls were initially kept at the same temperature. Having reached the desired rotation rate, the walls were then very slowly heated and cooled to the desired temperature values T_b and T_a, respectively.

This procedure allows one to state unequivocally the critical conditions of rotation rate and thermal forcing for the establishment of the different regimes. It is however completely artificial. Indeed, as noted by Fultz et al. (1959), different regimes can be reached for the same values of Ω and ΔT depending upon the way the steady values of these parameters are attained. This "hysteresis," making different states of motion correspond to the same values of Ω and ΔT, is a consequence of the nonlinear character of the equations governing the flow (Lorenz, 1962; Quinet, 1973b).

2.2. The Essential Distinction between Hadley and Rossby Flows

Before dealing with vacillating flows in more detail, one may recall that the experiments have shown that there are fundamental differences between the two classes of Hadley and Rossby flows, the most characteristic one being the process of heat transport. In the Hadley flow, this transport is accomplished by meridional circulation (in planes containing the axis of rotation) as at low latitudes in the atmosphere, while in all Rossby regimes the quasi-horizontal disturbances ensure the major part of this transport, as in the atmospheric westerlies. In laboratory experiments, Hadley circulations in particular are associated with slow rotation rates Ω. On the other hand, the vertical component of the earth's rotation is zero at the equator and increases poleward. Thus, in laboratory experiments as well as in the atmosphere, the flow tends to be three-dimensional if the vorticity of the absolute rotation is notably less than the vorticity of the relative motion. The comparison between the relative and entrainment vorticities is expressed by the Rossby number

$$(2.1) \qquad \mathrm{Ro} = \bar{u}/L\Omega = \bar{\zeta}/2\Omega$$

where $\bar{u}$ is the horizontal relative velocity scale, L the length scale, and $\bar{\zeta}$ the mean vertical component of the relative vorticity of the flow. This number also expresses the ratio between the relative acceleration and the Coriolis acceleration. Quasi-geostrophic motions, for which the Coriolis acceleration dominates, are consequently associated with the low values of Ro, that is to say with high rotation rates Ω. On the contrary, when these rotation rates are small, there exists an unbalanced part of the Coriolis force (the ageostrophic part) and the zonal flow is systematically deflected in a direction normal to the flow, forcing the fluid particles to travel along horizontal spirals. Vertical (ageostrophic) motions, imposed by continuity, then take place at the cold (descending motion) and warm (ascending motion) sources and allow the spiraling fluid particles to travel between the heat sources, ensuring in this way the closure of the heat transfer cycle.

At higher rotation rates, quasi-geostrophic flows become possible. In this case, the flow remains essentially two-dimensional (quasi-horizontal), the fluid particles approaching the cold source being able to progress later southward on the same horizontal surface as a consequence of the balance between the Coriolis forces and the horizontal pressure gradient forces. Consequently, the heat transfer is now accomplished by horizontal eddies, the fluid particles successively traveling quasi-horizontally from south to north and then from north to south. Since the thermal wind law holds for quasi-geostrophic flow, there is a strong correlation between the flows in the different layers. This is not so for Hadley flows where trade or antitrade winds prevail in the lower or upper layer respectively.

The sudden transition from the Hadley to the Rossby regime when the rotation rate is progressively increased suggests the intervention of some mechanism of dynamic instability. According to Bolin (1952) and following the theory of baroclinic (Charney, 1947; Eady, 1949) and barotropic (Kuo, 1949) instability, the dynamic instability condition would result from the increase of the wind shear (both horizontal and vertical) that might be associated with the increase of the rotation rate. But the observed restabilization of the symmetric Hadley regime at high thermal forcing (upper part of Fig. 1) then appears as paradoxical. Indeed, it can be expected that intense thermal forcing generates high horizontal temperature gradients within the fluid and consequently, according to the thermal wind relation, high vertical wind shear. The paradox was solved by Lorenz (1962) who pointed out that intense thermal forcing gives rise to intense upward motions of warm air near the outer rim and downward motion of cold air near the center (or the inner rim). Consequently, the static stability of the fluid increases with the thermal forcing so that finally the flow becomes baroclinically stable again and the symmetric Hadley regime is reestablished. In this context it has to be emphasized that baroclinic instability is considered as the single mechanism responsible for the transition from the Hadley to the steady Rossby regime. This has been indirectly confirmed by Lorenz (1962) who generated steady Rossby regimes in a model in which the process of barotropic instability was suppressed.

Let us also point out that, if vertical motion plays an essential role in realizing the critical conditions of static stability allowing the reappearance of Hadley flows, some differences can be anticipated according to the way in which the increase of static stability can be accomplished. In the Hadley regime, the cellular structure of the flow ensures a self-stabilization. This is not the case in Rossby flows where the motion is essentially quasi-horizontal. Accordingly, the reestablishment of a Hadley regime from a Rossby regime will presumably need a higher thermal forcing than the one compatible with a directly established Hadley flow. This hysteresis in the observation of Hadley flows according to the sequence of generated flows has indeed been observed by Fultz et al. (1959) and has been explained by Lorenz (1962).

The laboratory simulation of both Hadley and Rossby flows strongly suggests that, in spite of the obvious geometrical distortion, laboratory experiments are able to capture some of the very essential properties of the large-scale atmospheric motions. Interest will be more particularly focused here on Rossby vacillating flows which appear sufficiently simple as to remain "understandable" and, consequently, capable of improving our knowledge of the large-scale atmospheric motions. Indeed, the distinction between vacillation and nonperiodic flows seems to be more a question of degree than one of fundamental character, as already denoted by the fact that they both belong to the same class of flow. The observation of several sequences of upper air synoptic maps suggests that the evolution of the atmospheric flow pattern could be interpreted as some "perturbation" of a vacillation cycle. If this could actually be confirmed by a detailed analysis, much would be gained about the understanding of the behavior of the atmosphere.

2.3. The Characteristic Features of Vacillation

2.3.1. Vacillation in the Laboratory Flows. It has already been mentioned that Hide has been the first to recognize a new phenomenon, he called vacillation, in which the waves of the flow pattern are submitted to periodic fluctuations in amplitude, shape, and progression rate. This type of flow has also been extensively studied by Fultz and his collaborators. Figure 2 gives one example of the upper flow observed in a cylindrical annulus during a vacillation cycle whose period is 16.25 revolutions (one revolution simulating one day). The time interval between two successive photographs is four revolutions. The most peculiar feature is the periodic change in the horizontal plane of the orientation of the trough and ridge lines of the five-wave, while the zonal flow is mainly concentrated in a meandering jet stream. Moreover, the amplitude of the five-wave also exhibits fluctuations, increasing from the tilted five-wave pattern of Fig. 2a to the strong upper cyclonic circulation of Fig. 2c.

One of the essential results of laboratory experiments has been to show that vacillation is accompanied by a periodic variation in the heat transported from the warm to the cold source. A representation of these fluctuations can be given by measuring the temperature difference between the inlet and outlet water circulating at the cold bath of the annulus as represented in Fig. 3. Fultz *et al.* (1959) have also shown that the meridional profile of the zonal wind undergoes substantial fluctuations during the cycle. On the one hand, the position of the maximum zonal wind fluctuates in latitude, being successively near the outside and the inside wall. On the other hand, the intensity of the zonal wind varies simultaneously with the same period. These fluctuations in wind speed are such that the maximum zonal westerlies are attained near the time of the minimum heat transport at the cold source.

Obviously, many properties of the flow exhibit the same kind of periodic fluctuations. Thus large positive values of the zonally averaged relative northward eddy angular momentum transport $r^2[u'v']$, due to the fluctuations u', v' of the velocity components u ($u > 0$, westwind) and v ($v > 0$, southwind) from their zonal average $[u]$, $[v]$ along the latitude circle of radius r, are followed by smaller values. As a rule, the maximum transport is located

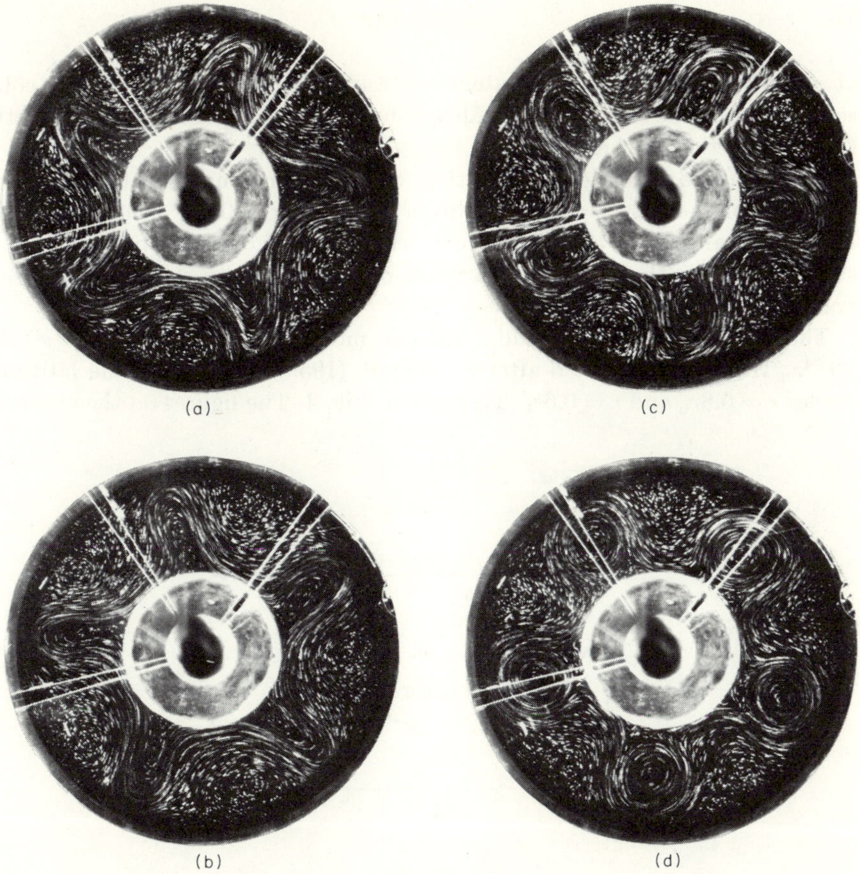

Fig. 2. Sequence of photographs showing a kinetic energy vacillation cycle. Notice the changes in the orientation of the trough axis from NW–SE in Fig. 2b to NE–SW in Fig. 2d and the change of amplitude of the wave between Fig. 2a and Fig. 2c. The period of the cycle is $16\frac{1}{4}$ revolutions; the time interval between two successive photographs is 4 revolutions (courtesy of D. Fultz).

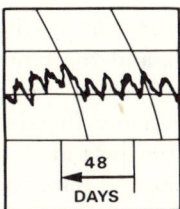

Fig. 3. Trace obtained by Fultz *et al.* (1959) of the temperature difference between inlet and outlet water at the cold source during vacillation. The period of the cycle is about 12 revolutions.

at middle latitude (approximately midway between the warm and cold sources) while the subsequent smaller values are slightly shifted to the north (in the direction of the cold source).

According to Fultz et al. negative values of the momentum transport occur when the waves are tilting from NW toward SE, a situation which arises generally when the troughs are deepening to generate upper cyclones, somewhat after the time of occurrence of the minimum value of the heat transport at the cold source.

The relative northward eddy angular momentum transport $r^2[u'v']$ in $10^{-4} r_b^2 \Omega^2$ units, estimated after Fultz et al. (1959) data, across the latitude circles $r = 0.8 r_b$ and $r = 0.6 r_b$, is shown in Fig. 4. The figures on the abscissa

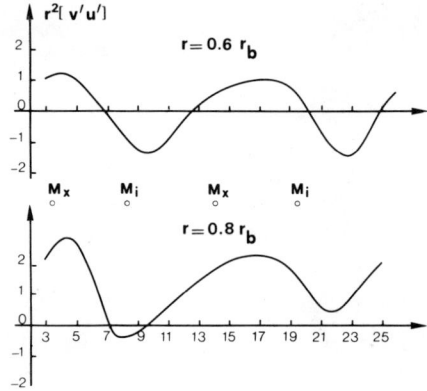

FIG. 4. Relative northward eddy momentum transport $r^2[u'v']$ in $10^{-4} r_b^2 \Omega^2$ units across latitude circle $r = 0.8 r_b$ and $r = 0.6 r_b$ during the vacillation cycle, Ω being the angular velocity of rotation and r_b the radius of the outer rim; the inner rim is at radius $0.4 r_b$. The small circles denote the time of occurrence of maximum (Mx) and minimum (Mi) heat transport to the cold source; numbers in abscissa are revolution numbers (After Fultz et al., 1959).

denote revolution numbers so that the vacillation period is about 12 days. The outer rim is at a radius r_b and the inner rim lies at $0.4 r_b$. The mean transport during one vacillation cycle at $0.8 r_b$ is definitely toward the north. The small circles denote respectively the time of occurrence of maximum (Mx) and minimum (Mi) heat transport at the cold source. Moreover, Pfeffer and Chiang (1967) noted that Fultz's data can also be used to demonstrate that the barotropic transfer rate between eddy kinetic energy and zonal kinetic energy undergoes substantial fluctuations in magnitude and also perhaps in sign during the cycle. Consequently, while baroclinic processes are necessarily associated with the thermal forcing, nevertheless vacillation involves also nonnegligible barotropic energy exchanges. Moreover, the cyclic increase and decrease of the transfer between the two types of kinetic energy proves that the zonal flow interacts with the eddies so that nonlinear mechanisms play an essential role in vacillating flows.

More recently, Pfeffer and his collaborators at Florida State University (1965), proceeded to very careful experiments, especially in the domain of vacillating flows. While recognizing that the difference is presumably rather one of degree than one of intrinsic nature, they were led to distinguish two kinds of vacillation.

A first one, of the kind presented in Fig. 2, is called a "*tilted trough vacillation*." Moreover, in order to draw attention to the very fact observed by Fultz et al. (1959) that the fluctuations of the tilt of trough and ridge lines are associated with fluctuations of the zonal and eddy kinetic energies, Pfeffer and Chiang (1967) suggest also calling this type of flow a "*kinetic energy vacillation*." But Pfeffer et al. (1965) and Pfeffer and Chiang (1967) also reported the existence of a different kind of vacillation, an example of which is given in Fig. 5. The essential difference from the case reported in

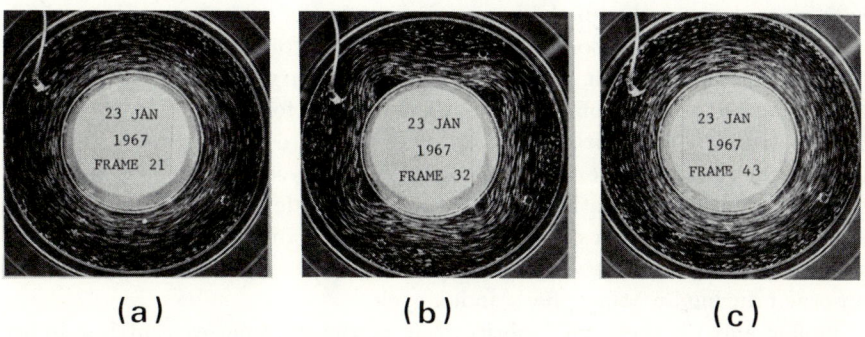

FIG. 5. Sequence of photographs showing extreme stages of a potential energy vacillation cycle. Notice the change in the amplitude of the wave without noticeable change of its tilt. The period of the cycle is 92 revolutions; the time interval between two successive photographs is 46 revolutions (courtesy of R. Pfeffer and W. Fowlis).

Fig. 2 lies in the fact that the trough and ridge lines no longer exhibit fluctuations in orientation during the cycle but that the wave amplitude undergoes major oscillations. The wave may even nearly vanish during the cycle so that the flow appears then as quasi-zonal (Figs. 5a and 5c). Therefore, these authors have proposed calling this type of vacillation "*amplitude vacillation*." As a matter of fact, amplitude vacillation is most common with high viscous fluids and appears at a lower rotation rate and higher thermal forcing than the tilted trough vacillation (Fowlis and Hide, 1965).

2.3.2. Meteorological Interest of Vacillation. The similarity of all the characteristics of vacillation with what is known about the atmosphere makes vacillating flows of considerable meteorological interest. The tilting of the planetary waves has indeed been recognized by Starr (1948) to play a fundamental role in the necessary northward transport of angular momentum of the atmosphere by the large-scale atmospheric eddies. The positive correlation between the fluctuations of northward and eastward velocity components required for a northward transport of angular momentum across latitude circles is realized when the troughs exhibit a SW to NE tilt as, for example, in Fig. 2a and Fig. 2d. Thus, in this respect, laboratory vacillating flows do simulate one of the most fundamental processes of the atmospheric large-scale turbulence. Figure 5 and, less obviously, Fig. 2 also suggest an analogy between vacillation and the recognized quasi-periodic redistribution of the angular momentum within the atmosphere giving rise to the so-called index cycles (Namias, 1954). In this respect, Fig. 5a or Fig. 2a would be typical of a high index, quasi-zonal circulation and Fig. 5b or Fig. 2c of a low index one. The estimates of the heat transport show that, in the atmosphere as well as in the laboratory flows, the low index configuration is associated with a maximum value of the heat transport to the cold source while a high index circulation corresponds to a minimum value of this transport. Finally, the period of vacillation has been observed to vary between large ranges, depending on the rotation rate Ω and the thermal forcing ΔT. Nevertheless, these periods are most often of the order (in the corresponding units, recalling that one day is equal to one revolution) of several weeks which is of the order of magnitude of the duration of the atmospheric index cycles (Namias, 1954). Moreover, the evolution of the meridional profile of the zonal wind during a vacillation cycle, as recalled in Section 2.3.1, is consistent with what is observed during an atmospheric index cycle.

Pfeffer and Chiang (1967) notice that in the extreme case of low index during an amplitude vacillation (Fig. 5b) the flow pattern closely looks like what is known as a blocking situation in synoptic meteorology. However, blocking situations do not generally repeat themselves around a whole hemisphere. Nevertheless configurations of three to five blocking highs are not uncommon, and have been reported by Pfeffer and Chiang (1967). The transition from a high index (quasi-zonal) circulation of the atmosphere to a configuration of a few blocking highs, regularly distributed around the hemisphere, is generally rather abrupt and does not reappear with a definite periodicity in time. According to the cases reported by Pfeffer *et al.* (1965), a particular feature of this major change of the whole atmospheric circulation of the Northern hemisphere would happen usually in winter, either in late December or in January, and be followed by an extreme maximum amount of zonal available potential energy. The energy study of this type of process

(Krueger et al. (1965)) has shown that it is characterized by a rapid adiabatic conversion from potential energy into kinetic energy. On this base, Pfeffer and Chiang (1967) have suggested also calling the amplitude vacillation a "*potential energy vacillation.*"

All these considerations lead to the idea that potential energy vacillation should essentially result from cyclic variation of the conversion rate between available potential energy and kinetic energy only (baroclinic process) while the appearance of kinetic energy vacillation necessarily implies the existence of substantial transfers between eddy and zonal kinetic energy (barotropic process). However, Pfeffer and Chiang (1967) recall that vacillation undoubtedly involves nonlinear effects, which make it impossible to isolate a single energy process, either barotropic or baroclinic, in the energy cycle of the flow. A detailed analysis of the energy cycle of potential and kinetic energy vacillation would therefore considerably improve knowledge of the phenomenon. This problem will be dealt with in particular in this review with the aid of numerical simulation.

3. A Model Atmosphere for the Study of Vacillation

3.1. *General Comments on the Theoretical Approach to the Study of Vacillation*

Theoretical studies (Davies, 1953, 1959; Kuo, 1954, 1957; Lorenz, 1953) have been devoted to the different types of flow observed in laboratory experiments in order to derive, in particular, the critical operational conditions of thermal forcing and rotation rate for the appearance of the different flow regimes. The Hadley regime and the conditions of transition from this regime to the Rossby regimes have especially been extensively discussed by the authors mentioned above.

More recently, Lorenz (1962) put forward the fruitful idea of using in this approach the spectral form of a set of thermohydrodynamical equations capable of describing the fluid motion. To simulate the laboratory flow, Lorenz uses a quasi-geostrophic, two-layer model of an inviscid and non-heat-conducting fluid, the friction and heating effects being taken into account in a roughly parameterized form. The spectral equations are obtained by expanding the dependent variables into their spectral components in a functional space and retaining in the equations a few scales of motion only.

This procedure seems specially suited here since, in most cases, the actual laboratory flow, and particularly vacillation, involves only very few waves. As the infinite number of degrees of freedom of the initial formulation is, in this way, reduced to only a few, the theoretical study turns out to be greatly simplified. This technique, however, does not suppress the essential nonlinear character of the equations.

The Lorenz (1962) determination of the critical conditions of the instability of the Hadley regime in a cylindrical model is in good qualitative agreement with the laboratory measurements in an annulus region, as reported by Fultz et al. (1959) (Fig. 1). The quantitative agreement, however, is rather poor. In order to obtain a better concordance between the theory and the experiments, Merilees (1968) used the correct cylindrical annulus geometry and added to Lorenz' model the effects of the viscosity and of the heat conductivity of the fluid as well as those of the lateral boundary. In this way, he demonstrated that these physical factors play an important role in the dynamics of the laboratory flows.

Nevertheless, and even surprisingly, Lorenz (1963) has simulated all types of laboratory regimes in his physically highly simplified model where, moreover, the channel geometry was used. This seems to support the hypothesis that the different regimes observed in laboratory experiments constitute a general characteristic of thermally driven rotating fluids. If the geometry of the experiment and the physical properties of the fluid influence the critical conditions of transition between different regimes or between different wavenumbers within the Rossby regimes, they do not critically alter the global mechanical behavior of the system. Consequently, and although we are aware that our simplification could appear as rather drastic, we shall use the channel type Lorenz' model for the study of vacillation. Needless to say, the aim is to capture only some of the essential, nonlinear aspects of the phenomenon. First the dynamics will be considered and then its energetics. Finally, the existence of vacillation in the atmosphere will be briefly presented.

3.2. The Tensor Formulation of Quasi-Hydrostatic Flows

3.2.1. General Characteristics of Tensor Calculus. Vacillation is possible with systems of various geometrical shapes. Therefore, it will be useful to express the equations of flow in the most general form, valid in any coordinate system. For this purpose, use will be made of tensor calculus.

Let us first specify a distinction between (i) a reference body R, which will be rigid and (ii) a coordinate system, rigidly attached to R (imbedded in R) or not. The following convention will prove to be convenient: *an index represented by a Greek letter takes the values 1, 2, and 3 while one represented by a Roman letter will take the values 1 and 2 only.*

The velocity of a material point P of coordinates $x^1(t)$, $x^2(t)$, $x^3(t)$, where t is time, with respect to a reference body R is given by

$$(3.1) \qquad V^\alpha = v^\alpha - r^\alpha \qquad (\alpha = 1, 2, 3)$$

where v^α is the velocity of P with respect to the coordinate system (x^1, x^2, x^3),

(3.2) $$v^\alpha = dx^\alpha(t)/dt \equiv \dot{x}^\alpha$$

while r^α is the velocity of R with respect to this coordinate system. Obviously, the velocity of all the fluid particles form a velocity field

(3.3) $$V^\alpha = V^\alpha(x^1, x^2, x^3, t)$$

It should be noted that V^α being defined with respect to R is a vector, independent of the choice of the coordinate system, while both v^α and r^α depend upon this choice.

The kinetic energy per unit mass with respect to R then assumes the form

(3.4) $$K = \tfrac{1}{2} g_{\alpha\beta} V^\alpha V^\beta = \tfrac{1}{2} V_\alpha V^\alpha$$

In (3.4), as everywhere else in this section, the usual summation convention of tensor calculus must be applied: if an index (dummy index) appears twice in an expression, once as a subscript and once as a superscript, it has to be summed for all allowed values of the index. The symmetric tensor $g_{\alpha\beta} = g_{\beta\alpha}$ is known as the metric tensor of the three-dimensional space, usually introduced from the expression of the square of the length δs of the vector δx^α:

(3.5) $$(\delta s)^2 = g_{\alpha\beta}(x^\gamma, t)\, \delta x^\alpha\, \delta x^\beta$$

the unit length being chosen. The components V_α and V^α are respectively the covariant and contravariant components of the vector **V** and are associated by the metric tensor according to

(3.6) $$V_\alpha = g_{\alpha\beta} V^\beta$$

a property which has already been used in (3.4). The covariant components V_α, v_α, r_α associated respectively with the contravariant components V^α, v^α, r^α are such that

(3.1') $$V_\alpha = v_\alpha - r_\alpha$$

For the sake of completeness, let us also recall that the so-called "physical components" of **V** are given by

(3.7) $$V^1\sqrt{g_{11}},\ V^2\sqrt{g_{22}},\ V^3\sqrt{g_{33}}$$

These are the components of the vector **V** referred to a Cartesian system coinciding with the axis of the basis vectors of the arbitrary coordinate system x^α. These basis vectors are defined at each point and are tangent to the coordinate curve along which x^α alone varies. It will also prove necessary to use the twice contravariant tensor $g^{\alpha\beta}$ defined by

(3.8) $$g^{\alpha\beta} = \operatorname{cof} g_{\alpha\beta}/G$$

where (cof $g_{\alpha\beta}$) denotes the cofactor of $g_{\alpha\beta}$ in the determinant

(3.9) $$G = |g_{\alpha\beta}|,$$

so that

(3.10) $$g_{\alpha\gamma} g^{\gamma\beta} = \delta_\alpha{}^\beta$$

where $\delta_\alpha{}^\beta$ is the Kronecker tensor

(3.11) $$\delta_\alpha{}^\beta = 1 \quad \text{if} \quad \alpha = \beta, \qquad \delta_\alpha{}^\beta = 0 \quad \text{if} \quad \alpha \neq \beta$$

With the usual notation of a partial derivative

(3.12) $$\partial_\alpha(\cdots) \equiv (\partial/\partial x^\alpha)(\cdots)$$

and considering K as a function of t, x^α, and $\dot{x}^\alpha$ ($\alpha = 1, 2, 3$), the equation of the fluid motion with respect to R may be written (Defrise, 1964)

(3.13) $$(d/dt)/(\partial K/\partial \dot{x}^\alpha) - \partial K/\partial x^\alpha = -2\omega_{\alpha\beta} V^\beta - \partial_\alpha \phi - \rho^{-1} \partial_\alpha p$$

where $d/dt\ (\cdots)$ is the derivative along the flow and $2\omega_{\alpha\beta} V^\beta$ is the Coriolis acceleration. When E is the velocity of R with respect to an absolute reference body (entrainment velocity) the vorticity tensor $2\omega_{\alpha\beta}$ of the absolute motion of R is given by

(3.14) $$2\omega_{\alpha\beta} = \partial_\alpha E_\beta - \partial_\beta E_\alpha = -2\omega_{\beta\alpha}$$

In (3.13) ϕ is the potential of the body forces, ρ the density of the fluid, and p the pressure. Actually, the right side of (3.13) should also include the viscous and turbulent stresses; these however will be disregarded here in accordance with the previous assumption concerning the viscosity of the fluid.

For the study of large-scale atmospheric motion it is convenient to use the pressure as the vertical coordinate so that the coordinate system (x^1, x^2, $x^3 \equiv p$) is not imbedded in R. Van Isacker (1963) adopts arbitrary "horizontal" coordinates x^i ($i = 1, 2$) at the mean sea level of the earth's surface and $x^3 = p$. Moreover he takes for the length of the vector (δx^1, δx^2, $\delta x^3 \equiv \delta p$) the approximate form

(3.15) $$(\delta s)^2 = g_{ij}(x^1, x^2)\, \delta x^i\, \delta x^j + \rho^{-2} g^{-2}(\delta p)^2$$

where g is the (constant) acceleration of gravity and $\rho = \rho(x^\alpha, t)$. The choice of the metric (3.15) implies three geometrical approximations: (1) the isobaric surfaces are considered as being horizontal for *measuring distances*; (2) the distance between two points (x^1, x^2, p), (x'^1, x'^2, p) of the same isobaric surface is supposed to be independent of the pressure p; and (3) for measuring the vertical distance δz, use is made of the hydrostatic equation

(3.16) $$\delta z = -(\rho g)^{-1} \delta p$$

along the ascending vertical ($x^1 = a$, $x^2 = b$) of each mean sea level point (a, b). These approximations are acceptable for the description of the large-scale flow of a shallow atmosphere in quasi-static motion. From a geometrical point of view, the splitting between horizontal and vertical coordinates in (3.15) implies that the three-dimensional space (x^1, x^2, p) is layered into isobaric sheets by the pressure parameter p.

Let us immediately note that, according to (3.15),

(3.17) $\quad\quad\quad g_{3i} = g_{i3} = 0 \quad \text{for} \quad i = 1, 2 \quad \text{and} \quad g_{33} = (\rho g)^{-2}$

(3.18) $\quad\quad\quad\quad\quad\quad G = g_{33} \gamma$

where γ is the determinant $|g_{ij}|$ so that

(3.19)
$$g^{3i} = g^{i3} = 0 \quad \text{for} \quad i = 1, 2 \quad \text{and} \quad g^{33} = 1/g_{33}$$
$$g^{11} = g_{22}/\gamma,\ g^{22} = g_{11}/\gamma,\ g^{12} = g^{21} = -g_{12}/\gamma$$

and

(3.20) $\quad\quad\quad\quad g_{ik} g^{kj} = \delta_i{}^j, \quad k = 1, 2$

Finally, the velocity of the earth (R) with respect to the coordinate system (x^1, x^2, p) is given by

(3.21) $\quad\quad\quad\quad\quad\quad r^1 = r^2 = 0$

the coordinates x^1, x^2 being imbedded in R, while

(3.22) $\quad\quad\quad\quad\quad\quad r^3 = \partial p/\partial t$

(x^1, x^2, z being kept constant) so that

(3.23) $\quad\quad\quad\quad V^1 = v^1,\ V^2 = v^2,\ V^3 = \omega - r^3$

with the usual notation

(3.24) $\quad\quad\quad\quad\quad\quad v^3 = dp/dt \equiv \omega$

Tensor calculus has been used by several authors in meteorology (see, for instance, Van Mieghem and Vandenplas, 1950; Langlois and Kwok, 1969). A systematic treatment of tensor calculus specially applied to atmospheric mechanics has been given by Defrise (1964).

3.2.2. *The Equations of Large-Scale Quasi-Static Flows.* According to (3.4) and (3.1)

(3.25) $\quad\quad\quad \partial K/\partial \dot{x}^\alpha = g_{\alpha\beta}(\dot{x}^\beta - r^\beta) = g_{\alpha\beta} V^\beta = V_\alpha$

(3.26)
$$\partial K/\partial x^\alpha = \tfrac{1}{2}(\dot{x}^\beta - r^\beta)(\dot{x}^\nu - r^\nu)\, \partial_\alpha g_{\beta\nu} - g_{\beta\nu}(\dot{x}^\nu - r^\nu)\, \partial_\alpha r^\beta$$
$$= \tfrac{1}{2} V^\beta V^\nu\, \partial_\alpha g_{\beta\nu} - V_\beta\, \partial_\alpha r^\beta$$

so that

(3.27) $$\frac{d}{dt}\left(\frac{\partial K}{\partial \dot{x}^\alpha}\right) - \frac{dK}{\partial x^\alpha} = \frac{dV_\alpha}{dt} - \frac{1}{2} V^\beta V^\nu \, \partial_\alpha g_{\beta\nu} + V_\beta \, \partial_\alpha r^\beta$$

The equation of motion (3.13) then assumes the form

(3.28) $\quad dV_\alpha/dt - \frac{1}{2} V^\beta V^\nu \, \partial_\alpha g_{\beta\nu} + V_\beta \, \partial_\alpha r^\beta + 2\omega_{\alpha\beta} V^\beta = -\partial_\alpha \phi - \rho^{-1} \, \partial_\alpha p$

or the equivalent, more usual form

(3.28') $\quad dV_\alpha/dt - V^\beta V_\nu \, \Gamma^\nu_{\alpha\beta} + V_\beta \, \partial_\alpha r^\beta + 2\omega_{\alpha\beta} V^\beta = -\partial_\alpha \phi - \rho^{-1} \, \partial_\alpha p$

with the Christoffel symbol

(3.29) $$\Gamma^\nu_{\alpha\beta} = \tfrac{1}{2} g^{\nu\varepsilon}(\partial_\alpha g_{\varepsilon\beta} + \partial_\beta g_{\alpha\varepsilon} - \partial_\varepsilon g_{\alpha\beta}).$$

Equation (3.28) can also be put in the more convenient form (Van Isacker, 1963; Phillips, 1965)

(3.29') $\quad \partial \mathbf{V}/\partial t + \nabla(\tfrac{1}{2} V_\beta V^\beta) - \mathbf{V} \times \text{curl}(\mathbf{V} + E) = -\nabla\phi - \rho^{-1}\nabla p$

free of Γ symbols and where $\times$ is the vector product and ∇ the usual "del" operator.

When (3.15) is taken into account, the equation of the horizontal motion deduced from (3.28) can be written

(3.30) $\quad dV_i/dt - \tfrac{1}{2} V^j V^k \, \partial_i g_{jk} - \tfrac{1}{2} V^3 V^3 \, \partial_i g_{33} + V^3 \, \partial_i r^3 + 2\omega_{i\alpha} V^\alpha = -\partial_i \phi$

for $i = 1, 2$.

The third equation of (3.28) is always replaced, in the study of large-scale atmospheric flow, by the hydrostatic approximation

(3.31) $$\partial_p \phi = -\rho^{-1}$$

which filters out the meteorologically nonsignificant compressibility waves (see, for instance, Holton, 1972). The usefulness of this approximation results from the fact that the period τ of the compressibility waves in the atmosphere can be much shorter ($\tau < 1$ sec) than the period of the large-scale Rossby waves ($\tau > 10^5$ sec). Accordingly, dealing explicitly with the compressibility waves in atmospheric models would necessitate using an integration time step which is about 10^5 times less than when the flow is considered to be hydrostatic. One may nevertheless recall that this approximation is justified only by the empirical fact that the large-scale motions of the atmosphere are quasi-static.

This simplification, although offering the greatest practical advantages, violates the conservation laws expressing the fundamental physical principles which are intrinsically contained in the original equation (3.28). Energetic consistency of approximate sets of hydrodynamic equations specially suited for studying the atmosphere has been discussed by Lorenz (1960), Van Isacker (1963), and Phillips (1965). As the diagnostic equation (3.31) replaces the equation of motion along the vertical, a consistent approximate kinetic energy is defined as the kinetic energy of the horizontal motion only. In this case, the theorem of the (approximate) kinetic energy holds provided (3.30) reduces to

$$dV_i/dt - \tfrac{1}{2} V^j V^k \, \partial_i g_{jk} + 2\omega_{ij} \, V^j = -\partial_i \phi \tag{3.32}$$

Note that the terms which have been suppressed in (3.30) may be considered as negligibly small from scale analysis considerations.

In analogy with (3.29'), (3.32) can also be put in the form

$$\partial V_i/\partial t + \partial_i(\tfrac{1}{2} V_j V^j) - e_{ij} V^j(\zeta + f) + \omega \, \partial_3 V_i = -\partial_i \phi \tag{3.33}$$

where ζ is the "vertical" component of the relative vorticity

$$\zeta = e^{ij} \, \partial_i V_j = (1/\sqrt{\gamma})(\partial_1 V_2 - \partial_2 V_1) \tag{3.34}$$

and f the vertical component of the vorticity of the earth's rotation

$$f = 2\omega_{12}/\sqrt{\gamma} \tag{3.35}$$

called the Coriolis parameter. The e^{ij} and e_{ij} tensors are defined by

$$e_{11} = e_{22} = e^{11} = e^{22} = 0$$
$$e_{12} = -e_{21} = \sqrt{\gamma}; \ e^{12} = -e^{21} = 1/\sqrt{\gamma} \tag{3.36}$$

and satisfy the relation

$$e^{ij} g_{jk} e^{kl} = -g^{il} \tag{3.37}$$

With (3.15) the continuity equation (Defrise, 1964) assumes the form

$$(1/\sqrt{\gamma}) \, \partial_i(\sqrt{\gamma} \, V^i) + \partial_3 \omega = 0 \tag{3.38}$$

where the first term on the left-hand side is the isobaric divergence of the velocity field (V^1, V^2),

$$\delta \equiv (1/\sqrt{\gamma}) \, \partial_i(\sqrt{\gamma} \, V^i) \tag{3.39}$$

Finally, the thermodynamic equation assumes the form

$$c_p \, dT/dt - \rho^{-1} \, dp/dt = H \tag{3.40}$$

where T is the absolute temperature, c_p the specific heat of the fluid at constant pressure, and H the rate of heating per unit mass.

With the ideal gas state equation

$$p = \rho RT, \tag{3.41}$$

where R is the specific gas constant for dry air, Eqs. (3.31), (3.33), (3.38), and (3.40) form a closed differential set in the six unknown functions V^1, V^2, ω, ρ, ϕ, and T and determine, at least in principle, the evolution of an inviscid dry atmosphere once suitable initial and boundary conditions are prescribed.

The fluid used in laboratory experiments (generally water) can be considered as being incompressible so that ρ is a function of T only and is expressed by the thermal expansion law

$$(3.41') \qquad \rho = \rho_0[1 - \varepsilon(T - T_0)]$$

where ε is the coefficient of thermal expansion of the liquid and ρ_0 its density at temperature T_0. Consequently, except for using (3.41') instead of (3.41), the large-scale atmospheric flow and the geophysical laboratory flows are governed by analogous sets of equations. In the remaining part of this paper, we shall limit ourself to the atmospheric case.

3.3. The Equations of the Two-Layer Balanced Model

3.3.1. The Equations of Quasi-Geostrophic Flow. We shall now establish the invariant (tensorial) form of the equations of a quasi-geostrophic two-layer model. The third equation of motion having been replaced by the diagnostic hydrostatic equation (3.31), we are concerned here only with the two equations (3.33) of the horizontal motion.

Let us first recall the invariant form of the Laplacian

$$(3.42) \qquad \nabla^2 A = (1/\sqrt{\gamma})\, \partial_i(\gamma^{1/2} g^{ij}\, \partial_j A)$$

and of the Jacobian

$$(3.43) \qquad J(A, B) = e^{ij}\, \partial_i A \cdot \partial_j B$$

in the two-dimensional space (x^1, x^2), A and B being two scalars. Van Isacker (1963) considers also the invariant differential operator

$$(3.44) \qquad Q(A, B) = g^{ij}\, \partial_i A \cdot \partial_j B$$

equivalent to $\nabla A \cdot \nabla B$ in the usual "del" operator notation, and finds it convenient to introduce the operator S defined by

$$(3.45) \qquad S(A, B) = (1/\sqrt{\gamma})\, \partial_i(\gamma^{1/2} g^{ij} A\, \partial_j B) = A\nabla^2 B + Q(A, B)$$

which is also the divergence of $g^{ij} A\, \partial_j B$ in (x^1, x^2)-space.

Following Helmholtz' theorem, the isobaric velocity field can be represented with the aid of two functions ψ and χ, namely

$$(3.46)[1] \qquad V^i = -e^{ij}\, \partial_j \psi + g^{ij}\, \partial_j \chi$$

---

[1] The reversed convention regarding the sign of χ is also used.

so that the covariant components of **V** are

(3.47) $$V_i = g_{ij} V^j = -e_{ij} g^{jk} \partial_k \psi + \partial_i \chi$$

ψ is the stream function and χ the velocity potential of the isobaric wind $\mathbf{V} \equiv (V^1, V^2)$.

Recalling the definitions (3.34) and (3.39) of the isobaric vorticity ζ and divergence δ respectively, we have

(3.48) $$\zeta = \nabla^2 \psi$$

(3.49) $$\delta = \nabla^2 \chi$$

so that, according to (3.46), ψ expresses the nondivergent, rotational part of the wind and χ its divergent, irrotational part.

Operating with $e^{ji} \partial_j(\cdots)$ and $(\gamma^{1/2})^{-1} \partial_j(\gamma^{1/2} g^{ji} \cdots)$ on the equations of the horizontal motion (3.33) we obtain a prognostic equation for ζ and δ respectively, the so-called isobaric vorticity and divergence equations. Formally, the unknown functions V_1 and V_2 are now replaced by ζ and χ, a substitution which proves useful in dynamic meteorology.

At this stage, it is essential to mention two attitudes in the mathematical modeling of the atmosphere. One is based on the empirical fact that, in a shallow atmosphere, the large-scale wind field is nearly determined by the geopotential field ϕ. Assuming that this condition strictly holds leads to replacing the equation of the isobaric divergence of the wind by a diagnostic equation, the so-called balance equation, between the geopotential field ϕ and the stream function ψ only. From a physical point of view, the balance equation ensures a permanent adjustment between the ϕ- and the ψ-fields and, consequently, radically suppresses the gravity inertial waves. Such models are usually called *balanced models*. It then becomes possible to use numerical integration time steps of the order of 10^4 sec which are sufficient to adequately describe the evolution of the remaining Rossby waves (periods of the order of 10^5 sec).

The other attitude consists of maintaining the isobaric divergence equation. But, for the atmosphere, the integration time step is then necessarily of the order of 10^2 sec, which seems extremely painstaking as compared with the period of the Rossby waves. However, because gravity waves are supposed to be important in some circumstances, in particular when the air flows over mountains, this procedure is now widely used in forecasting models which are then referred to as *primitive equation models*. Finally, it should be noted that J. Van Isacker (personal communication, 1971) recently developed a new type of atmospheric model in which the gravity inertial waves are filtered out without imposing a strict condition of balance between the ϕ- and

ψ-fields. The main advantage of Van Isacker's procedure is the avoidance of the often difficult problem of the resolution of a nonlinear balance equation, though allowing integration time steps of the same order or magnitude as in balanced models.

For our purpose, a balanced model seems quite appropriate since vacillation appears as a pure Rossby type flow. As the balance equation results from various simplifications introduced in the original equation of the divergence of the wind (see, for example, Van Mieghem, 1973), some restrictions are also imposed on the other equations of the model in order to maintain energetic consistency. Moreover, different types of balanced models can be formulated, corresponding to different balance equations and energetically consistent models (Lorenz, 1960; Van Isacker, 1963; Phillips, 1965). The simplest form of the balance equation shall be used here, namely the quasi-geostrophic balance equation

$$(3.50) \qquad \nabla^2 \phi = S(f, \psi)$$

the differential operator S having been defined in (3.45). This choice is compatible with the quasi-geostrophic nature of Rossby flows, as discussed in Section 2.2, and offers the substantial advantage of providing a balance equation easily tractable from the mathematical point of view.

Since the equation of the isobaric divergence of the wind is replaced by a diagnostic balance equation, so that $\partial_t \delta \equiv 0$, the kinetic energy budget has to deal with the kinetic energy of the nondivergent flow only. This restriction is analogous to the exclusion of the kinetic energy of the vertical motion, when the equation of the vertical wind is replaced by the diagnostic hydrostatic equation. With this new definition of the kinetic energy, the total energy of a nondissipative insulated quasi-geostrophic model will be conserved provided the relative vorticity equation assumes the form (Van Isacker, 1963)

$$(3.51) \qquad (\partial/\partial t)\zeta = -J(\psi, \eta) - S(f, \chi)$$

where the absolute vorticity η is given by

$$(3.52) \qquad \eta = \zeta + f$$

Writing

$$(3.53) \qquad X(p) = -\int_0^p \chi(p')\, dp'$$

the continuity equation (3.38) transforms into

$$(3.54) \qquad \omega = \nabla^2 X$$

when the upper boundary condition

(3.55) $$\omega = 0 \quad \text{for} \quad p = 0$$

is introduced.

Taking (3.46) into account, the thermodynamic equation (3.40) may be given the form

(3.56) $$\partial \theta / \partial t = -J(\psi, \theta) - S(\theta, \chi) - \partial_p(\theta \nabla^2 \chi) + (H/c_p)(p_0/p)^\kappa$$

where θ is the potential temperature

(3.57) $$\theta = T(p_0/p)^\kappa = -(1/R)(p_0/p)^\kappa \, \partial \phi / \partial \ln p$$

with $p_0 = 100$ cbar and $\kappa = R/c_p$. Using (3.57), the balance equation (3.50) transforms into

(3.58) $$c_p p_0^{-\kappa} \nabla^2 \theta = -(\partial/\partial p^\kappa) S(f, \psi)$$

so that the system of the two prognostic equations (3.51) and (3.56) and the diagnostic balance equation (3.58) forms a closed differential set in the three unknown functions ψ, χ (or X), and θ. The diagnostic equations (3.57), (3.41), and (3.54) then allow one to compute ϕ, ρ, and ω, and thus complete the description of the flow.

3.3.2. The Equations of the Two-Layer Model. For the numerical integration of the system (3.51), (3.56), (3.58), the atmosphere is subdivided into isobaric layers. Obviously, it is necessary that this partitioning not violate the law of conservation of total energy. Here the procedure followed by Lorenz (1960) in the case of a two-layer model is recalled. Its vertical structure is schematically represented in Fig. 6. The three isobaric surfaces of the model are denoted by p_0, p_2, and p_4; p_0 refers to the surface pressure and will be put equal to 100 cbar, $p_2 = p_0/2$, while at the uppermost level $p_4 = 0$. Each of the functions ψ, χ, and θ is replaced by two functions $\psi_1, \psi_3; \chi_1, \chi_3; \theta_1, \theta_3$

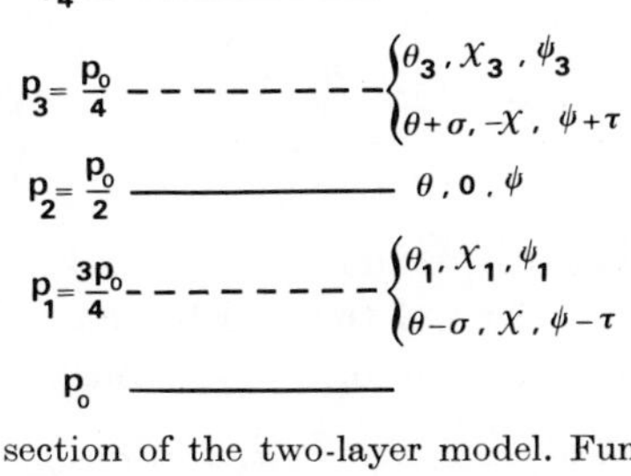

$p_4 = 0$ ——————————

$p_3 = \dfrac{p_0}{4}$ — — — — — — $\begin{cases} \theta_3, X_3, \psi_3 \\ \theta+\sigma, -X, \psi+\tau \end{cases}$

$p_2 = \dfrac{p_0}{2}$ —————————— $\theta, 0, \psi$

$p_1 = \dfrac{3p_0}{4}$ — — — — — — $\begin{cases} \theta_1, X_1, \psi_1 \\ \theta-\sigma, X, \psi-\tau \end{cases}$

p_0 ——————————

FIG. 6. Vertical cross section of the two-layer model. Functions θ_i, χ_i, and ψ_i are defined at odd levels p_i with $i = 1, 3$; their expression in terms of the variables θ, σ, χ, ψ, and τ are also reported.

defined at the isobaric levels $p_1 = 3p_0/4$ and $p_2 = p_0/4$ respectively and depending on the two space coordinates x^1 and x^2 only (not on p). For convenience, the subscript of any variable will refer to the isobaric level where this variable is defined. Note that all the dependent variables are now scalars so that no confusion is possible concerning the meaning of this subscript with respect to those used in tensor calculus.

With the usual boundary condition for a flat earth

(3.59) $$\omega = 0 \quad \text{for} \quad p = p_0$$

and the upper limit condition (3.55), we have, by virtue of (3.54),

(3.60) $$\nabla^2 X = 0 \quad \text{at} \quad p = p_0 \quad \text{and} \quad p = p_4 = 0$$

By virtue of the continuity equation (3.53), the function X_2 defined at the surface p_2 separating the layers is given by

$$\nabla^2 X_4 - \nabla^2 X_2 = \nabla^2 \int_0^{p_0/2} \chi_3(x^1, x^2) \, dp'$$

so that taking (3.60) into account,

(3.61) $$\nabla^2 X_2 = -\tfrac{1}{2} p_0 \nabla^2 \chi_3$$

In the same way,

(3.62) $$\nabla^2 X_2 - \nabla^2 X_0 = \nabla^2 X_2 = \tfrac{1}{2} p_0 \nabla^2 \chi_1$$

Consequently, the continuity of ω in p_2 implies

(3.63) $$\chi \equiv \chi_1 = -\chi_3$$

The finite difference form of (3.51) and (3.56) is obtained by replacing the derivatives with respect to p by finite differences, thus

(3.64) $$(\partial/\partial t)\nabla^2 \psi_j = -J(\psi_j, \eta_j) - S(f, \chi_j)$$

and

(3.65) $$(\partial/\partial t)\theta_j = -J(\psi_j, \theta_j) - S(\theta_j, \chi_j)$$
$$- \frac{\theta_{j-1} \nabla^2 X_{j-1} - \theta_{j+1} \nabla^2 X_{j+1}}{p_{j-1} - p_{j+1}} + \frac{H_j}{c_p} \left(\frac{p_0}{p_j}\right)^\kappa$$

for $j = 1$ and 3, H_j being the heating rate in the layer j, while some rule is necessary to define θ_2 in terms of the two thermal variables θ_1 and θ_3. In the same way the finite difference form of (3.58) can be written

(3.66) $$c_p p_0^{-\kappa} \nabla^2 \theta_2 = -(p_1^\kappa - p_3^\kappa)^{-1} S(f, \psi_1 - \psi_3)$$

which is the balance equation for the two-layer model.

3.3.3. The Energetics of the Two-Layer Model. The aim of this section is to ensure the energetic consistency of the formulation of the two-layer model. Hence, it will deal with the conservative, insulated model only.

In an atmosphere in strict hydrostatic equilibrium, the gravitation potential energy P and the internal energy I are proportional (Margules, 1903). Therefore, in large-scale atmospheric energetics, there is no reason to deal separately with these two quantities which are always associated in the definition of the total potential energy

$$(3.67) \qquad P + I = c_p \int T \, dm = c_p g^{-1} p_0^{-\kappa} \int p^\kappa \theta \, dp \, d\Sigma$$

where dm is the mass element of the fluid and $d\Sigma$ an element of horizontal area, the integration being extended to the system under consideration. In the two-layer model, (3.67) is approximated by

$$(3.68) \qquad P + I = c_p g^{-1} p_0^{-\kappa} \int [p_1^\kappa \theta_1 (p_0 - p_2) + p_3^\kappa \theta_3 (p_2 - p_4)] \, d\Sigma$$

According to (3.4), the kinetic energy of the nondivergent horizontal motion is given by

$$(3.69) \qquad K = \tfrac{1}{2} g^{-1} \int g_{ij} V_\psi^i V_\psi^j \, dp \, d\Sigma,$$

where the subscript ψ indicates that only the rotational part of the wind is taken into account. Using (3.46), we have

$$(3.70) \qquad K = \tfrac{1}{2} g^{-1} \int g_{ij} e^{im} e^{jn} \partial_m \psi \, \partial_n \psi \, dp \, d\Sigma$$

or, with (3.37) and (3.44),

$$(3.71) \qquad K = \tfrac{1}{2} g^{-1} \int Q(\psi, \psi) \, dp \, d\Sigma$$

In the two-layer model (3.71) assumes the form

$$(3.72) \qquad K = \tfrac{1}{2} g^{-1} \int [Q(\psi_1, \psi_1)(p_0 - p_2) + Q(\psi_3, \psi_3)(p_2 - p_4)] \, d\Sigma$$

Before expressing the energy budgets, it is perhaps worth recalling that when the system is closed (no net exchange of mass with the surroundings) and mechanically and thermally insulated (no net work and heat flux through the boundary) the horizontal average of the (two-dimensional) divergence of any vector of the flow field vanishes. The property is easily obtained in mathematical form from Stokes' theorem. Physically it expresses that the net flux of any flow quantity across the lateral boundary is zero. On the other hand, the horizontal average of a Jacobian is easily seen to be zero too. Indeed, the Jacobian always involves a stream function ψ and, recalling (3.46), assumes the form

(3.73) $$J(\psi, \xi) = V_\psi^i \partial_i \xi$$

where ξ is a flow variable. Such a term expresses the transport of ξ by the nondivergent wind and simply redistributes ξ on the given isobaric surface. Obviously such a process cannot contribute to modify the average value of ξ on the surface.

According to (3.67) and (3.65), the budget of $P + I$ for an insulated model then assumes the form

(3.74) $$(\partial/\partial t)(P + I) = c_p g^{-1} p_0^{-\kappa}(p_1^\kappa - p_3^\kappa) \int X_2 \nabla^2 \theta_2 \, d\Sigma$$

In the same way, by virtue of (3.72) and (3.64),

(3.75) $$\partial K/\partial t = g^{-1} \tfrac{1}{2} p_0 \int S(f, \chi)(\psi_1 - \psi_3) \, d\Sigma$$

(3.76) $$= g^{-1} \int X_2 S(f, \psi_1 - \psi_3) \, d\Sigma$$

Thus, from (3.74) and (3.76), the balance equation (3.66) ensures the conservation of the total energy of the insulated nondissipative two-layer model. Note that this condition leaves θ_2 arbitrary.

Of considerable interest in atmospheric energetics also is the concept of available potential energy, originally introduced by Margules (1903) and which represents the (small) portion of the total potential energy capable of being transformed into kinetic energy. Conventionally, the available potential energy is defined as the excess of the total potential energy above its value obtained by an adiabatic rearrangement of the mass leading to a stably stratified atmosphere with horizontal isobaric (or isentropic) surfaces. Van Mieghem (1957) has shown that this reference state is the one of minimum total potential energy. An approximate form of the available potential energy, presenting the main advantage of being expressible only in terms of the current state of the fluid, has been derived by Lorenz (1955).

In the two-layer model considered, the potential temperature is defined at a few levels only, so that a continuous redistribution of the mass cannot be envisaged. Nevertheless, it is still possible to define a state of minimum potential energy.

Introducing, with Lorenz (1960), the following new variables θ and σ:

(3.77) $$\theta = \tfrac{1}{2}(\theta_1 + \theta_3), \; \sigma = \tfrac{1}{2}(\theta_3 - \theta_1)$$

so that θ is the vertically averaged potential temperature and σ a measure of the static stability, the total potential energy (3.68) can be written

(3.78) $$P + I = c_p g^{-1} p_0 \int (a\theta - b\sigma) \, d\Sigma$$

with $a = [(3/4)^\kappa + (1/4)^\kappa]/2$ and $b = [(3/4)^\kappa - (1/4)^\kappa]/2$. Now, Eq. (3.65) with $H_j = 0$ implies the conservation of the mean square potential temperature

(3.79) $$\tfrac{1}{2}\overline{(\theta_1^2 + \theta_3^2)} = \overline{\theta^2 + \sigma^2}$$

where a bar denotes a horizontal average, provided (Lorenz, 1960)

(3.80) $$\theta_2 = \tfrac{1}{2}(\theta_1 + \theta_3) = \theta$$

Recalling that the mean square potential temperature $\theta^2 + \sigma^2$ is conserved and that θ is also an invariant for adiabatic flow, the quantity

(3.81) $$\bar{\sigma}^2 + \overline{\sigma'^2} + \overline{\theta'^2} = \sigma_m^2$$

with $\theta = \bar{\theta} + \theta'$ and $\sigma = \bar{\sigma} + \sigma'$, is also conserved. Accordingly, $\bar{\sigma}$ has a maximum value σ_m when $\sigma' = \theta' = 0$, i.e. when the isobaric surfaces (where the temperature field is defined) are also isentropic. Clearly, in this case, the total potential energy (3.78) reaches its minimum value attainable by adiabatic process. The available potential energy in the two-layer model is then given by

(3.82) $$A = c_p g^{-1} p_0 \int [(a\theta - b\sigma) - (a\bar{\theta} - b\sigma_m)] \, d\Sigma = c_p g^{-1} p_0 b \int (\sigma_m - \sigma) \, d\Sigma$$

or, taking (3.81) into account,

(3.83) $$A = c_p g^{-1} p_0 b \int (\overline{\theta'^2} + \overline{\sigma'^2})/(\sigma_m - \bar{\sigma}) \, d\Sigma$$

Obviously, from (3.82) and (3.81), the budget of A assumes the same form as the budget of $P + I$ so that the adiabatic two-layer model conserves also the sum of the (so defined) available potential energy and of the kinetic energy. Note that this condition imposes the definition of θ_2 according to (3.80).

For the sake of completeness, let us also mention that the energetic consistency of less simplified models allows one to define (Lorenz, 1960) ψ_2 as follows:

(3.84) $$\psi_2 = \tfrac{1}{2}(\psi_1 + \psi_3)$$

It is then convenient to introduce, together with (3.77) defining θ and σ, the new variables

(3.85) $$\psi = \tfrac{1}{2}(\psi_1 + \psi_3), \quad \tau = \tfrac{1}{2}(\psi_3 - \psi_1)$$

so that ψ is the stream function at the mean level p_2 and τ the stream function of the vertical wind shear.

3.3.4. The Basic Physical Processes of the Model.
With the new variables ψ, τ, θ, and σ, the balance equation (3.66) assumes the form

$$bc_p \nabla^2 \theta = S(f, \tau) \qquad (3.86)$$

When the vorticity equations (3.64) are added for $j = 1$ and 3, we obtain, taking (3.63) into account,

$$(\partial/\partial t)\nabla^2 \psi = -\tfrac{1}{2}\{J(\psi_1, \eta_1) + J(\psi_3, \eta_3)\} \qquad (3.87)$$

$$= -J(\psi, \nabla^2 \psi + f) - J(\tau, \nabla^2 \tau) \qquad (3.88)$$

Recalling (3.73), the local rate of change, (3.87), of the vertically averaged relative vorticity $\nabla^2 \psi$ is given by the mean absolute vorticity advected by the *nondivergent* wind in the two layers. Thus, Eq. (3.87) governs the local *barotropic* change of relative vorticity. Considering (3.88), $\partial_t \nabla^2 \psi$ results from an advection of mean absolute vorticity $\nabla^2 \psi + f$ by the nondivergent mean flow and an advection of shear vorticity $\nabla^2 \tau$ by the shear flow. Obviously, the horizontal mean of $\nabla^2 \psi$ is left unaltered by these processes.

On the other hand, subtracting (3.64) for $j = 1$ from (3.64) for $j = 3$, the vertical shear vorticity equation assumes the forms

$$(\partial/\partial t)\nabla^2 \tau = \tfrac{1}{2}[J(\psi_1, \eta_1) - J(\psi_3, \eta_3)] + S(f, \chi) \qquad (3.89)$$

$$= -[J(\tau, \nabla^2 \psi + f) + J(\psi, \nabla^2 \tau)] + S(f, \chi) \qquad (3.90)$$

The two terms in square brackets in (3.89) represent the difference between the advection of absolute vorticity in the upper (ψ_3) and the lower (ψ_1) layers. The third term in (3.89) represents the generation rate

$$S(f, \chi) = f\nabla^2 \chi + Q(f, \chi) \qquad (3.91)$$

of shear vorticity by the divergent wind (χ).

By virtue of the balance equation (3.86), the vertical wind shear (τ) is related to the isobaric temperature gradients at the mean level p_2. For instance, in the particularly simple case of constant f, the balance equation (3.86) leads to the thermal wind equation

$$-e^{ij} \partial_j \tau = V_s^{\,i} = -(bc_p/f)e^{ij} \partial_j \theta \qquad (3.92)$$

where $V_s^{\,i}$ denotes the vertical shear wind. Thus Eq. (3.89) expresses the local baroclinic change of relative vorticity. From (3.90) the differential advection term (the square bracket term) consists of an advection of mean absolute vorticity by the shear wind and an advection of shear vorticity by the mean wind. These processes, as well as those appearing in (3.88), simply transport vorticity from place to place, and do not alter the horizontally averaged values of $\nabla^2\psi$ and $\nabla^2\tau$. The actual source of vorticity in the model is represented by the S term in (3.90), involving the divergent wind potential χ. This term is associated with the (baroclinic) conversion processes between potential and kinetic energy [see Eq. (3.75)]. Thus, the change of shear vorticity is made up of two contributions: (i) an initial one, represented by the square bracket of (3.89), which locally modifies, but does not change in the mean, the shear vorticity of the flow; and (ii) a source term, the S term in (3.89), which represents the influence of the conversions between kinetic and potential energy and which alters, both locally and in the mean, the shear vorticity of the flow.

With the new variables (3.77), the adiabatic form of the thermal equations (3.65) becomes

$$(3.93) \qquad \partial\theta/\partial t = -J(\psi, \theta) - J(\tau, \sigma) + S(\sigma, \chi)$$

and

$$(3.94) \qquad \partial\sigma/\partial t = -J(\tau, \theta) - J(\psi, \sigma) + Q(\theta, \chi)$$

In both Eqs. (3.93) and (3.94) the first two terms on the right side represent the local rate of temperature change due to heat advection by the non-divergent wind. These terms are similar to the vorticity advection terms in (3.88) and (3.90), with θ substituted by $\nabla^2\psi + f$ and σ by $\nabla^2\tau$. The third terms S and Q express through the continuity equation (3.54), the adiabatic temperature changes resulting from vertical motions.

3.3.5. Friction and Heating Effects.

3.3.5.1. *Friction.* It has been recognized (Lorenz, 1962) that friction plays an essential role in the dynamics of models intended to simulate laboratory flows. In order to take frictional effects into account, Lorenz introduces a drag at the surface p_0 proportional to the wind in the lower layer and also a drag at the surface p_2 proportional to the difference between the winds in the upper and lower layers. If the coefficients of proportionality, i.e. the coefficients of friction, are $2F$ and F' at p_0 and p_2 respectively, the additional terms on the right side of (3.88) and (3.90) assume, respectively, the forms

$$(3.95) \qquad -F\nabla^2\psi + F\nabla^2\tau$$

and

$$(3.96) \qquad F\nabla^2\psi - (F + 2F')\nabla^2\tau$$

3.3.5.2. *Heating.* The parameterization of the heating will also be performed following Lorenz' procedure (1962). The heat exchange between the underlying surface and the lower layer is assumed proportional to the difference between the preassigned temperature of the surface θ^* and the temperature of the lower layer. Similarly, the heat exchange between the lower and the upper layers is taken proportional to the difference between the temperature of these layers. If the heating coefficients are denoted by $2H$ and H' for the lower and upper layer respectively, the additional terms on the right side of (3.93) and (3.94) are, respectively

(3.97) $\qquad S_1 = (1/c_p)\{a_1 H\theta^* - (H + H' - a_2 H')\theta + (H - a_3 H')\sigma\}$

and

(3.98) $\qquad S_2 = (1/c_p)\{-a_1 H\theta^* + (H + a_3 H')\theta - (H + H' + a_2 H')\sigma\}$

with

$$a_1 = (4/3)^\kappa \simeq 1.085, \ a_2 = (3^\kappa + 3^{-\kappa})/2 \simeq 1.29,$$

and

$$a_3 = (3^\kappa - 3^{-\kappa})/2 \simeq 0.81$$

3.3.5.3. Evaluation of the friction and heating coefficients in the atmosphere. It may be of interest to evaluate the order of magnitude of F and H in the atmosphere. The atmospheric coefficient of friction F^* is usually introduced by the following expression

(3.99) $\qquad\qquad\qquad F^* g(\partial/\partial p)(\rho |\mathbf{V}| \mathbf{V})$

for the frictional drag. There is a large dispersion of the values of F^* depending upon whether it is directly measured or computed by an indirect way or adjusted in order to improve the results in atmospheric models. Obviously, there is also a large divergence within each set of estimations based on a given method. Roughly speaking, the values of F^* reported in various papers range from 10^{-3} to 10^{-2} (Stessel, 1969). For comparison, with the formulation adopted here, we write

(3.100) $\qquad\qquad\qquad 2F|\mathbf{V}_1| = F^* g \rho_2 |\mathbf{V}_2|^2/p_0/2$

where the subscripts 1 and 2 refer to levels p_1 and p_2, respectively. Midlatitude typical values (Statistiques Quinquennales, 1971) for the wind speed and the air density

$$|\mathbf{V}_1| \simeq 20 \text{ m sec}^{-1}, \ |\mathbf{V}_2| \simeq 30 \text{ m sec}^{-1}, \ \rho_2 \simeq 0.7 \text{ kg m}^{-3}, \ g \simeq 10 \text{ m sec}^{-2}$$

lead to

(3.101) $\qquad\qquad\qquad 3 \times 10^{-6} \leq F \leq 3 \times 10^{-5} \text{ sec}^{-1}$

The coefficient of heating is considered to be of the order of magnitude (see, for instance, Döös, 1969)

(3.102) $$H/c_p \simeq 10^{-5} \sec^{-1}$$

Consequently, in order to reduce the number of parameters in the model, the approximation

(3.103) $$F = H/c_p$$

does not seem unreasonable. Furthermore, we shall assume, with Lorenz (1962), that

(3.104) $$F = 2F' = H/c_p = 2H'/c_p$$

which corresponds to a vertical decrease of friction and heating as $(p/p_0)^2$. Hence, the friction and heating effects are represented with the aid of a single coefficient F. As a consequence of the commodity hypothesis (3.103) and (3.104) it seems unnecessary, and even somewhat misleading, to maintain the more or less sophisticated refinement introduced in (3.97) and (3.98) and therefore just the numerical values

(3.105) $$a_1 = 1, a_2 = a_3 = 0$$

will be used that still warrant that the direction of the heat transfers be correctly represented.

4. The Spectral Dynamics and Energetics of the Model

4.1. The Spectral Representation

The spectral form of the equations of the model is obtained by representing each dependent variable by a series using a complete set of orthonormal functions F_i. The particular choice of these functions depends on the geometry of the system. We consider here an infinite channel consisting of periodically repeating "units" of width πL and of fundamental length $2\pi L$. Obviously it is then sufficient to study the dynamics and energetics of a single "unit." If $x^1 = x$ and $x^2 = y$ are rectangular coordinates respectively oriented along (from W to E) and across (from S to N) the channel, the fundamental plane domain is defined by

(4.1) $$0 \leq x/L \leq 2\pi, \quad 0 \leq y/L \leq \pi$$

and the components of the metric tensor are simply

(4.2) $$g_{11} = g_{22} = 1, \quad g_{12} = g_{21} = 0$$

so that

(4.3) $$\gamma = 1$$

and

(4.4) $$g^{11} = g^{22} = 1, \quad g^{12} = g^{21} = 0$$

A suitable enumerable set of functions F_i orthogonal on the domain (4.1) is

(4.5)
$$\begin{aligned}\varphi_{0,0} &= 1 \\ \varphi_{m,0} &= \sqrt{2}\,\cos(my/L) \\ \varphi_{m,n} &= 2\sin(my/L)\cos(nx/L) \\ \varphi'_{m,n} &= 2\sin(my/L)\sin(nx/L)\end{aligned}$$

m and n being integers ($m, n = 1, 2, \ldots$). These functions are such that

$$\int_0^{2\pi L} dx \int_0^{\pi L} dy\, F_i\, F_j = 2\pi^2 L^2\, \delta_i^{\ j}$$

where $\delta_i^{\ j}$ is the Kronecker symbol (3.11). Thus each scale of the fields of motion is characterized by a pair of integers (m, n), where m refers to the meridional wave shape or *mode* and n to the zonal wavenumber. The decomposition of the fields into their spectral components not only allows a clear and easy identification of each scale of motion but also provides the possibility of further simplifying the model equations by discarding, in the development of the dependent variables, all but a small number of scales.

Vacillation has been identified as consisting of a cyclic modification of a pattern including a single longitudinal wave n. Consequently, we shall use a set of functions F_i including only one single zonal wavenumber n ($n \neq 0$). Moreover, it was shown that the wave motion cycle is accompanied by a concomitant variation of the zonal flow. Consequently vacillation is definitely a nonlinear phenomenon where the interaction of the wave with the zonal flow plays an essential role. The minimum set of basic functions to be used in a channel for describing this kind of interaction consists of the five functions $\varphi_{2,0}$, $\varphi_{1,n}$, $\varphi'_{1,n}$, $\varphi_{2,n}$, and $\varphi'_{2,n}$ with $n \neq 0$. Obviously we also need the function $\varphi_{0,0}$ associated with the horizontal mean of the dependent variables ($\psi, \tau, \theta, \ldots$). On the other hand, the function $\varphi_{1,0}$ is indispensable to model the laboratory thermal forcing condition. Thus, our ordered minimum set of basic functions F_i is

(4.6) $\{F_0 \equiv \varphi_{0,0},\ F_1 \equiv \varphi_{1,0},\ F_2 \equiv \varphi_{2,0},$
$F_3 \equiv \varphi_{1,n},\ F_4 \equiv \varphi'_{1,n},\ F_5 \equiv \varphi_{2,n},\ F_6 \equiv \varphi'_{2,n}\}$

Consequently, the dependent variables involve (at most) seven spectral components: (i) a component (0, 0) equal to the horizontal average of the dependent variable considered; (ii) two zonal components, (1, 0) and (2, 0), of the first and second mode respectively; and (iii) a wave component n of the first two modes (1, n) and (2, n). Then the ψ, τ, and θ fields assume the forms

(4.7)
$$\psi(x, y, t) = L^2 f \sum_{m=1}^{2} \{\psi_{m,0}(t)\varphi_{m,0}(y) + \psi_{m,n}(t)\varphi_{m,n}(x, y) + \psi'_{m,n}(t)\varphi'_{m,n}(x, y)\}$$

(4.8)
$$\tau(x, y, t) = L^2 f \sum_{m=1}^{2} \{\tau_{m,0}(t)\varphi_{m,0}(y) + \tau_{m,n}(t)\varphi_{m,n}(x, y) + \tau'_{m,n}(t)\varphi'_{m,n}(x, y)\}$$

(4.9) $$\theta(x, y, t) = BL^2 f \Big\{ \theta_{0,0}(t)\varphi_{0,0} + \sum_{m=1}^{2} [\theta_{m,0}(t)\varphi_{m,0}(y)$$
$$+ \theta_{m,n}(t)\varphi_{m,n}(x, y) + \theta'_{m,n}(t)\varphi'_{m,n}(x, y)] \Big\}$$

or, making use of (4.6) to establish the correspondence between the subscript i and the subscripts (m, n),

(4.7') $$\psi(x, y, t) = L^2 f \sum_{i=1}^{6} \psi_i(t) F_i(x, y)$$

(4.8') $$\tau(x, y, t) = L^2 f \sum_{i=1}^{6} \tau_i(t) F_i(x, y)$$

and

(4.9') $$\theta(x, y, t) = BL^2 f \sum_{i=0}^{6} \theta_i(t) F_i(x, y)$$

The spectral coefficients ψ_i, τ_i, θ_i, $i = 1, 2, \ldots, 6$, and θ_0 are functions of time t only and become the new variables. The factors $L^2 f$ and $BL^2 f$ with $B = f/b \cdot c_p$ where $b = [(3/4)^\kappa - (1/4)^\kappa]/2$ allow one to deal with dimensionless spectral coefficients by using L and f^{-1} as units of length and time respectively.

It is possible to restrict the static stability σ so that it varies only with time without breaking down the recognized fundamental structure necessary to simulate vacillating flows. Accordingly, and for the sake of simplicity, σ is specified only by a horizontal nondimensional mean component σ_0:

(4.10) $$\sigma(t) = BL^2 f \sigma_0(t)$$

Obviously, the equation of the static stability (3.94) must remain compatible with the hypothesis (4.10). Therefore (3.94) should be averaged horizontally,

(4.11') $$\partial \bar{\sigma}/\partial t = \partial \sigma/\partial t = -\overline{J(\tau, \theta)} - \overline{J(\psi, \sigma)} + \overline{Q(\theta, \chi)} + \overline{S_2}$$

a horizontal bar denoting a horizontal average over the domain (4.1). In (4.11'), S_2 is given by (3.98) with the values (3.105) for a_1, a_2, and a_3. Recalling that the system is closed and mechanically insulated, the horizontal average of the Jacobians and of the divergence $\nabla(\theta\nabla\chi)$ coming from the Q term vanish in (4.11') which reduces to

(4.11) $$\partial\sigma/\partial t = -\overline{\theta\nabla^2\chi} + \overline{S_2}$$

Then the velocity potential χ appears in the equation of the model under its Laplacian form only, so that it suffices to use

(4.12) $$\nabla^2\chi = f \sum_{m=1}^{2} \{\omega_{m,0}(t)\varphi_{m,0}(y) + \omega_{m,n}(t)\varphi_{m,n}(x,y) + \omega'_{m,n}(t)\varphi_{m,n}(x,y)\}$$

(4.12') $$= f \sum_{i=1}^{6} \omega_i(t) F_i(x,y)$$

where, again, the spectral coefficients $\omega_i(t)$ are dimensionless.

In order to simulate both the laboratory conditions and the major component of the earth's thermal forcing on the atmosphere, we use for θ^*:

(4.13) $$\theta^*(x,y) = BL^2 f\{\theta^*_{0,0}\varphi_{0,0} + \theta^*_{1,0}\varphi_{1,0}(y)\}$$

We consider this forcing to be steady so that $\theta^*_{0,0}$ and $\theta^*_{1,0}$ are simply constants.

The amplitude $A_{m,n}$ and the phase $\mu_{m,n}$ of the wave (m,n) of mode m and wavenumber n of any field C will be associated with the spectral coefficients $C_{m,n}$ and $C'_{m,n}$ in the following way:

(4.14) $$\sin(my/L)[C_{m,n}(t)\cos(nx/L) + C'_{m,n}(t)\sin(nx/L)]$$
$$= \sin(my/L) A_{m,n}(t) \cos[(nx/L) - \mu_{m,n}(t)]$$

so that

(4.15) $$A_{m,n} = |(C^2_{m,n} + C'^2_{m,n})^{1/2}|$$

where the vertical bars denote the absolute value and

(4.16) $$\mu_{m,n} = \tan^{-1}(C'_{m,n}/C_{m,n})$$

Consequently, $\mu_{m,n}$ varies counterclockwise $(d\mu_{m,n}/dt > 0)$ in the (A, μ) plane when the wave progresses eastward $(dx/dt > 0)$. It is then clear that the sign of the expression

(4.17) $$A_{1,n} A_{2,n} \sin(\mu_{2,n} - \mu_{1,n}) = C_{1,n} C'_{2,n} - C'_{1,n} C_{2,n}$$

or, using (4.6),

(4.17') $$A_{1,n} A_{2,n} \sin(\mu_{2,n} - \mu_{1,n}) = C_3 C_6 - C_4 C_5$$

allows one to determine the relative position of the two modes of the wave n in terms of their spectral components $C_{1,n}$, $C'_{1,n}$, $C_{2,n}$, and $C'_{2,n}$. Indeed, for eastward moving waves, the wave $(2, n)$ lags behind the wave $(1, n)$ if $\mu_{2,n} - \mu_{1,n} > \pi$ so that $\sin(\mu_{2,n} - \mu_{1,n}) < 0$ or, equivalently, if $C_{1,n} C'_{2,n} - C'_{1,n} C_{2,n}$ is negative.

4.2. The Spectral Equations

With a constant Coriolis parameter f and the static stability restricted according to (4.10), there remain only two differential operators, namely ∇^2 and J, in the equations of the model. The basic functions F_i [Eq. (4.5)] being the eigenfunctions of the two-dimensional Laplacian in Cartesian coordinates, we simply have

$$(4.18) \qquad L^2 \nabla^2 F_i = -a_i^2 F_i$$

where the eigenvalue $-a_i^2$ stands for $-(m^2 + n^2)$ when F_i is either $\varphi_{m,n}$ or $\varphi'_{m,n}$. The Jacobian of two basic functions

$$J(F_j, F_k) = \partial_x F_j \, \partial_y F_k - \partial_x F_k \, \partial_y F_j = -J(F_k, F_j)$$

is represented by a limited series using the basic functions (4.6) only,

$$(4.19) \qquad L^2 J(F_j, F_k) = \sum_{i=0}^{6} c_{ijk} F_i$$

where, owing to the orthogonality of the F_i's on the domain (4.1),

$$(4.20) \qquad c_{ijk} = (1/2\pi^2 L^2) \int_0^{2\pi L} dx \int_0^{\pi L} dy \, J(F_j, F_k) \cdot F_i$$

The interaction coefficients c_{ijk} are completely antisymmetric:

$$(4.21) \qquad c_{ijk} = c_{jki} = c_{kij} = -c_{jik} = -c_{ikj} = -c_{kji}$$

For the basic set (4.6) the only nonzero coefficients c_{ijk} can be deduced from

$$(4.22) \qquad c_{134}/5 = c_{256}/4 = c_{236}/8 = c_{254}/8 = -(8\sqrt{2}/15\pi) \cdot n$$

Figure 7 schematically describes the associated scale interactions allowed in the model. The upper connections represent the self-interaction of each

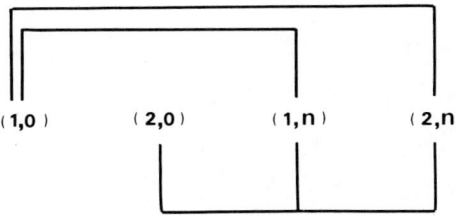

FIG. 7. Schematic representation of the scale interactions included in the model. The upper connections represent the self-interaction of wave $(1, n)$ or $(2, n)$ with the zonal component $(1, 0)$; the lower connections represent the triple interaction between scales $(2, 0)$, $(1, n)$, and $(2, n)$. In each parenthesis the first integer refers to the mode and, the second one to the zonal wavenumber.

component $(1, n)$ or $(2, n)$ with the zonal component of the first mode $(1, 0)$. The lower connections represent the interactions between the three scales $(2, 0)$, $(1, n)$, and $(2, n)$.

When the expressions (4.7)–(4.10) are introduced into the equations for ψ, τ, θ, and σ, we obtain (Lorenz, 1963) the spectral form of the equations of the model

$$(4.23) \quad \dot{\psi}_i = \sum_{j<k=1}^{6} a_i^{-2}(a_j^2 - a_k^2) c_{ijk}(\psi_j \psi_k + \tau_j \tau_k) - k(\psi_i - \tau_i)$$

$$(4.24) \quad \dot{\tau}_i = \sum_{j<k=1}^{6} a_i^{-2}(a_j^2 - a_k^2) c_{ijk}(\tau_j \psi_k + \tau_k \psi_j) - a_i^{-2} \omega_i + k\psi_i - 2k\tau_i$$

$$(4.25) \quad \dot{\theta}_i = \sum_{j<k=1}^{6} c_{ijk}(\theta_j \psi_k - \theta_k \psi_j) + \sigma_0 \omega_i - k(\theta_i - \sigma_i) + k\theta_i^*$$

$$(4.26) \quad \dot{\sigma}_0 = -\sum_{i=1}^{6} \theta_i \omega_i - 2k\sigma_0 + k(\theta_0 - \theta_0^*)$$

where a dot denotes a time derivative with respect to the nondimensional time $t_0 = ft$ and where $k = Ff^{-1}$. Taking into account (4.8) and (4.9) the balance equation (3.86) reduces simply to

$$(4.27) \quad \theta_i = \tau_i \quad \text{if} \quad i \neq 0$$

The system (4.23)–(4.27) involves the 26 unknowns ψ_i, τ_i, θ_i, ω_i, $i = 1, 2, \ldots, 6$, θ_0, and σ_0. The balance equation (4.27) allows one to eliminate six unknowns θ_i or τ_i ($i = 1, 2, \ldots, 6$). Moreover, using (4.27) it is still possible to eliminate the six unknowns ω_i, $i = 1, 2, \ldots, 6$ from (4.24) and (4.25), so that there finally remains a set of 14 prognostic equations in ψ_i, τ_i, (or θ_i) $i = 1, 2, \ldots, 6$, θ_0, and σ_0. If

$$(4.28) \quad \begin{aligned} &\alpha \equiv -8\sqrt{2}\, n/(15\pi), \; \beta \equiv n^2/(n^2+1) \\ &\beta' \equiv (n^2+3)/(n^2+4), \; \beta'' \equiv (n^2-3)/(n^2+4) \end{aligned}$$

these equations can be written

$$(4.29) \quad \dot{\psi}_1 = \phantom{-6\alpha\{\psi,\tau\}_{3,6} + 6\alpha\{\psi,\tau\}_{4,5}} \quad -k(\psi_1 - \tau_1)$$

$$(4.30) \quad \dot{\psi}_2 = -6\alpha\{\psi, \tau\}_{3,6} + 6\alpha\{\psi, \tau\}_{4,5} \quad -k(\psi_2 - \tau_2)$$

$$(4.31) \quad \dot{\psi}_3 = 5\alpha\beta\{\psi, \tau\}_{1,4} + 8\alpha\beta\{\psi, \tau\}_{2,6} \quad -k(\psi_3 - \tau_3)$$

$$(4.32) \quad \dot{\psi}_4 = -5\alpha\beta\{\psi, \tau\}_{1,3} - 8\alpha\beta\{\psi, \tau\}_{2,5} \quad -k(\psi_4 - \tau_4)$$

$$(4.33) \quad \dot{\psi}_5 = 4\alpha\beta'\{\psi, \tau\}_{1,6} + 8\alpha\beta''\{\psi, \tau\}_{2,4} \quad -k(\psi_5 - \tau_5)$$

$$(4.34) \quad \dot{\psi}_6 = -4\alpha\beta'\{\psi, \tau\}_{1,5} - 8\alpha\beta''\{\psi, \tau\}_{2,3} \quad -k(\psi_6 - \tau_6)$$

$$(4.35) \quad \dot{\tau}_1 = (1+\sigma_0)^{-1}\{ \quad 5\alpha(\tau,\psi)_{3,4} \quad +4\alpha(\tau,\psi)_{5,6} \\ + k(\theta_1^* - \theta_1) + k\sigma_0(\psi_1 - 2\tau_1)\}$$

(4.36) $\dot{\tau}_2 = (1 + 4\sigma_0)^{-1}\{-24\alpha\sigma_0[\tau, \psi]_{3,6} + 8\alpha(\tau, \psi)_{3,6} + 24\alpha\sigma_0[\tau, \psi]_{4,5}$
$- 8\alpha(\tau, \psi)_{4,5} - k\theta_2 + 4k\sigma_0(\psi_2 - 2\tau_2)\}$

(4.37) $\dot{\tau}_3 = [1 + (1 + n^2)\sigma_0]^{-1}$
$\{5n^2\alpha\sigma_0[\tau, \psi]_{1,4} - 5\alpha(\tau, \psi)_{1,4} + 8n^2\alpha\sigma_0[\tau, \psi]_{2,6}$
$- 8\alpha(\tau, \psi)_{2,6} - k\theta_3 + k\sigma_0(1 + n^2)(\psi_3 - 2\tau_3)\}$

(4.38) $\dot{\tau}_4 = [1 + (1 + n^2)\sigma_0]^{-1}$
$\{-5n^2\alpha\sigma_0[\tau, \psi]_{1,3} + 5\alpha(\tau, \psi)_{1,3} - 8n^2\alpha\sigma_0[\tau, \psi]_{2,5}$
$+ 8\alpha(\tau, \psi)_{2,5} - k\theta_4 + k\sigma_0(1 + n^2)(\psi_4 - 2\tau_4)\}$

(4.39) $\dot{\tau}_5 = [1 + (4 + n^2)\sigma_0]^{-1}$
$\{4(n^2 + 3)\alpha\sigma_0[\tau, \psi]_{1,6} - 4\alpha(\tau, \psi)_{1,6} + 8(n^2 - 3)\alpha\sigma_0[\tau, \psi]_{2,4}$
$- 8\alpha(\tau, \psi)_{2,4} - k\theta_5 + k\sigma_0(4 + n^2)(\psi_5 - 2\tau_5)\}$

(4.40) $\dot{\tau}_6 = [1 + (4 + n^2)\sigma_0]^{-1}$
$\{-4(n^2 + 3)\alpha\sigma_0[\tau, \psi]_{1,5} + 4\alpha(\tau, \psi)_{1,5} - 8(n^2 - 3)\alpha\sigma_0[\tau, \psi]_{2,3}$
$+ 8\alpha(\tau, \psi)_{2,3} - k\theta_6 + k\sigma_0(4 + n^2)(\psi_6 - 2\tau_6)\}$

(4.41) $\dot{\theta}_0 = \qquad\qquad\qquad\qquad\qquad\qquad k(\theta_0^* - \theta_0) + k\sigma_0$

(4.42) $\dot{\sigma}_0 = -\sum_{i=1}^{6} \theta_i \omega_i \qquad\qquad + k(\theta_0^* - \theta_0) - 2k\sigma_0$

where the following notations are used:

(4.43) $\{\psi, \tau\}_{i,j} \equiv \psi_i \psi_j + \tau_i \tau_j$

(4.44) $[\psi, \tau]_{i,j} \equiv \psi_i \tau_j + \psi_j \tau_i$

(4.45) $(\psi, \tau)_{i,j} \equiv \psi_i \tau_j - \psi_j \tau_i$

In (4.42) the ω_i's still have to be replaced by their values extracted from (4.24) and (4.25)

(4.46) $\omega_1 = (1 + \sigma_0)^{-1}\{5\alpha(\tau_4\psi_3 - \tau_3\psi_4) + 4\alpha(\tau_6\psi_5 - \tau_5\psi_6)$
$+ k(\psi_1 - \tau_1) - k\theta_1^*\}$

(4.47) $\omega_2 = (1 + 4\sigma_0)^{-1}\{8\alpha(\tau_6\psi_3 - 7\tau_3\psi_6) - 8\alpha(\tau_5\psi_4 - 7\tau_4\psi_5)$
$+ 4k(\psi_2 - \tau_2)\}$

(4.48) $\omega_3 = [1 + (1 + n^2)\sigma_0]^{-1}\{-5\alpha[\tau_4\psi_1 - (2n^2 + 1)\tau_1\psi_4]$
$- 8\alpha[\tau_6\psi_2 - (2n^2 + 1)\tau_2\psi_6] + (1 + n^2)k(\psi_3 - \tau_3)\}$

(4.49) $\omega_4 = [1 + (1 + n^2)\sigma_0]^{-1}\{5\alpha[\tau_3\psi_1 - (2n^2 + 1)\tau_1\psi_3]$
$+ 8\alpha[\tau_5\psi_2 - (2n^2 + 1)\tau_2\psi_5] + (1 + n^2)k(\psi_4 - \tau_4)\}$

(4.50) $\quad \omega_5 = [1 + (4+n^2)\sigma_0]^{-1}\{-4\alpha[\tau_6\psi_1 - (2n^2+7)\tau_1\psi_6]$
$\qquad\qquad - 8\alpha[7\tau_4\psi_2 - (2n^2+1)\tau_2\psi_4] + (4+n^2)k(\psi_5 - \tau_5)\}$

(4.51) $\quad \omega_6 = [1 + (4+n^2)\sigma_0]^{-1}\{4\alpha[\tau_5\psi_1 - (2n^2+7)\tau_1\psi_5]$
$\qquad\qquad + 8\alpha[7\tau_3\psi_2 - (2n^2+1)\tau_2\psi_3] + (4+n^2)k(\psi_6 - \tau_6)\}$

Note that according to (4.29) the waves $(1, n)$ and $(2, n)$ do not interact with the mean zonal flow of the first mode as a consequence of the vanishing coefficients $a_j^2 - a_k^2$ in this equation. Note also (Lorenz, 1963) that the equations of the model are invariant for a change of sign of the variables of the second mode (subscripts 2, 5, and 6).

Before dealing with the energetics of the model and in order to avoid any ambiguity, some dynamical properties of the field patterns will be presented in the next section.

4.3. Some Dynamical Properties of the Field of Motion

4.3.1. The Tilt of Wave n.
Let us consider a field C represented by

(4.52) $$C(x, y, t) = \sum_{i=0}^{6} Sc_i F_i$$

where S is a scale factor making the Fourier coefficients c_i dimensionless. The equation $\partial C/\partial x = 0$ of the trough and ridge lines of this field assumes the form

(4.53) $$\text{tg}\,\frac{nx}{L} = \frac{c_4 + 2c_6 \cos y/L}{c_3 + 2c_5 \cos y/L}$$

The slope with respect to the W–E direction of the trough and ridge lines is then given by

(4.54) $$\frac{\delta y}{\delta x} = n\frac{[c_3 + 2c_5 \cos(y/L)]^2 + [c_4 + 2c_5 \cos(y/L)]^2}{2(c_4 c_5 - c_3 c_6)\sin(y/L)}$$

Consequently, the change with time of the c_i coefficients generally not only gives rise to the progression of the wave but also to a modification of its shape. Note that the sign of the slope of the wave pattern does not change over the interval $0 \leq y \leq \pi L$ and is determined, according to (4.54) and (4.17'), by the relative position of modes 1 and 2 of the wave n. For instance, if the wave $(2, n)$ lags behind the wave $(1, n)$, then $c_4 c_5 - c_3 c_6 > 0$ so that $\delta y/\delta x > 0$ and the flow pattern exhibits a SW–NE tilt.

4.3.2. The S–N Transport of W–E Momentum.
Let us denote a zonal average by square brackets and the deviation along a latitude circle with

respect to the zonal average by primes. In particular, we have for the zonal u and meridional v wind components, respectively,

(4.55) $$V^1 \equiv u = [u] + u', \qquad V^2 \equiv v = [v] + v'$$

where obviously

(4.56) $$[u'] = [v'] = 0$$

The mean S–N transport rate of zonal momentum (W–E momentum) across a latitude circle in a layer of pressure depth of one unit is then given by

(4.57) $$[vu] = [v][u] + [v'u']$$

The first term $[v][u]$ on the right-hand side of (4.57) is associated with the existence of a mean meridional circulation $[v]$ at the considered latitude. The eddy momentum transport $[v'u']$ is due to the covariance between the fluctuations of v and u along a latitude circle and results from the eddies.

Recalling the definition (3.46) of V^i and taking (4.2)–(4.4) into account, the nondivergent wind at the mean level p_2 is simply given by

(4.58) $$u_\psi = -\partial_y \psi = -L^2 f \sum_{i=1}^{6} \psi_i \, \partial_y F_i$$

and

(4.59) $$v_\psi = \partial_x \psi = L^2 f \sum_{i=1}^{6} \psi_i \, \partial_x F_i$$

With the basic functions (4.6) we have

(4.60) $$[u_\psi] = -L^2 f \sum_{i=1}^{2} \psi_i \, \partial_y F_i$$

and

(4.61) $$[v_\psi] = 0$$

the zonal average of terms linear in $\sin(nx/L)$ or $\cos(nx/L)$ being equal to zero. This yields for the fluctuations:

(4.62) $$u' = -L^2 f \sum_{i=3}^{6} \psi_i \, \partial_y F_i$$

(4.63) $$v' = L^2 f \sum_{i=1}^{6} \psi_i \, \partial_x F_i$$

Taking (4.58)–(4.63) into account, we obtain for the mean S–N transport rate of zonal momentum, (4.57), across a latitude circle by the nondivergent wind at the mean level:

(4.64) $$[vu] = [v'u'] = 4nL^2 f^2 \sin^3(y/L)(\psi'_{1,n} \psi_{2,n} - \psi_{1,n} \psi'_{2,n})$$

Consequently, according to (4.17), the transport of W–E momentum is northward if the $(2, n)$ wave of the ψ-field lags behind its $(1, n)$ wave. As a consequence of (4.54), this transport is accomplished when the wave tilts from S–W toward N–E ($\delta y/\delta x > 0$), a classical result of the theory of momentum transport (Starr, 1948). From (4.64), $[v'u']$ is symmetric with respect to the central latitude, where it reaches its extremum value.

The convergence of the mean meridional momentum transport at a given latitude is given by

(4.65) $\quad -(\partial/\partial y)[v'u'] = -12nLf^2 \sin^2(y/L)\cos(y/L)(\psi'_{1,n}\psi_{2,n} - \psi_{1,n}\psi'_{2,n})$

Let us consider the ψ-field at a particular time so that $\psi'_{1,n}\psi_{2,n} - \psi_{1,n}\psi'_{2,n}$ is fixed and supposed to be positive. From (4.64) it can be seen that, in this case, the transport is toward the north along the whole width of the channel and maximum at its center. The convergence of momentum being positive to the north and negative to the south of the center of the channel [Eq. (4.65)], the zonal wind tends to increase northward and to be concentrated into a jet at the maximum convergence latitude, y_0, easily obtained from (4.65)

(4.66) $\quad\quad\quad y_0 = L \cos^{-1}(-1/\sqrt{3}), \quad\quad \pi/2 < y_0 < \pi$

When the time flows, $\psi'_{1,n}\psi_{2,n} - \psi_{1,n}\psi'_{2,n}$ varies and, consequently, so does the latitude of the jet. If $\psi'_{1,n}\psi_{2,n} - \psi_{1,n}\psi'_{2,n}$ keeps its sign, the maximum zonal wind remains in the northern region while it will travel across the center of the channel when $\psi'_{1,n}\psi_{2,n} - \psi_{1,n}\psi'_{2,n}$ changes its sign.

Obviously, the x-dependence of the momentum transport, vu, related to the presence of the disturbances, creates a longitudinally meandering jet. Let us also note that, according to (4.64), the transport increases with n, i.e. with decreasing wavelength.

4.3.3. The Northward Sensible Heat Transport. The northward sensible heat transport per unit mass across a latitude circle at the mean level is proportional to the covariance

(4.67) $\quad\quad\quad\quad\quad\quad [v\theta] = [v'\theta']$

In a manner similar to that followed in the preceding subsection, we find

(4.68)

$[v'\theta'] = BL^3 f^2 2n\{\sin^2(y/L)(\psi'_{1,n}\theta_{1,n} - \psi_{1,n}\theta'_{1,n})$
$\quad\quad + \sin(y/L)\sin(2y/L)[(\psi'_{1,n}\theta_{2,n} - \psi_{1,n}\theta'_{2,n}) + (\psi'_{2,n}\theta_{1,n} - \psi_{2,n}\theta'_{1,n})]$
$\quad\quad + \sin^2(2y/L)(\psi'_{2,n}\theta_{2,n} - \psi_{2,n}\theta'_{2,n})\}$

The correlations between the waves (1, n) of the ψ- and θ-fields and between the waves (2, n) of the same fields the [terms in $\sin^2 y/L$ and $\sin^2 2y/L$ in (4.68)] determine a systematic northward or southward heat transport at all latitudes. According to (4.68) and (4.17) the transport is northward when the wave (m, n), with $m = 1$ or 2, of the temperature field lags behind the same wave of the stream field. In contrast, the heat transport associated with the correlations between two different modes of the wave n of the ψ- and θ-fields [the term in $\sin y/L \sin 2y/L$ in (4.68)] changes sign at the central latitude $y = \pi L/2$.

4.4. The Spectral Energetics of the Model

4.4.1. The Global Forms. The total potential energy (3.78) being a linear function of θ and σ, we immediately obtain from (4.9) and (4.10)

(4.69) $$P + I = cb^{-1}\{a\theta_0 - b\sigma_0\}$$

where the scale factor c is given by

(4.70) $$c = 2\pi^2 L^4 p_0 g^{-1} f^2$$

The kinetic energy (3.72) can be given the form

(4.71) $$K = \frac{1}{2} c \sum_{i=1}^{6} a_i^2 (\psi_i^2 + \tau_i^2)$$

which can be partitioned into the zonal kinetic energy

(4.72) $$K_Z = \frac{1}{2} c \sum_{i=1}^{2} a_i^2 (\psi_i^2 + \tau_i^2)$$

and the eddy kinetic energy

(4.73) $$K_E = \frac{1}{2} c \sum_{i=3}^{6} a_i^2 (\psi_i^2 + \tau_i^2)$$

We shall also consider each term of (4.71) as the spectral component i of K.

The available potential energy (3.83) is given by

(4.74) $$A = [c/(\sigma_0 + \sigma_{0m})] \sum_{i=1}^{6} \theta_i^2$$

where, according to (3.81),

(4.75) $$\sigma_{0m}^2 = \sigma_0^2 + \sum_{i=1}^{6} \theta_i^2$$

Each term of (4.74) can be considered as the spectral component i of A while the classical zonal A_Z and eddy A_E available potential energy are given by

(4.76) $$A_Z = [c/(\sigma_0 + \sigma_{0m})] \sum_{i=1}^{2} \theta_i^2$$

and

(4.77) $$A_E = [c/(\sigma_0 + \sigma_{0m})] \sum_{i=3}^{6} \theta_i^2$$

When use is made of the equations of the model (4.29)–(4.42), the budget of the energy forms (4.69), (4.72), (4.73), (4.76), and (4.77) can be written

(4.78) $\partial(P+I)/\partial t = -(C_Z + C_E) \qquad\qquad\qquad + G$

(4.79) $\partial A_Z/\partial t = -C_Z \qquad -C_A - C_{AR} \qquad + G_Z + G_{ZR}$

(4.80) $\partial A_E/\partial t = \qquad\quad -C_E + C_A + C_{AR} \qquad + G_E + G_{ER}$

(4.81) $\partial K_Z/\partial t = \quad C_Z \qquad\qquad\qquad -C_K \qquad\qquad - D_Z$

(4.82) $\partial K_E/\partial t = \qquad\quad C_E \qquad\qquad +C_K \qquad\qquad - D_E$

The terms appearing on the right sides of (4.78)–(4.82) can be given the following interpretation:

(4.83) $$C_Z = -cf \sum_{i=1}^{2} \theta_i \omega_i$$

is the conversion rate of A_Z into K_Z.

(4.84) $$C_E = -cf \sum_{i=3}^{6} \theta_i \omega_i$$

is the conversion rate of A_E into K_E.

(4.85) $$C_A = -[2cf/(\sigma_0 + \sigma_{0m})] \sum_{i=1}^{2} \theta_i \sum_{j<k=1}^{6} c_{ijk}(\theta_j \psi_k - \theta_k \psi_j)$$

(4.85′) $$= +[2cf/(\sigma_0 + \sigma_{0m})] \sum_{i=3}^{6} \theta_i \sum_{j<k=1}^{6} c_{ijk}(\theta_j \psi_k - \theta_k \psi_j)$$

is the transfer rate of A_Z into A_E resulting from sensible heat advection.

(4.86) $$C_{AR} = -[cf/(\sigma_0 + \sigma_{0m})^2] \sum_{i=1}^{2} \theta_i \sum_{j=1}^{6} \theta_j(\theta_i \omega_j - \theta_j \omega_i)$$

(4.86′) $$= +[cf/(\sigma_0 + \sigma_{0m})^2] \sum_{i=3}^{6} \theta_i \sum_{j=1}^{6} \theta_j(\theta_i \omega_j - \theta_j \omega_i)$$

is the transfer rate of A_Z to A_E due to the evolution of σ_0.

(4.87) $$G_Z = [2cfk/(\sigma_0 + \sigma_{0m})] \sum_{i=1}^{2} \theta_i(\theta_i{}^* - \theta_i)$$

is the generation rate of A_Z by external heating.

(4.88) $$G_E = [2cfk/(\sigma_0 + \sigma_{0m})] \sum_{i=3}^{6} \theta_i(\theta_i{}^* - \theta_i)$$

is the generation rate of A_E by external heating.

(4.89)
$$G_{ZR} = \frac{cfk}{\sigma_{0m}(\sigma_0 + \sigma_{0m})} \left\{ (\sigma_0{}^* - \theta_0) + 2\sigma_0 + \frac{1}{\sigma_0 + \sigma_{0m}} \sum_{j=1}^{6} \theta_j(\theta_j - \theta_j{}^*) \right\} \sum_{j=1}^{2} \theta_i{}^2$$

is the generation rate of A_Z due to nonadiabatic effects on σ_0 and the reference state (σ_{0m}),

(4.90)
$$G_{ER} = \frac{cfk}{\sigma_{0m}(\sigma_0 + \sigma_{0m})} \left\{ (\theta_0{}^* - \theta_0) + 2\sigma_0 + \frac{1}{\sigma_0 + \sigma_{0m}} \sum_{j=1}^{6} \theta_j(\theta_j - \theta_j{}^*) \right\} \sum_{j=3}^{6} \theta_i{}^2$$

is the generation rate of A_E due to nonadiabatic effects on σ_0 and the reference state (σ_{0m}).

(4.91) $$C_K = -cf \sum_{i=1}^{2} \sum_{j<k=1}^{6} (a_j{}^2 - a_k{}^2) c_{ijk}$$
$$\times \{\psi_i \psi_j \psi_k + \psi_i \tau_j \tau_k + \psi_j \tau_i \tau_k + \psi_k \tau_i \tau_j\}$$

(4.91′) $$= cf \sum_{i=3}^{6} \sum_{j<k=1}^{6} (a_j{}^2 - a_k{}^2) c_{ijk}$$
$$\times \{\psi_i \psi_j \psi_k + \psi_i \tau_j \tau_k + \psi_j \tau_i \tau_k + \psi_k \tau_i \tau_j\}$$

is the transfer rate of K_Z into K_E.

(4.92) $$D_Z = cfk \sum_{i=1}^{2} a_i{}^2 \{(\psi_i - \tau_i)^2 + \tau_i{}^2\}$$

is the dissipation rate of K_Z by friction.

(4.93) $$D_E = cfk \sum_{i=3}^{6} a_i{}^2 \{(\psi_i - \tau_i)^2 + \tau_i{}^2\}$$

is the dissipation rate of K_E by friction, and finally,

(4.94) $$G = cfkb^{-1}\{(a+b)(\theta_0{}^* - \theta_0) + (a+2b)\sigma_0\}$$

is the generation rate of total potential energy by external heating.

The energy structure of large-scale quasi-static flows as expressed by (4.79)–(4.82) is well known in meteorology (Lorenz, 1955). One can simply point out that the spectral equations allow one to assess easily the role of the time changes of σ_0 and σ_{0m} on the conversion and generation rates of available potential energy, while this effect is generally overlooked in most studies of atmospheric energetics. Basically there are two possible sources of eddy kinetic energy. The first one consists of a baroclinic conversion C_E of eddy available potential energy into eddy kinetic energy. Note however that this process necessarily implies a previous transfer C_A from zonal into eddy available potential energy or an external generation G_E of eddy available potential energy which establish along the latitude circles the temperature fluctuations needed for the conversion of potential into kinetic energy (Van Mieghem, 1952). The second possible source is a direct barotropic transfer C_K of kinetic energy from the zonal flow toward the eddy flow.

4.4.2. The Individual Forms. For convenience, we shall use throughout this section the following notation:

(4.95) $\quad \varepsilon \equiv c_{134} = -(8\sqrt{2}/3\pi)n; \ \varepsilon' \equiv c_{156} = -(32\sqrt{2}/15\pi)n;$

$$\varepsilon'' \equiv c_{236} = c_{254} = -(64\sqrt{2}/15\pi)n$$

We shall consider here the contributions to the energy budgets (4.83)–(4.94) of each of the interactions involved in the model and which are schematically shown in Fig. 7. In order to avoid any confusion, let us nevertheless recall that there is no self-interaction of the components $(1, n)$ or $(2, n)$ of the ψ-field with its component $(1, 0)$.

4.4.2.1. The barotropic transfers of kinetic energy. The single barotropic transfers of kinetic energy allowed in the model concern the scales $i = (2, 0)$, $j = (1, n)$, and $k = (2, n)$. According to (4.91) and (4.91'), they assume the form

(4.96) $\quad (\partial K/\partial t)_i^{\text{BT}} = cf\varepsilon''(a_j^2 - a_k^2)T$

(4.97) $\quad (\partial K/\partial t)_j^{\text{BT}} = cf\varepsilon''(a_k^2 - a_i^2)T$

(4.98) $\quad (\partial K/\partial t)_k^{\text{BT}} = cf\varepsilon''(a_i^2 - a_j^2)T$

where the superscript BT recalls that we are considering only the barotropic changes of kinetic energy. The coefficients

(4.99) $\quad a_i^2 = 4, \quad a_j^2 = 1 + n^2, \quad a_k^2 = 4 + n^2$

occur twice with opposite sign so that the total kinetic energy is conserved. The expression for T is

$$(4.100) \quad T = \psi_{2,0}(\psi_{1,n}\psi'_{2,n} - \psi'_{1,n}\psi_{2,n}) + \psi_{2,0}(\tau_{1,n}\tau'_{2,n} - \tau'_{1,n}\tau_{2,n}) \\ + \tau_{2,0}(\tau_{1,n}\psi'_{2,n} - \tau'_{1,n}\psi_{2,n}) + \tau_{2,0}(\psi_{1,n}\tau'_{2,n} - \psi'_{1,n}\tau_{2,n})$$

For given values of c, ε'', a_i^2, a_j^2, and a_k^2, the sign of T governs the direction of the barotropic kinetic energy transfers between the components i, j, and k of the kinetic energy as in a purely barotropic flow (Fjørtoft, 1953). Taking into account the values of a_i^2, a_j^2, and a_k^2 from (4.99) and recalling the negative value (4.95) of ε'', the kinetic energy is transferred from scale $(1, n)$ toward the scales $(2, 0)$ and $(2, n)$ when $T > 0$. From (4.100), the sign of T is determined by the zonal flow of the second mode ($\psi_{2,0}$ and $\tau_{2,0}$) and the relative position of waves $(1, n)$ and $(2, n)$ of the ψ- and τ-fields. For instance, with $\psi_{2,0} > 0$, the first group of terms in T contributes positively provided the wave $(1, n)$ of the ψ-field lags behind its wave $(2, n)$ with, according to (4.17), a maximum contribution when the phase difference between these waves reaches the value $\pi/2$. Obviously each group of terms in T may receive a similar interpretation. Nevertheless, the sign of the whole expression T, resulting from the contributions of the four groups of terms, cannot be *a priori* stated in a simple way.

4.4.2.2. The baroclinic transfers of available potential energy. Let us first consider (see Fig. 7) the baroclinic transfers between the zonal available potential energy of the first mode $A_{1,0}$ and the eddy available potential energy of the first $A_{1,n}$ or second $A_{2,n}$ mode. According to (4.85) and (4.85'), these are given by

$$(4.101) \quad (\partial A/\partial t)^*_{1,0} = [2cf/(\sigma_0 + \sigma_{0m})]\varepsilon\theta_{1,0}(\theta_{1,n}\psi'_{1,n} - \theta'_{1,n}\psi_{1,n}) \\ \doteq -(\partial A/\partial t)^*_{1,n}$$

or

$$(4.102) \quad (\partial A/\partial t)^*_{1,0} = [2cf/(\sigma_0 + \sigma_{0m})]\varepsilon'\theta_{1,0}(\theta_{2,n}\psi'_{2,n} - \theta'_{2,n}\psi_{2,n}) \\ = -(\partial A/\partial t)^*_{2,n}$$

where the asterisk indicates that only the contribution of the transfer between $A_{1,0}$ and $A_{1,n}$ or $A_{2,n}$ to the value of $A_{1,0}$ is considered. Recalling that ε and ε' have negative values, (4.95), available potential energy is transferred from scale $(1, 0)$ toward scale $(1, n)$ [or $(2, n)$] if the wave $(1, n)$ [or $(2, n)$] of the temperature field lags behind the wave $(1, n)$ [or $(2, n)$] of the stream field. Here, as everywhere else in this discussion, it is assumed that $\theta_{1,0} > 0$, i.e. the zonal temperature field of the first mode is such that the temperature decreases northwards. It is well known from linearized perturbation theory (see, for instance, Thompson, 1961) that the growth of baroclinic disturbances imposes such a configuration of the stream and temperature fields. Thus, in a purely zonal flow, the transfer of available potential energy from scale $(1, 0)$ toward scale $(1, n)$ [or $(2, n)$] appears as a necessary step in the conversion of the initially exclusively zonal available potential energy into kinetic energy of the disturbance. According to (4.68) this transfer implies meridional heat transport.

For the additional interactions between scales $(2, 0)$, $(1, n)$, and $(2, n)$ we get, from (4.85) and (4.85′),

(4.103) $\qquad (\partial A/\partial t)^*_{2, 0} = [2cf/(\sigma_0 + \sigma_{0m})]\varepsilon''\{\quad T_1 + T_2 \quad\quad\}$

(4.104) $\qquad (\partial A/\partial t)^*_{1, n} = [2cf/(\sigma_0 + \sigma_{0m})]\varepsilon''\{-T_1 \quad\quad + T_3\}$

(4.105) $\qquad (\partial A/\partial t)^*_{2, n} = [2cf/(\sigma_0 + \sigma_{0m})]\varepsilon''\{\quad\quad - T_2 - T_3\}$

with

(4.106) $\qquad T_1 = \theta_{2, 0}\{\theta_{1, n}\psi'_{2, n} - \theta'_{1, n}\psi_{2, n}\}$

(4.107) $\qquad T_2 = \theta_{2, 0}\{\theta_{2, n}\psi'_{1, n} - \theta'_{2, n}\psi_{1, n}\}$

(4.108) $\qquad T_3 = \psi_{2, 0}\{\theta_{1, n}\theta'_{2, n} - \theta'_{1, n}\theta_{2, n}\}$

Obviously, these transfers conserve the total available potential energy and each T_i can easily be given an interpretation in terms of the flow pattern. Let us emphasize the completely different structure of the transfers (4.103)–(4.105) of available potential energy as compared with the barotropic transfers (4.96)–(4.99) of kinetic energy between different scales of motion. In the present case, there are three separate contributions T_1, T_2, and T_3, each of them appearing twice with an opposite sign. Consequently, and contrary to the barotropic transfers of kinetic energy, the same scale can simultaneously receive available potential energy from another scale and feed the third one.

4.4.2.3. The baroclinic conversions between available potential energy and kinetic energy. The interactions schematically shown in Fig. 7 lead to concomitant changes in different components of K and A during energy conversions. As for the transfers between different spectral components of available potential energy, so the set of conversions between available potential energy and kinetic energy can be partitioned into two subsets. The first one concerns the zonal component of the first mode $(1, 0)$ and one single eddy component $(1, n)$ or $(2, n)$ of K and A. Using (4.83) and (4.84) and substituting for ω_i its values given by (4.46)–(4.51), we obtain for the adiabatic frictionless baroclinic (BC) rate of change of kinetic energy the following expressions:

(4.109) $\quad (\partial K/\partial t)_{1,0}^{BC} = cf\varepsilon(1+\sigma_0)^{-1}\theta_{1,0}(\theta_{1,n}\psi'_{1,n} - \theta'_{1,n}\psi_{1,n})$
$\quad\quad\quad = -(\partial A/\partial t)_{1,0}^{BC}$

(4.110) $\quad (\partial K/\partial t)_{1,n}^{BC} = -cf\varepsilon[1+(1+n^2)\sigma_0]^{-1}(2n^2+1)$
$\quad\quad\quad \times \theta_{1,0}(\theta_{1,n}\psi'_{1,n} - \theta'_{1,n}\psi_{1,n})$
$\quad\quad\quad = (\partial A/\partial t)_{1,n}^{BC}$

resulting from the interaction between scales (1, 0) and (1, n) and

(4.111) $\quad (\partial K/\partial t)_{1,0}^{BC} = cf\varepsilon'(1+\sigma_0)^{-1}\theta_{1,0}(\theta_{2,n}\psi'_{2,n} - \theta'_{2,n}\psi_{2,n})$
$\quad\quad\quad = -(\partial A/\partial t)_{1,0}^{BC}$

(4.112) $\quad (\partial K/\partial t)_{2,n}^{BC} = -cf\varepsilon'[1+(4+n^2)\sigma_0]^{-1}\theta_{1,0}(\theta_{2,n}\psi'_{2,n} - \theta'_{2,n}\psi_{2,n})$
$\quad\quad\quad = -(\partial A/\partial t)_{2,n}^{BC}$

resulting from the interaction between scales (1, 0) and (2, n). The meaning of Eqs. (4.109)–(4.112) should be well understood. Equations (4.109) and (4.111) express conversion rates between $A_{1,0}$ and $K_{1,0}$. However, each of these processes does not operate individually but is necessarily coupled with another conversion taking place respectively [Eqs. (4.110) and (4.112)] between the components (1, n) and (2, n) of A and K. Let us also emphasize that according to (4.109)–(4.112), an increase of kinetic energy at one scale implies a decrease of kinetic energy at the associated scale. In fact the following conservation law governs the baroclinic kinetic energy production in this case:

(4.113) $\quad [2(m^2+n^2)-1][1+(m^2+n^2)\sigma_0]^{-1}\,\partial E_{1,0}/\partial t$
$\quad\quad\quad + (1+\sigma_0)^{-1}\,\partial E_{m,n}/\partial t = 0$

where E is either K or A and $m = 1$ [Eqs. (4.109), (4.110)] or 2 [Eqs. (4.111), 4.112)]. Equation (4.113) is a particular case of a more general conservation law governing the baroclinic energy conversions in the model (Quinet, 1973a). Let us also mention that, according to (4.109)–(4.112) and (4.68), the growth of baroclinic disturbances implies a poleward transportation of heat.

The second subset of associated energy conversions involves the zonal components of the second mode $i = (2, 0)$ and the two eddy components $j = (1, n)$ and $k = (2, n)$ of A and K and results from the interactions between scales $(2, 0)$, $(1, n)$, and $(2, n)$ (see Fig. 7). Following (4.83) and (4.84), these conversion rates assume the form

$$
\begin{aligned}
(4.114) \quad (\partial K/\partial t)^{BC}_{2,0} &= cf\varepsilon'(1+4\sigma_0)^{-1} \\
&\quad \times \{(a_k^2 + a_i^2 - a_j^2)T_1 + (a_j^2 + a_i^2 - a_k^2)T_2\} \\
&= -(\partial A/\partial t)^{BC}_{2,0}
\end{aligned}
$$

$$
\begin{aligned}
(4.115) \quad (\partial K/\partial t)^{BC}_{1,n} &= cf\varepsilon''[1 + (1+n^2)\sigma_0]^{-1} \\
&\quad \times \{-(a_k^2 + a_j^2 - a_i^2)T_1 + (a_j^2 + a_i^2 - a_k^2)T_3\} \\
&= -(\partial A/\partial t)^{BC}_{1,n}
\end{aligned}
$$

$$
\begin{aligned}
(4.116) \quad (\partial K/\partial t)^{BC}_{2,n} &= cf\varepsilon''[1 + (4+n^2)\sigma_0]^{-1} \\
&\quad \times \{-(a_k^2 + a_j^2 - a_i^2)T_2 - (a_k^2 + a_i^2 - a_j^2)T_3\} \\
&= -(\partial A/\partial t)^{BC}_{2,n}
\end{aligned}
$$

where T_1, T_2, and T_3 have been defined in (4.106)–(4.108). In this case, the following conservation law governs the baroclinic production of kinetic energy within the three nonlinearly associated components of K.

$$
\begin{aligned}
(4.117) \quad & [1+(1+n^2)\sigma_0]^{-1}[1+(4+n^2)\sigma_0]^{-1}(a_k^2+a_j^2-a_i^2)(\partial K/\partial t)_i \\
& +[1+(4+n^2)\sigma_0]^{-1}(1+4\sigma_0)^{-1}(a_k^2+a_i^2-a_j^2)(\partial K/\partial t)_j \\
& +(1+4\sigma_0)^{-1}[1+(1+n^2)\sigma_0]^{-1}(a_i^2+a_j^2-a_k^2)(\partial K/\partial t)_k = 0
\end{aligned}
$$

Obviously the same relation could be written with A substituted for K.

4.5. Comments on the Various Physical Processes

A by-product of Sections 4.3 and 4.4 is the provision of a common basis to the various possible interpretations of the time evolution of the model. Different, physically equivalent points of view have been adopted, all of them being classical. Thus, we can conceive the need for a poleward transport of heat (4.67) as the driving mechanism of the flow, and it becomes apparent that this transport can be accomplished by disturbances superimposed on the zonal flow (4.68). Generally speaking, following (4.64), the eddies are then able to accomplish also a meridional transport of zonal momentum which favors [Eq. (4.65)] the generation of narrow jets. On the other hand, if the point of view of the energetics is adopted, the transfers of zonal kinetic energy toward eddy kinetic energy [(4.96)–(4.98)] or the conversions of potential into kinetic energy [(4.109)–(4.112) and (4.114)–(4.116)] will appear as the energy sources of the eddy motion. Clearly, this point of view invites the concept of dynamic instability and the meridional momentum or heat transports appear then as consequences of the instability of the zonal flow. It should be kept in mind that these transports can progressively (re)build new instability conditions. For instance, the momentum convergence associated with the growth of baroclinic disturbances could lead to barotropically unstable zonal wind profiles.

All the energy processes can receive an interpretation in terms of flow configuration through (4.17) and (4.54). The description of the link between dynamic and energetic processes is simple as long as only one zonal scale and one eddy scale are envisaged [(4.101), (4.102); (4.109)–(4.112)]. The energetic interpretation of the effects of interactions between three scales of motion is less simple except for the barotropic kinetic energy transfers [(4.96)–(4.98)]. For the baroclinic case, the main interest in formulas (4.103)–(4.105), and (4.114)–(4.116) is to provide processes relating only two scales of motion as identified by the T_1, T_2, and T_3 terms.

5. The Numerical Study of Vacillation

5.1. The Numerical Procedure

The model presented in the previous sections has been used by Lorenz (1963) to discuss some aspects of vacillating flows. In particular, Lorenz has shown that vacillation arises when the instability conditions of the steady Rossby regime prevail. We shall be concerned here with a detailed description of the properties of vacillation, presenting first its dynamical characteristics and then paying more attention to its energetics.

In the numerical model, the single longitudinal wave of the first two modes has been chosen, more or less arbitrarily, to be at wavenumber $n = 3$. Accordingly the dependent variables of the model have, at most, the spectral components (0, 0), (1, 0), (2, 0), (1, 3), and (2, 3). The numerical integration of the equations has been performed by the Runge–Kutta method which possesses sufficient numerical stability (Young, 1968) as to allow the identification of periodic flows within reasonable limits of confidence. It should be recalled that Fultz' experiments are performed in such a way that the fluid is brought first in solid rotation with the vessel; then the temperature contrast is very slowly established. Since our purpose has never been to reproduce the laboratory conditions of appearance of vacillation (obviously the geometry of the model is not suited to this), the thermal forcing and the rotation rate are both directly applied with their final values. The equations are such that if no wave is initially present, no wave can develop in the system [see Eqs. (4.29)–(4.42)]. Consequently, in order to allow the generation of disturbances, weak waves are initially superimposed on a weak zonal flow. This numerical expedient simulates the random disturbances inevitably present in the laboratory fluid when the experiment is started.

We have seen that the equations of the model are invariant for a change of sign of the variables of the second mode, the variables of the first mode being left unchanged. Consequently, when one solution is obtained, a second solution always exists which is similar to the first, except for the sign of the variables of the second mode. The initial conditions of integration decide which of the solutions develops.

In order to choose adequately the numerical experiments to be performed for simulating the two kinds of vacillation delineated by Pfeffer and Chiang (1967), it is useful to recall the specific characteristics of potential energy and kinetic energy vacillations. It has been seen that the tilt of the waves undergoes only minor oscillations in a potential energy vacillation (see Section 2). Accordingly, the barotropic process of momentum transport (4.64) can be considered as being of secondary importance in this kind of flow. However, a "tilted trough" vacillation necessarily implies considerable fluctuations of the meridional transport of west to east momentum and consequently a marked intervention of a nonsteady barotropic mechanism. Now it can quite safely be stated that the dominant mechanism of generation of the waves is of baroclinic character. On the other hand, it has been noted (Quinet, 1973b) that the higher the rotation rate, the relatively greater becomes the influence of the barotropic kinetic energy transfers. Thus, at a given thermal forcing, and assuming that with this thermal forcing both kinds of vacillation are possible, the potential energy vacillation should precede the kinetic energy vacillation when the rotation rate is gradually increased.

The origin of the temperature scale is chosen so that $\theta_{0,0}^* = 0$ and the value $\theta_{1,0}^* = 0.25$ is used. With the wave $n = 3$ for the first and second modes, this thermal forcing warrants the possibility of generating different kinds of vacillating flows (Quinet, 1973b). Following the arguments of the above discussion, we shall examine the kind of flow generated at a "low" rotation rate and determine if it compares reasonably well with a potential energy vacillation. Then it will be seen that kinetic energy vacillation occurs at higher rotation rates.

Let us recall that, according to the definition of k ($k = Ff^{-1}$), the rotation rate is proportional to k^{-1}.

5.2. The Potential Energy Vacillation

5.2.1. The Dynamics of Potential Energy Vacillation.

The values of $\psi_{1,0}$, $\psi_{2,0}$, $\psi_{1,3}$, $\psi'_{1,3}$, $\psi_{2,3}$, $\psi'_{2,3}$, $\theta_{1,0}$, $\theta_{2,0}$, $\theta_{1,3}$, $\theta'_{1,3}$, $\theta_{2,3}$, $\theta'_{2,3}$, and σ_0 computed for $\theta_{1,0}^* = 0.25$ and $k = 0.2925$ are listed in Table I for every five nondimensional time units $t_0 = ft$. The waves are given by their amplitude and their phase difference with respect to the wave (1, 3) of the ψ-field. Basically, the flow is viewed in a coordinate system moving with the (1, 3)

TABLE I. Numerical integration of the model for $k = 0.2925$, $\theta_{1,0} = 0.25$[a]

t_0	$\psi_{1,0}$	$\psi_{2,0}$	$\psi_{1,3}$	$\psi_{2,3}$	$\theta_{1,0}$	$\theta_{2,0}$	$\theta_{1,3}$	$\theta_{2,3}$	σ_0
0	0.2191	0.0147	0.0537	0.0221, 41°	0.2196	—0.0	0.0222, −52°	0.0095, 11°	0.0778
5	0.2199	0.0136	0.0526	0.0222, 38°	0.2203	−0.0006	0.0218, −53°	0.0095, 7°	0.0772
10	0.2204	0.0117	0.0530	0.0210, 34°	0.2206	−0.0013	0.0221, −53°	0.0091, 4°	0.0770
15	0.2204	0.0100	0.0549	0.0190, 32°	0.2203	−0.0017	0.0229, −53°	0.0083, 3°	0.0773
20	0.2201	0.0087	0.0573	0.0170, 32°	0.2197	−0.0017	0.0238, −52°	0.0074, 4°	0.0779
25	0.2194	0.0083	0.0594	0.0153, 34°	0.2189	−0.0014	0.0246, −51°	0.0067, 6°	0.0787
30	0.2187	0.0086	0.0607	0.0145, 38°	0.2182	−0.0010	0.0250, −51°	0.0064, 9°	0.0793
35	0.2181	0.0096	0.0611	0.0148, 41°	0.2179	−0.0005	0.0251, −50°	0.0065, 12°	0.0796
40	0.2179	0.0111	0.0605	0.0160, 44°	0.2178	−0.0001	0.0248, −50°	0.0070, 15°	0.0796
45	0.2179	0.0128	0.0589	0.0179, 45°	0.2181	0.0002	0.0242, −51°	0.0078, 15°	0.0793
50	0.2183	0.0141	0.0568	0.0200, 45°	0.2187	0.0003	0.0233, −51°	0.0087, 14°	0.0788
55	0.2189	0.0145	0.0545	0.0216, 42°	0.2194	0.0	0.0225, −52°	0.0093, 11°	0.0780
56	0.2191	0.0145	0.0541	0.0218, 41°	0.2196	−0.0001	0.0224, −52°	0.0094, 11°	0.0779

[a] Variables are observed in a coordinate system moving with wave (1.3) of the ψ-field.

152   A. QUINET

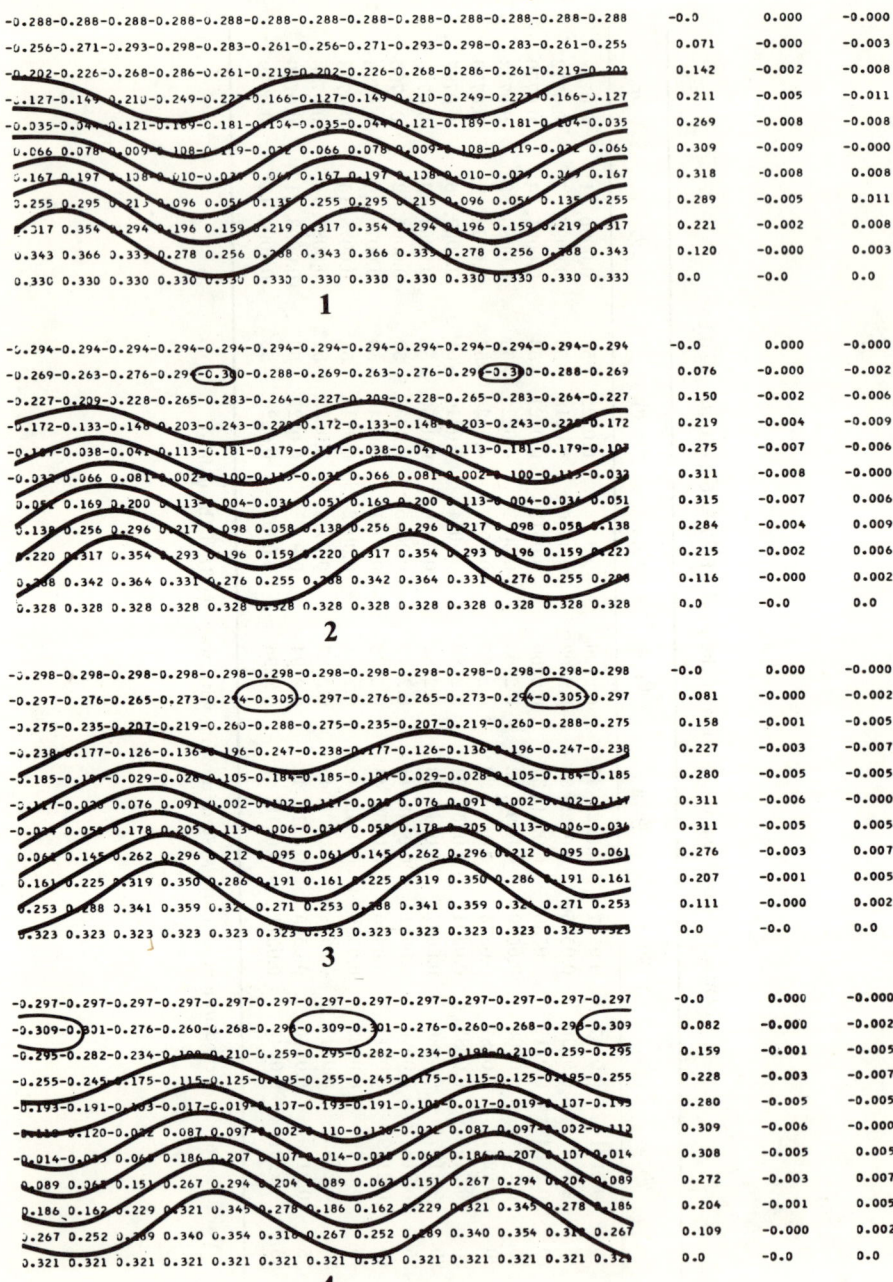

FIG. 8. See facing page for legend.

A NUMERICAL STUDY OF VACILLATION

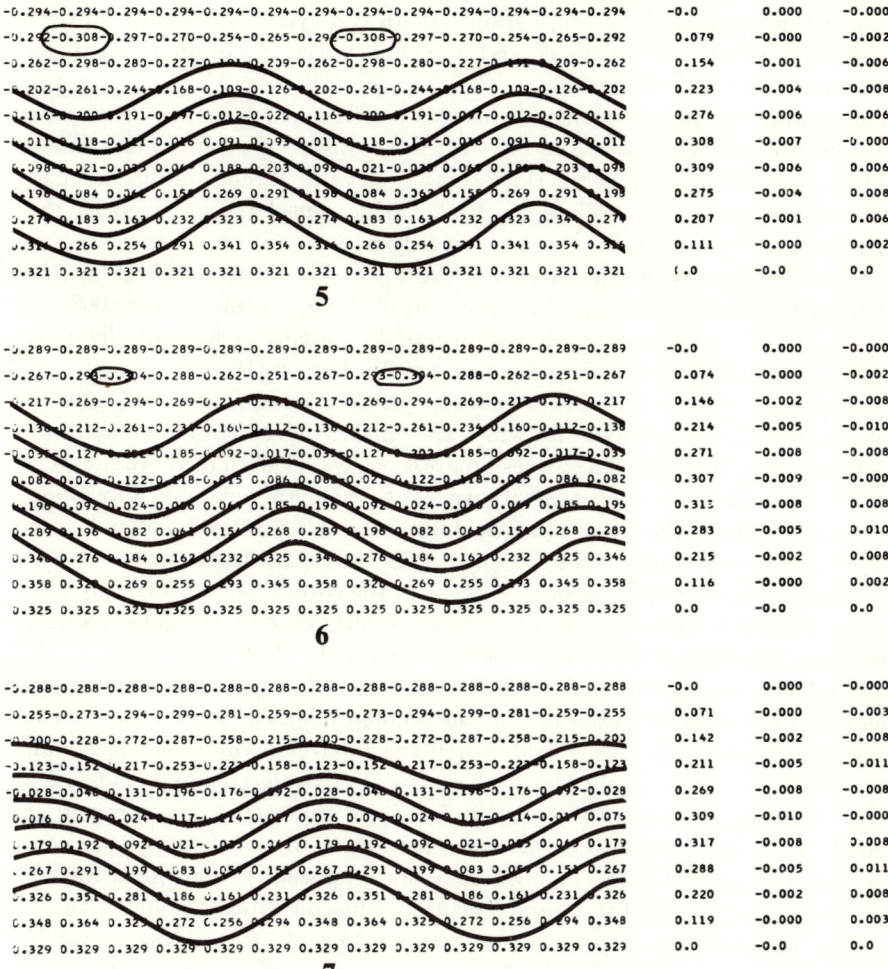

FIG. 8. Synoptic representation of the ψ-field during the potential energy vacillation observed for $k = 0.2925$, $\theta^*_{1,0} = 0.25$. Spacing between isolines is 0.1 in $L^2 f$ units. The time interval between two successive maps is 9 nondimensional time units $t_0 = ft$. The zonal wind in Lf units, the zonally averaged northward W–E momentum transport in $L^2 f^2$ units, and momentum transport convergence in Lf^2 units are respectively printed in the three columns at the right of the maps.

wave of the ψ-field. The periodic character of the flow can be illustrated by comparing the first and last row of Table I, so that the period T of the phenomenon is 56 nondimensional time units. It should, however, be mentioned that in a fixed reference frame, there is a systematic phase shift on each wave component of the ψ- and θ-fields between the beginning and the end of the period. Accordingly, there is some ambiguity regarding the definition of the period of the flow as this period depends on the coordinate system chosen. When only the physical mechanisms of evolution are taken into consideration, this distinction seems irrelevant. Hence, and in order to avoid ambiguity, the vacillation period will be considered as given by the period of a nonwavelike component, namely $\psi_{2,0}$.[2]

Table I shows weak fluctuations of the $\psi_{1,0}$ component ($0.2179 \leq \psi_{1,0} \leq 0.2204$) and also rather limited fluctuations of the $\psi_{2,0}$ component ($0.0083 \leq \psi_{2,0} \leq 0.0147$). Such a periodic flow where $\psi_{2,0}$ oscillates without changing sign has been called *unsymmetric vacillation* by Lorenz (1963).

The amplitudes of the waves exhibit cyclic variations, these being more pronounced for the ψ-waves, for which they may reach 50 % of the amplitude of the (2, 3) component. The waves progress one wavelength in approximately $8t_0$, and their relative positions with respect to each other undergo only minor alterations during the cycle. The θ-field lags behind the ψ-field by 50 to 53 deg for the first mode and by 28 to 31 deg for the second mode. As expected, this configuration points to a baroclinic origin of the eddies ((4.109)–(4.112)). The phase difference between the two modes of the wave of the ψ-field varies between 32 and 45 deg, the (2, 3) wave preceding the (1, 3) wave. Accordingly [(4.54)], the wave pattern tilts SE–NW and there is a systematic southward momentum transport (4.64) during the whole cycle. The amplitudes of the wave (1, 3) of the ψ- and θ-fields reach their extreme values nearly simultaneously, as do the amplitudes of the wave (2, 3) of the ψ- and θ-fields. As a rule, the lowest values of the amplitude of the waves (1, 3) are attained at nearly the same time as the maximum amplitude of waves (2, 3), and conversely. Consequently, the fluctuation of the amplitude of the global wave 3 [waves (1, 3) + (2, 3)] of the ψ and θ fields is reduced to about its lowest value.

The synoptic representation of the stream field at the mean level is given in Fig. 8. The spacing of stream lines is 0.10 nondimensional units and the time interval between two successive maps is 9 time units, the uppermost map corresponding to the first row of Table I. Except for an eastward translation, the last map is a repetition of the first. The zonal wind, the zonally averaged eddy momentum transport, and momentum transport convergence, expressed respectively in Lf, L^2f^2, and Lf^2 units, at latitude $y/L = 18°n$, with $n = 0$,

---

[2] It should also be noted that the small phase shift could be due to the numerical integration scheme.

1, ..., 10, corresponding to the different printed rows of the ψ-values, are listed on the right side of the maps. Clearly the tilt of the trough and ridge lines is practically constant during the cycle. The zonal wind U being given by

(5.1) $\qquad U = (Lf)(\sqrt{2}\bar{\psi}_{1,0}\sin(y/L) + 2\sqrt{2}\bar{\psi}_{2,0}\sin(2y/L))$

its maximum remains south of the center of the channel since $\bar{\psi}_{2,0} > 0$. The maximum zonal wind oscillates in the narrow latitude belt $y/L = 76°$ at $t = 0$ (or $t = T$) and $y/L = 82°$ at $t = T/2$, corresponding respectively to the maximum (0.0147) and minimum (0.0084) values of $\bar{\psi}_{2,0}$. The small variations of the zonal wind profile are represented in Fig. 9 where the figures on

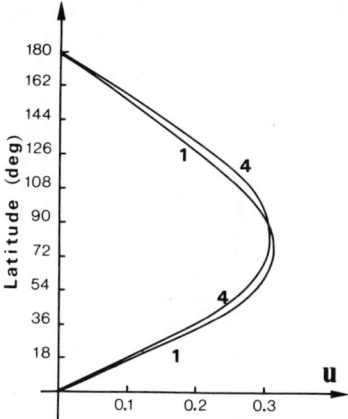

FIG. 9. Meridional profile of the zonal wind corresponding to maps 1 and 4 of Fig. 8. Latitudes are given in degrees according to (4.1).

the curves refer to the corresponding maps in Fig. 8. The largest zonal wind maximum is observed on map 1 (and 7) when this maximum occupies its lowest latitude (76°), and the smallest maximum appears on map 4 where it reaches its highest latitude (82°).

Owing to the presence of the second mode of wave 3, a distinction should be made when comparing the flow patterns in the northern and southern parts of the different maps. It appears then that in the northern part, the troughs deepen from map 1 to map 4, where well defined closed lows are observed. In this respect, map 1 could be considered as characteristic of a high index circulation and map 4 as corresponding to a low index circulation. This comparison is consistent with the fluctuations of the zonal wind, the closed cyclones being associated with relatively weak jets. The southern part undergoes a similar pulsation, the most intense ridge being observed on map 1 and then decaying down to map 4. A detailed comparison would be needed in order to establish definitely the analogy of the flow presented here with the potential energy vacillation emphasized by Pfeffer and Chiang (1967). Let it be simply stated that the nearly steady character of the wave tilt during its amplitude fluctuation and the fluctuations of the intensity of the zonal jet during the cycle are important characteristics common to both phenomena.

The southward transport of west to east momentum reported in the second column of Fig. 8 is evidently associated with the NW–SE tilt of the waves. The maximum southward transport occurs on map 1 and then decays, approximately down to map 4, where the northern cyclone has developed. The momentum transport during the cycle is given for the three latitudes $y/L = 54°$, $90°$, and $126°$, from bottom to top in Fig. 10a, where the figures

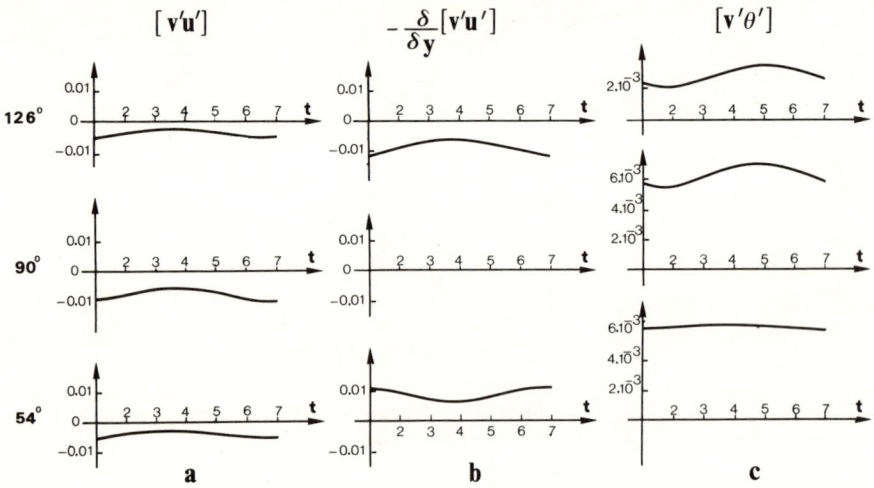

FIG. 10. Time evolution of the zonally averaged (a) northward W–E eddy momentum transport $[v'u']$ in L^2f^2 units; (b) W–E eddy momentum transport convergence $-\partial[v'u']/\partial y$ in Lf^2 units; and (c) eddy heat transport $[v'\theta']$ in BL^3f^2 units, at the three latitudes $y/L = 54°$, $90°$, and $126°$, during the potential energy vacillation observed at $k = 0.2925$, $\theta^*_{1,0} = 0.25$. The figures along the x-axis refer to the maps of Fig. 8.

along the x-axis refer to the maps of Fig. 8. As imposed by (4.64), there is maximum southward momentum transport at the central latitude and identical weaker values at the two other latitudes.

The third column of Fig. 8 indicates that there is momentum divergence north of the center of the channel and, as a consequence of (4.65), momentum convergence in the south. Accordingly, the southward export rate of west to east momentum leads to a reduction of the zonal flow in the northern part and creates a zonal jet in the southern part. Figure 10b gives the time variation of $-\partial[v'u']/\partial y$ during the cycle. The maximum values of the low latitude momentum transport convergence and of its high latitude divergence arise at time $t = 0$ (and $t = T$), when the zonal wind maximum occupies its lowest latitude $y/L = 76°$ and reaches its largest value. At time $t = T/2$, the low latitude convergence is minimum, corresponding to the weakest jet at the highest latitude $y/L = 82°$. Accordingly, the zonal jet fluctuations are strongly correlated with the momentum transport and its convergence.

The heat transport at the three above mentioned latitudes is given in Fig. 10c. It is toward the north at all latitudes and during the whole cycle. The largest values occur in the region of maximum zonal wind. They are approximately the same at 54° and 90° latitude, and roughly half as much at $y/L = 126°$. The heat transport is nearly constant at low latitude during the cycle but exhibits fluctuations of approximately the same amplitudes at the central and high latitudes, where the fluctuations are in phase. The maximum heat transport corresponds to map 5, i.e. somewhat later than the time of occurrence of low index circulation, when the zonal flow is minimum. It should be noted that the evolution of $[v'\theta']$ does not show any asymmetry with respect to the extreme values while Pfeffer *et al.* (1965) have reported the existence of such asymmetries in the potential energy vacillation.

The occurrence in time of the extreme values of U, $[v'u']$, $-\partial[v'u']/\partial y$, and $[v'\theta']$ is summarized in Table II, where M stands for "maximum" and m for "minimum." Since, at a given latitude, these quantities do not change their sign during the cycle, the modulus only has been considered for labeling the extreme values.

TABLE II. Distribution in time of the occurrence of extreme values (M = maximum, m = minimum) of zonal wind u, zonally averaged relative S–N eddy momentum transport $[v'u']$ and momentum transport convergence $-\partial[v'u']/\partial y$ and zonally averaged northward heat transport $[v'\theta']$ during the potential energy vacillation cycle observed at $k = 0.2925$, $\theta^*_{1,0} = 0.25$

Maps	1	2	3	4	5	6	7
u	M			m			M
$[v'u']$	M			m			M
$-\partial[v'u']/\partial y$	M			m			M
$[v'\theta']$		m			M		

5.2.2. The Energetics of the Potential Energy Vacillation.

Figure 11 shows the evolution of the available potential energy A and the kinetic energy K, expressed in $c = 2\pi^2 p_0 g^{-1} L^4 f^2$ units, K being approximately one half of A

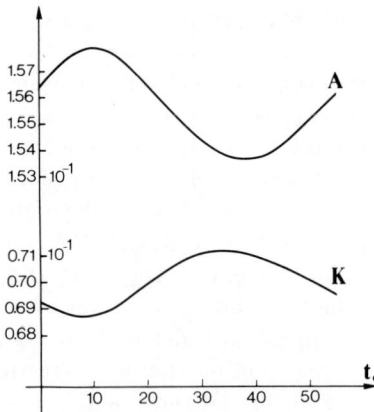

Fig. 11. Time evolution of the available potential energy A and of the kinetic energy K, expressed in $2\pi^2 p_0 g^{-1} L^4 f^2$ units, during the potential energy vacillation observed at $k = 0.2925$, $\theta_{1,0}^* = 0.25$.

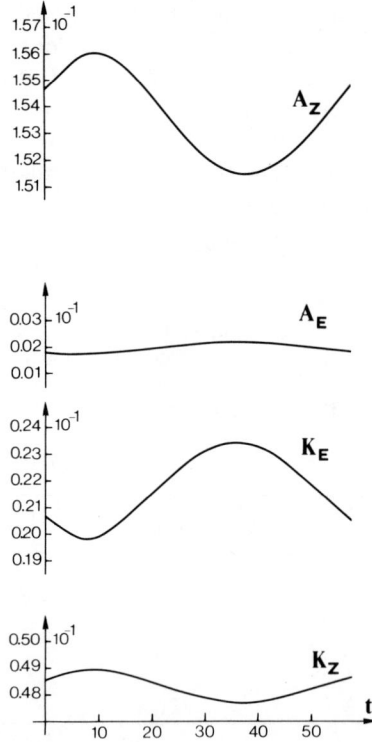

Fig. 12. Time evolution of the zonal A_Z and eddy A_E available potential energy and of the zonal K_Z and eddy K_E kinetic energy, expressed in $2\pi^2 p_0 g^{-1} L^4 f^2$ units, during the potential energy vacillation observed at $k = 0.2925$, $\theta_{1,0}^* = 0.25$.

during the cycle. The striking feature is the nearly perfect phase opposition between the two curves although nonadiabatic processes are taking place. The fluctuations of A are of greater amplitude than those of K.

Figure 12 gives the graphic representation of A_Z, A_E, K_E, and K_Z during the cycle. A_Z and K_E undergo the major oscillations while A_E is nearly constant. As a consequence of the very intense zonal thermal forcing, A_Z is two orders of magnitude greater than A_E; K_Z is approximately one third of A_Z and K_E one order of magnitude greater than A_E. In opposition to the curves of A and K, the A_Z and K_Z curves or the A_E and K_E curves are nearly in phase, so that any increase of A_Z (A_E) is accompanied by a simultaneous increase of K_Z (K_E).

In order to investigate the nature of the energetics of the flow, Fig. 13 gives the evolution of the various terms of the energy budgets (4.79)–(4.82). Clearly C_Z is the dominant conversion term, so that the zonal kinetic energy is essentially maintained against dissipation by conversion from A_Z. In lesser proportion A_Z is also transferred into A_E, by C_A and C_{AR}, which, in turn, is converted ($C_E > 0$) into K_E at an important rate. Accordingly, there must be a considerable generation of A_Z by heating, G_Z, feeding both C_Z and $C_A + C_{AR}$. Evidently, without any longitudinally distributed heat source, $G_E < 0$ [see (4.88)]. The kinetic energy transfers are toward the zonal flow ($C_K < 0$) but the contribution of this term is two orders of magnitude less than C_Z so that, as expected, the barotropic kinetic energy transfers are extremely weak.

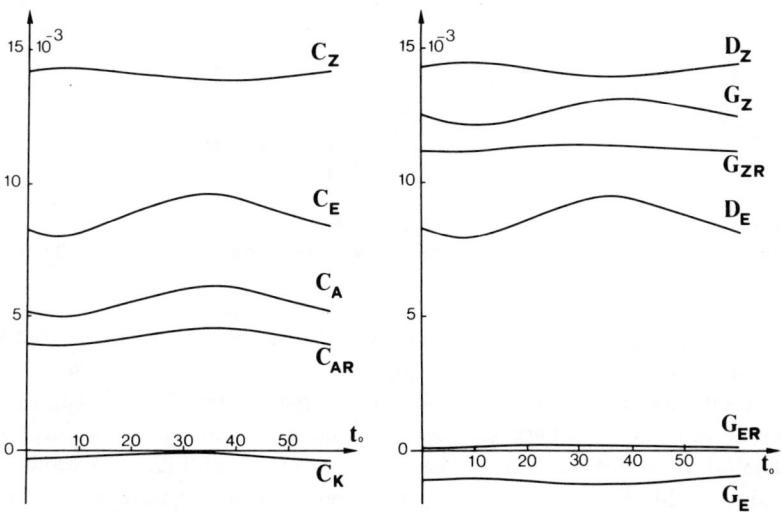

FIG. 13. Time evolution of the different components of the energy budgets [see Eqs. (4.79)–(4.82) for definitions], expressed in $2\pi^2 p_0 g^{-1} L^4 f^3$ units, during the potential energy vacillation observed at $k = 0.2925$, $\theta_{1,0}^* = 0.25$.

G_Z, C_A, C_{AR}, C_E, and D_E are nearly in phase but in phase opposition with C_Z and D_Z. Thus, weak values of G_Z are associated with high values of C_Z and vice versa, a situation which leads to large fluctuations of A_Z. These are however partly reduced by $C_A + C_{AR}$. In contrast, neglecting the fluctuations of G_E, the nearly perfect correlation between $C_A + C_{AR}$ and C_E tends to reduce to zero the fluctuations of A_E. Finally, the fluctuations of $C_A + C_{AR}$ being more intense and in phase opposition with those of C_Z, the zonal available potential energy A_Z and kinetic energy K_Z increase or decrease simultaneously while the evolution of K_E is essentially governed by C_E.

Figure 14 shows the evolution of T, defined in (4.100), which governs the barotropic kinetic energy transfers arising between scales (2, 0), (1, 3), and (2, 3), and the evolution of T_1, T_2, T_3, defined in (4.106)–(4.108), which

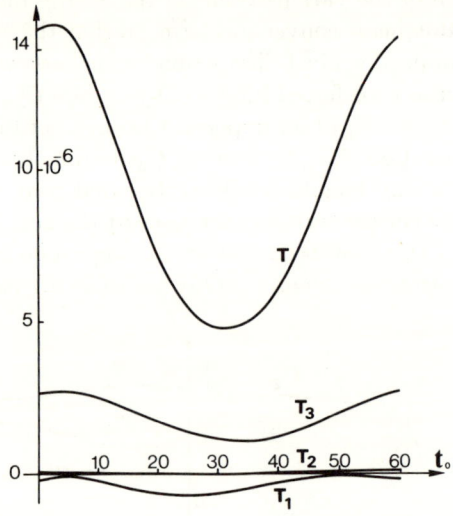

FIG. 14. Time evolution of T, T_1, T_2, and T_3 [see Eqs. (4.100) and (4.106)–(4.108) for definitions] during the potential energy vacillation observed at $k = 0.2925$, $\theta^*_{1,0} = 0.25$.

governs the available potential energy transfers and conversions between the same scales. The essential fact to note is the weakness of these terms as compared with the global conversion and transfer rates. Thus, it can be considered that the nonlinear character of the potential energy vacillation is restricted to the interactions of the two modes (1, 3) and (2, 3) of waves with the zonal field (1, 0). Since these self-interactions do not influence the barotropic kinetic energy transfers (see Section 4.4.2), the potential energy vacillation results only from the baroclinic transfers and conversions of available potential energy. The strong correlation between C_A, C_Z, and C_E

appears then to result from the fact that only the self-interaction terms (4.101), (4.102), and (4.109)–(4.112) come into play in the energy processes.

The term T being positive, there is a permanent transfer from $K_{1,3}$ toward $K_{2,0}$ and $K_{1,3}$. On the other hand, $A_{1,3}$ is transferred toward $A_{2,0}$ ($T_1 < 0$) but chiefly toward $A_{2,3}$ ($T_3 > 0$) while the transfers between $A_{2,0}$ and $A_{2,3}$ are extremely small ($T_2 \simeq 0$). Let us finally note that, regarding the role of the (2, 0), (1, 3), and (2, 3) interactions, the barotropic processes, although very weak in comparison with the self-interaction contributions, are the dominant ones.

Figure 15 describes schematically the energy cycle of the potential energy vacillation; the values indicated on the arrows represent mean values over one period.

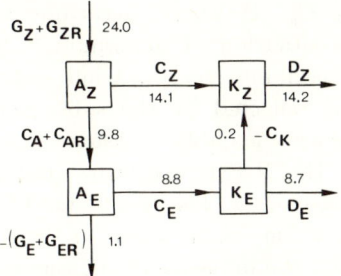

Fig. 15. Energy diagram averaged over one period of the potential energy vacillation observed at $k = 0.2925$, $\theta_{1,0}^* = 0.25$. Energies are expressed in $c = 2\pi^2 p_0 g^{-1} L^4 f^2$ units, the conversion and transfer rates are expressed in cf units.

One of the major deficiencies of the simulated flow with respect to laboratory potential energy vacillation is the symmetric character of the heat transport during the cycle and the relatively weak fluctuations of the amplitudes of the waves. Once it has been recognized that potential energy vacillation involves essentially baroclinic energy exchanges, it is possible to obtain intense potential energy vacillation at a high rotation rate by simply not including the barotropic energy exchanges in the model. Such a model, with the correct cylindrical annular geometry, has been dealt with by Merilees (1972). If high amplitude potential energy vacillation can indeed be simulated in this way, the heat transport nevertheless remains symmetric during the cycle. The introduction of viscosity and heat conductivity into the numerical model would improve the results. It has indeed been observed that amplitude vacillation is most common with highly viscous fluids (Fowlis and Hide, 1965). However, whether these physical factors will introduce in the model the specific effects capable of leading to asymmetries in the heat transfer cycle remains an open question.

5.2.3. A Physical Interpretation of the Potential Energy Vacillation.

We will now try to give a simple physical interpretation of the flow. Taking into account the results of subsection 5.2.2, one need only deal with the dominant baroclinic processes related to the self-interactions of the waves with the zonal thermal field of the first mode $\theta_{1,0}$.

The generation of A_Z consists of a cooling of cold air in the north and a warming of warm air in the south. Accordingly, $\theta_{1,0}^*$ being steady, G_Z is a minimum when the zonal temperature contrast between the southern and northern part of the fluid is the largest ($\theta_{1,0}$ a maximum). In this case, the intensity of the direct Hadley cell generated by the N-S thermal contrast into the fluid, i.e. the conversion rate C_Z of A_Z into K_Z is maximum (see Table I and Fig. 13). The associated relative excess of heat transport by an intense Hadley cell allows a reduction of the eddies which are then generated at their minimum rate C_E. However, according to the linear baroclinic instability theory, this situation of maximum N-S temperature contrast and, consequently, of maximum vertical wind shear, favors the generation of disturbances. This is reflected by the subsequent increase of C_E. The amplifying disturbances are then able to assume a supplementary part of the heat transport. Hence, the N-S temperature contrast into the fluid is reduced, entailing a decrease of C_Z. The rate C_E (and C_A) goes on growing until it reaches its maximum value, which occurs at the time of minimum C_Z.

In this situation, under the influence of the thermal forcing, G_Z attains its maximum value since $\theta_{1,0}$ is a minimum (see Table I and Fig. 13). After the time of minimum value of $\theta_{1,0}$, and in accordance with linear baroclinic instability theory, the baroclinic generation rate of the eddies C_E decreases. The associated deficit of heat transport is then taken up by an intensification of the Hadley cell (C_Z) which is, however, insufficient to avoid the reappearance of a strong thermal contrast between the southern and northern part of the fluid. When $\theta_{1,0}$ is maximum, G_Z and C_E are minimum while C_Z is maximum (see Table I and Fig. 13) and the cycle starts again.

The most striking aspect of the potential energy vacillation energetics is that the maximum baroclinic generation rate of the eddies C_E appears at the minimum N-S thermal gradient. This shows how the nonlinear mechanism of the interplay between the zonal fields and the eddies can lead to situations radically different from those suggested by the theory of linear perturbations.

5.3. The Kinetic Energy Vacillation

5.3.1. The Dynamics of the Kinetic Energy Vacillation.

We now describe the results of a numerical experiment performed with the same thermal forcing $\theta_{1,0}^* = 0.25$ as in the previous case, but at a higher rotation rate, $k = 0.275$. Table III lists the values of the flow variables for every five time steps in a

TABLE III. Numerical integration of the model for $k = 0.275$, $\theta^*_{1,0} = 0.25$[a]

t_0	$\psi_{1,0}$	$\psi_{2,0}$	$\psi_{1,3}$	$\psi_{2,3}$	$\theta_{1,0}$	$\theta_{2,0}$	$\theta_{1,3}$	$\theta_{2,3}$	σ_0
0	0.2186	−0.0294	0.0458	0.0336, −122°	0.2188	−0.0053	0.0188, −55°	0.0136, −156°	0.0782
5	0.2203	−0.0225	0.0339	0.0357, −151°	0.2223	−0.0014	0.0138, −58°	0.0146, 175°	0.0746
10	0.2230	−0.0084	0.0366	0.0296, −177°	0.2246	0.0035	0.0156, −57°	0.0120, 150°	0.0726
15	0.2241	0.0030	0.0470	0.0211, 165°	0.2239	0.0067	0.0206, −56°	0.0080, 135°	0.0738
20	0.2225	0.0125	0.0572	0.0156, 126°	0.2206	0.0073	0.0245, −54°	0.0052, 95°	0.0774
25	0.2192	0.0248	0.0570	0.0249, 77°	0.2176	0.0065	0.0237, −53°	0.0095, 41°	0.0798
30	0.2184	0.0296	0.0462	0.0337, 58°	0.2186	0.0054	0.0190, −55°	0.0136, 24°	0.0784
35	0.2201	0.0227	0.0339	0.0359, 29°	0.2222	0.0015	0.0137, −58°	0.0147, −5°	0.0747
40	0.2230	0.0085	0.0365	0.0296, 3°	0.2246	−0.0035	0.0155, −57°	0.0120, −30°	0.0726
45	0.2241	−0.0031	0.0469	0.0212, −15°	0.2239	−0.0067	0.0205, −56°	0.0080, −45°	0.0737
50	0.2226	−0.0126	0.0570	0.0158, −54°	0.2207	−0.0073	0.0245, −54°	0.0052, −85°	0.0773
55	0.2197	−0.0230	0.0581	0.0227, −98°	0.2179	−0.0067	0.0243, −53°	0.0085, −134°	0.0797
60	0.2185	−0.0296	0.0458	0.0338, −122°	0.2187	−0.0054	0.0188, −55°	0.0136, −156°	0.0783

[a] Variables are observed in a coordinate system moving with wave (1, 3) of the ψ-field.

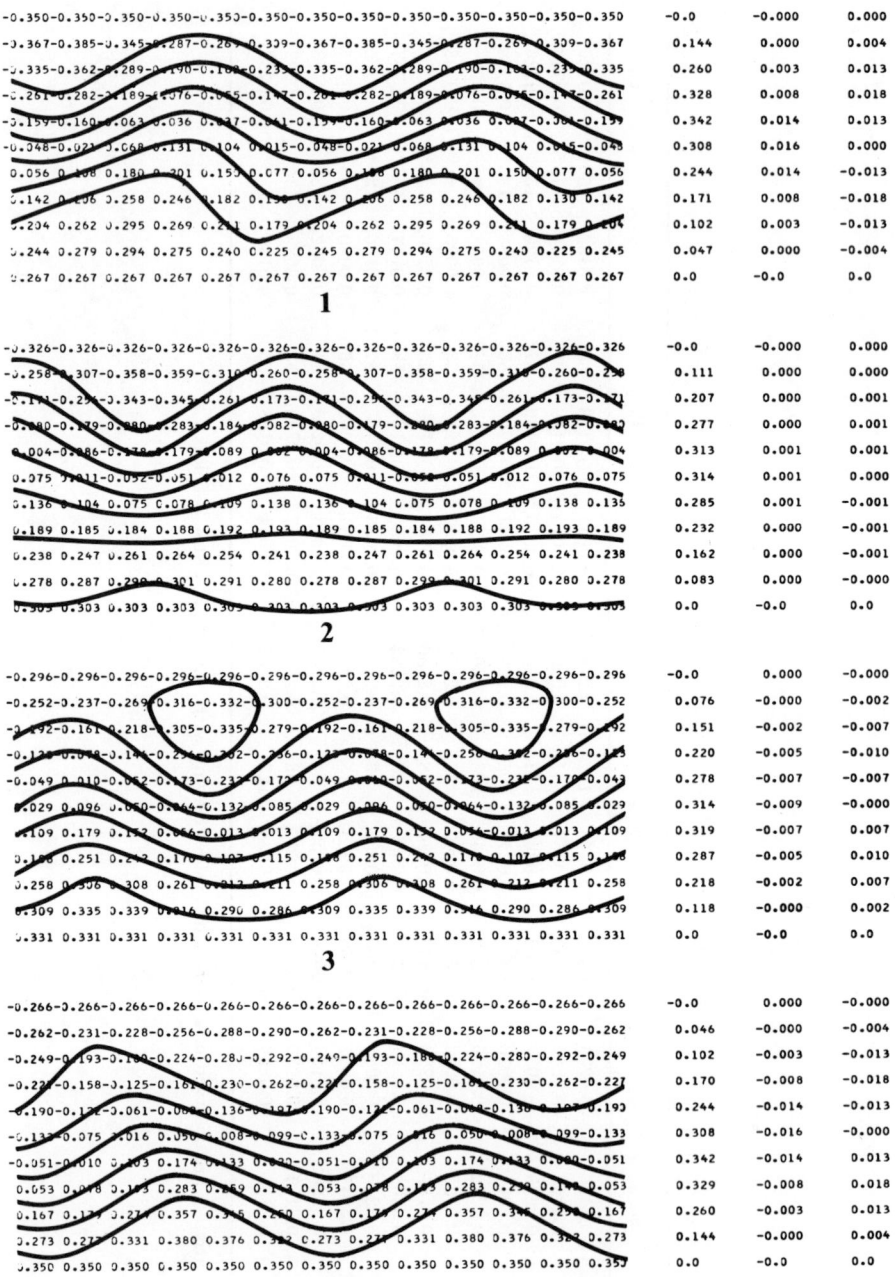

Fig. 16. See facing page for legend

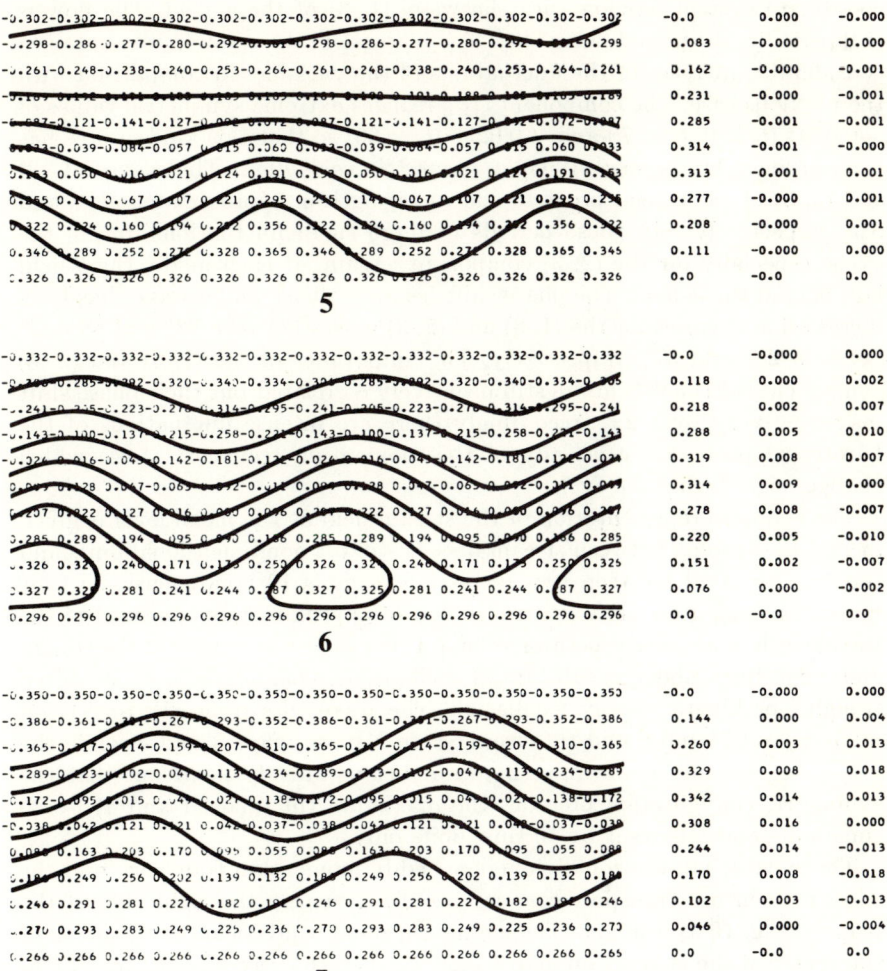

Fig. 16. Synoptic representation of the ψ-field during the kinetic energy vacillation observed at $k = 0.275$, $\theta_1^*{}_0 = 0.25$. Spacement between isolines is 0.1 in $L^2 f$ units. The time interval between two successive maps is 10 nondimensional time units $t_0 = ft$. The zonal wind in Lf units, the zonally averaged S–N momentum transport in $L^2 f^2$ units, and the momentum transport convergence in Lf^2 units respectively, are, printed on the three columns at the right of the maps.

coordinate system moving with the wave (1, 3) of the ψ-field. The flow is still periodic, the period being $60t_0$. The increase of the rotation rate entails a considerable increase of the fluctuations of the variables in comparison with the previous case. The component $\psi_{2,0}$ reaches extreme symmetric values of -0.0294 ($t_0 = 0$, $t_0 = T$) and $+0.0296$ ($t_0 = 30$) so that this kind of flow has been termed by Lorenz (1963) a *symmetric vacillation*. The waves (1, 3) and (2, 3) of the ψ-field progress one wavelength in approximately $7\frac{1}{3}t_0$ and $8\frac{1}{2}t_0$ respectively, but these periods undergo small fluctuations during the cycle, especially for the (2, 3) component. Again, at each mode, the θ-field lags behind the ψ-field. The phase shift between the ψ- and θ-waves has been increased and varies for the (1, 3) and (2, 3) waves between 53° and 58° and between 28° and 37°, respectively. The amplitudes of the (1, 3) and (2, 3) components of the ψ-field are still negatively correlated but their phase shift may now vary over 2π. Accordingly there are intense fluctuations of the barotropic processes, especially of the momentum transport (4.64) which changes sign during the cycle.

The synoptic representation of the stream field at the mean level is given in Fig. 16. Spacing between stream lines is still 0.1 nondimensional units and the time interval between two successive maps is 10 nondimensional time units. Except for an eastward translation of approximately one sixth of the wavelength, map 7 is a repetition of map 1. It can be seen now that the trough and ridge lines undergo substantial oscillations, characteristic of a "tilted trough" or kinetic energy vacillation. The wave tilts from SW to NE on maps 1 (and 7) and 6 and tilts from SE to NW on maps 3 and 4 while the waves are symmetric on maps 2 and 5. Let us already point out that the symmetric configuration precedes the maximum tilt (maps 1 and 4) by 20 time units and follows it by 10 time units only.

The zonal wind, the zonally averaged eddy momentum transport, and the eddy momentum transport convergence are given on the right side of the maps of Fig. 16 as in Fig. 8. The maximum zonal wind occurs to the north of the center of the channel on map 1 (and 7) when the waves are tilting northeast and when there is a northward transport of W–E momentum through the whole width of the channel. At this time there is momentum transport convergence to the north of the center of the channel and momentum transport divergence to the south. The situation is reversed on map 4, the southward momentum transport converging south of the center of the channel where the maximum zonal wind is located. Between these two situations, but not midway, the momentum transport and the momentum transport convergence vanish at all latitudes (maps 2 and 5).

The negative values of the momentum transport observed after map 2 are associated with the deepening of the waves, generating high latitude cyclones which are established on map 3, somewhat before the time of the maximum zonal jet of map 4. Similarly, positive values of the momentum transport are associated with amplifying low latitude highs.

The zonal flow profile is given in Fig. 17 where the figures on the different curves refer to the corresponding map of Fig. 16. The extreme latitudes of the maximum zonal wind are respectively 112° on map 1 (and 7) and 67° on map 4, when the waves reach their maximum tilt. At these times, the zonal flow is barotropically unstable for small linear perturbations, the zonal wind profile presenting an inflection point on Fig. 17. This is not without suggesting an explanation for the rapid reestablishment of the symmetric configuration of maps 2 and 5 after the time of maximum tilt.

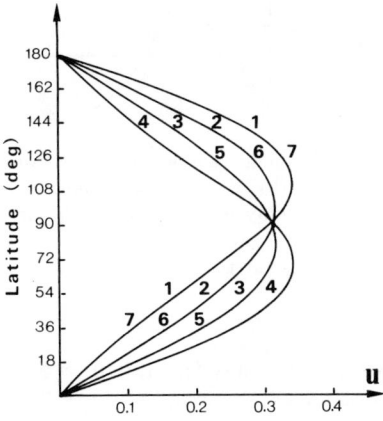

FIG. 17. Meridional profile of the zonal wind corresponding to maps 1 to 7 of Fig. 16. Latitudes are given in degrees according to (4.1); the inflection point is at 63° and 117° on curves 1 and 4, respectively.

The fast evolution after the occurrence of an unstable zonal wind profile is particularly well represented in Fig. 18 which is the analog of Fig. 10. Northward or southward eddy momentum transport is simultaneously observed at the three latitudes of 54°, 90°, and 126° while high latitude momentum transport convergence is balanced by low latitude divergence, and conversely. As noted by Fultz, the maximum value of $[u'\,v']$ is located at the middle latitude. In the model considered here, this circumstance is necessarily a consequence of (4.64).

There is considerable asymmetry on each of the curves of Figs. 18a and 18b, the extreme values, corresponding to maps 1 and 4, being slowly reached and then rapidly left. This allows one to give the following interpretation of the momentum evolution during the cycle: the momentum transport convergence associated with the tilting of the waves tends to remove the jet from the center of the channel, where the thermal forcing would locate it.

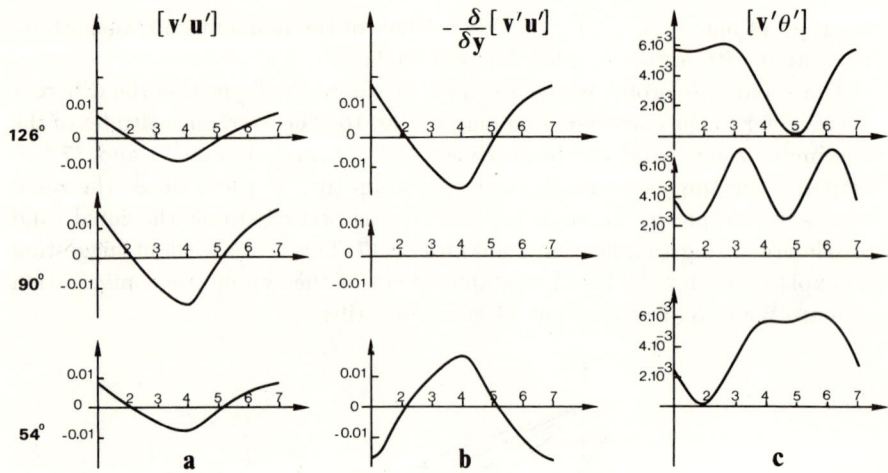

Fig. 18. Time evolution of the zonally averaged (a) northward W–E eddy momentum transport $[v'u']$ in L^2f^2 units; (b) W–E eddy momentum transport convergence $-\partial[v'u']/\partial y$ in Lf^2 units; and (c) eddy heat transport $[v'\theta']$ in BL^3f^2 units, at the three latitudes $y/L = 54°$, $90°$, and $126°$ during the kinetic energy vacillation observed at $k = 0.275$, $\theta_{1,0} = 0.25$. The figures on the x-axis refer to the corresponding maps of Fig. 16.

Before the maximum wind reaches the latitude of maximum momentum convergence the mechanism of barotropic instability comes into play and rapidly redistributes the momentum within the zonal flow, releasing barotropic instability.

The heat transport shown in Fig. 18c exhibits important differences at the three latitudes and differs radically from the previous case (Fig. 10c). The curves are out of phase at the two extreme latitudes, but the amounts of heat transported across these latitudes during the complete cycle are now of equal importance. There are two strong maxima and minima at the central latitude, while there exist only one well defined maximum and one well defined minimum, separated by a weak secondary minimum and maximum, on the other curves. Thus, at latitudes of 54° and 126°, the heat transport slowly increases and then rapidly decays. The maximum high latitude heat transport occurs near the time of appearance of the closed cyclones at those latitudes. An analogous situation occurs at low latitude, at the time of appearance of the closed highs. It will be established in the next section that, as suggested by these correlations, the cellular circulations arise from intense transformation rates of eddy available potential energy into eddy kinetic energy. Consistent with Fultz' observation the minima of heat transport at 54° and 126° latitude occur near the time of maximum zonal wind in the northern and southern part of the channel, respectively, when the momentum transport and the momentum transport convergence through the whole width vanish. Likewise, as in laboratory experiments, the minimum heat transport at a given latitude follows the minimum momentum convergence at this latitude. As for the potential energy vacillation, the relative positions of the times of occurrence of the extreme values of U, $[v'u']$, $-\partial[v'u']/\partial y$, and $[v'\theta']$ are schematically shown in Table IV. M stands for

TABLE IV. Distribution in time of the occurrence of extreme values (M = maximum, m = minimum) of the zonal wind U, the zonally averaged S–N momentum transport $[v'u']$ and momentum transport convergence $-\partial[v'u']/\partial y$, and zonally averaged heat transport $[v'\theta']$ during the kinetic energy vacillation cycle observed at $k = 0.275$, $\theta^*_{1,0} = 0.25$ [a]

Maps	1	2	3	4	5	6	7
U	M(N)			M(S)			M(N)
$[v'u']$	M	0	<0	m	0	>0	M
$-\partial[v'u']/\partial y$	M(N) m(S)	0	<0 >0	m(N) M(S)	0	>0 <0	M(N) m(S)
$[v'\theta']$			m(S)	M(N)	m(N)	M(S)	

[a] The region of location of the extreme value, north (N) or south (S) of the center of the channel, is also indicated.

positive maximum and m stands for (negative) minimum; N (north) and S (south) indicate the region of location of the associated extreme value with respect to the center of the channel.

5.3.2. *The Energetics of the Kinetic Energy Vacillation.* Figure 19 shows the evolution of A and K expressed in nondimensional units. Once again, any increase of A is accompanied by a simultaneous decrease of K, and vice versa. The orders of magnitude of A and K are respectively the same as in Fig. 11 but, noting the scale differences along the y-axis in Figs. 19 and 11, the fluctuations are now increased and amount to approximately 10 % of the total values. There is no noticeable asymmetry in the curves.

Figure 20, the analog of Fig. 12, now shows the presence of small phase shifts between A_Z, A_E, K_E, and K_Z. The curve of K_Z is asymmetric with respect to the extreme values. As discussed in the previous section, this characteristic can be ascribed to the barotropic instability conditions prevailing at certain times (maps 1 and 4) in the fluid. On the contrary, the curve of K_E appears symmetric with respect to the extreme values. There is also some asymmetry on the curve of A_E. This will be discussed later.

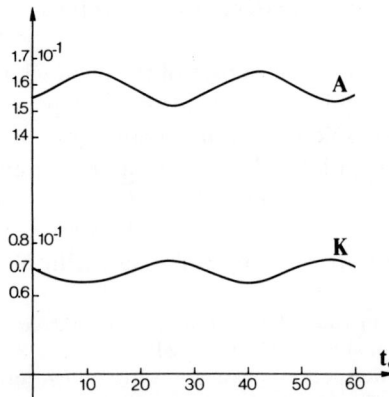

Fig. 19. Time evolution of the available potential energy A and of the kinetic energy K, expressed in $2\pi^2 p_0 g^{-1} L^4 f^2$ units, during the kinetic energy vacillation observed at $k = 0.275$, $\theta_{1,0}^* = 0.25$.

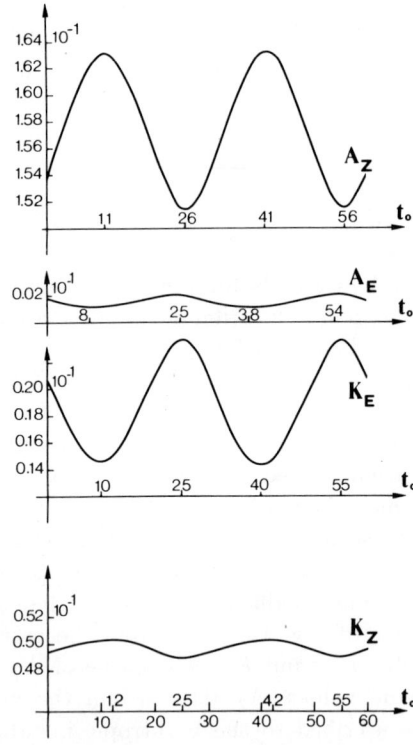

Fig. 20. Time evolution of the zonal A_Z and eddy A_E available potential energy and of the zonal K_Z and eddy K_E kinetic energy, expressed in $2\pi^2 p_0 g^{-1} L^4 f^2$ units, during the kinetic energy vacillation observed at $k = 0.275$, $\theta_{1,0}^* = 0.25$.

The conversion rate C_Z is still the dominant term in Fig. 21, which corresponds to Fig. 13. Each curve now exhibits two sequences of maximum and minimum, the first sequence corresponding to $-0.0294 \leq \psi_{2,0} \leq +0.0296$ and the second one to $0.0296 \geq \psi_{2,0} \geq -0.0296$. Except for an increase of the fluctuations, there is great qualitative similarity between Fig. 13 and Fig. 21 regarding the behavior of G_Z, C_Z, C_E, C_A, C_{AR}, D_Z, and D_E. Accordingly, as in the potential energy vacillation, so the self-interactions of the waves (1, 3) and (2, 3) with the zonal thermal field (1, 0) remain the dominant contributions to *baroclinic* energy processes in the kinetic energy vacillation.

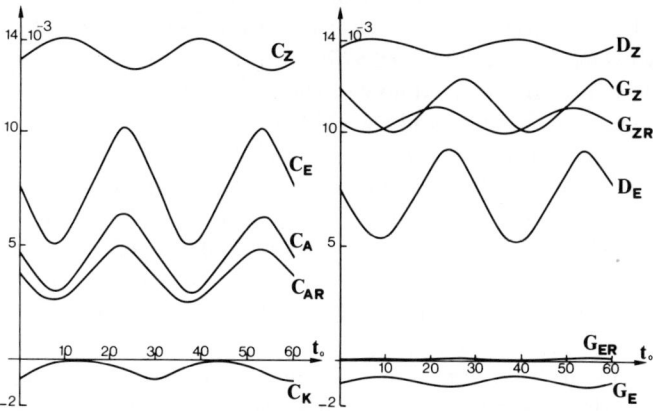

FIG. 21. Time evolution of the different components of the energy budgets [see Eqs. (4.79)–(4.82) for definitions], expressed in $2\pi^2 p_0 g^{-1} L^4 f^3$ units, during the kinetic energy vacillation observed at $k = 0.275$, $\theta^*_{1,0} = 0.25$.

In contrast, the barotropic energy exchanges C_K now play an important role. C_K being negative, this term represents a barotropic stabilization of the eddies. In this respect, we note that, paradoxically, the maximum stabilization (minimum values of C_K) occurs at time 30 corresponding to the curve labeled 4 in Fig. 17, when the zonal flow is definitely barotropically unstable for a small perturbation.

To understand this paradox, the nonlinear character of the flow should be invoked by considering the threefold kinetic energy transfers defined in (4.96)–(4.98) where, according to C_K, T should be positive. Let us first recall that the only energy source in the model is at scale (1, 0), namely $\theta^*_{1,0}$. The kinetic energy at scale (2, 0) has then necessarily to come via the [(2, 0), (1, 3), (2, 3)] interactions since there is no possibility of a direct supply of $K_{2,0}$ from $K_{1,0}$. Now, it has already been mentioned, and it is also observed in this case, that the baroclinic transfers associated with the [(2, 0), (1, 3), (2, 3)] interactions are quite small. Accordingly, the major source of kinetic energy

$K_{2,0}$ is a barotropic energy transfer from $K_{1,3}$. This actually is the reason why T is positive; in this sense, the flow is barotropically stable. Thus the rate of increase or decrease of $K_{2,0}$ or, considered at the mean level, of $\psi_{2,0}$ is mainly governed by the respective increase or decrease of T.

It is easily seen that the linear condition of barotropic instability $d^2U/dy^2 = 0$ assumes the form

(5.2) $$|\psi_{1,0}/16\psi_{2,0}| < 1$$

so that the relative increase of $\psi_{2,0}$ leads to barotropic instability. The maximum value of $\psi_{2,0}$ is precisely observed when $K_{2,0}$ is fed at the maximum rate by the eddies. After that time, we do not observe a transfer of $K_{2,0}$ towards K_E, but a rapid *reduction* of the kinetic energy transfer from the eddy flow toward the zonal flow of the second mode. This transfer nearly falls to zero in 10 time units and then slowly increases again to reach its new maximum value 20 time units later.

The evolution of K_Z nevertheless remains dominated by C_Z as indicated by the strong positive correlation between C_Z and K_Z. The curves C_Z and C_K being also nearly in phase, the relative deficit of conversion of A_Z into K_Z tends to be compensated by an increase of the barotropic transfer from K_E towards K_Z. As a result the fluctuations of K_Z are much weaker than the fluctuations of A_Z.

In contrast with C_K, the baroclinic energy transfer and conversion are symmetric with respect to their extreme values. Surprisingly it is observed that the maximum generation rate of the eddies C_E occurs near the time of maximum static stability. The fluctuations of $\tau_{2,0}$ being of the same order of magnitude as the fluctuations of $\tau_{1,0}$, the vertical zonal wind shear oscillation is controlled chiefly by $\tau_{2,0}$. It can be observed that the maximum values of C_E are attained near the time of extreme (positive or negative) values of $\tau_{2,0}$. On the other hand, the time of minimum C_E nearly coincides with that of zero $\tau_{2,0}$. Accordingly, and owing to its nonlinear nature, the baroclinic behavior of the flow is subjected to the antagonistic influences of an increase of static stability, proceeding with an intensification of the vertical zonal wind shear associated with the increase of $\tau_{2,0}$. Presumably such an interplay between two factors is responsible for the symmetric character of the baroclinic process curves, contrasting with the strong asymmetry of C_K. Let us also mention that, as for the potential energy vacillation, the maximum value of C_E occurs when $\theta_{1,0}$ is relatively small, and vice versa.

Figure 22 shows the evolution of T [(4.100)] and T_1, T_2, and T_3 [(4.106)–(4.108)] during the vacillation cycle. With respect to the analogue, Fig. 14, all these terms have been increased, the term T remaining one order of magnitude greater than the others. Accordingly as long as only the influence of the [(2, 0), (1, 3), (2, 3)] interactions is considered, the barotropic kinetic energy

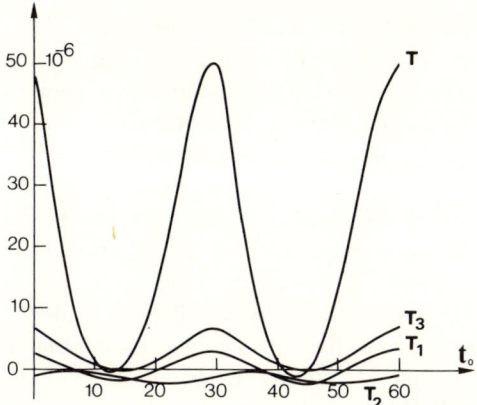

Fig. 22. Time evolution of T, T_1, T_2, and T_3 [see Eqs. (4.100) and (4.106)–(4.108) for definitions] during the kinetic energy vacillation observed at $k = 0.275$, $\theta^*_{1,0} = 0.25$.

transfers are still the dominant energetic processes. In contrast with the self-interaction terms, the terms T and particularly T_1, T_2, and T_3 are extremely weak. Nevertheless, their contributions are now sufficient to introduce the observed shifts between the curves of A_Z, A_E, K_E, and K_Z of Fig. 20. Owing to the weak values of A_E, indicating that the dominant self-interaction transfers from A_Z are nearly entirely converted into K_E, the influence of T_1, T_2, and T_3 is most marked on the A_E curve which exhibits noticeable asymmetries.

Figure 23, where the numbers on the arrows represent mean values over one period, schematically describes the energy cycle of the flow. Actually, there is little difference between Fig. 15 and Fig. 23. This shows how the averaging process can smooth important aspects of the energy cycle since we know that the role of the C_K term is quite different in both cases.

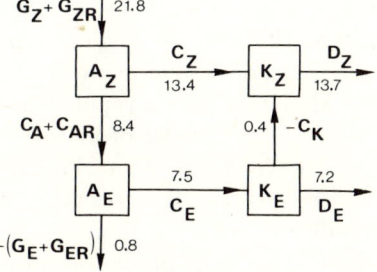

Fig. 23. Energy diagram averaged over one period of the kinetic vacillation observed at $k = 0.275$, $\theta^*_{1,0} = 0.25$. Energies are given in $c = 2\pi^2 p_0 g^{-1} L^4 f^2$ units; the conversion and transfer rates are expressed in cf units.

The generated flow exhibits remarkable similarity to the laboratory kinetic energy vacillation. This confirms indirectly that the neglected effects of viscosity and thermal conductivity should be of minor importance in this kind of flow. We have seen that this does not seem to be entirely true for the potential energy vacillation.

5.3.3. A Physical Interpretation of the Kinetic Energy Vacillation. The great similarity between Figs. 13 and 21 suggests that we may use our description of the potential energy vacillation as a starting point for the interpretation of the kinetic energy vacillation. Indeed, recalling that, as in the potential energy vacillation case, the maximum (minimum) value of C_E occurs near the time of minimum (maximum) value of $\tau_{1,0} = \theta_{1,0}$, the kinetic energy vacillation heat transport cycle has an important component similar to the potential energy vacillation heat transport cycle. Nevertheless, differences now occur under the influence of the intense fluctuations of the (2, 0) variables. Thus, the physical interpretation of the existence and influences of these fluctuations becomes the keystone of the interpretation of the kinetic energy vacillation.

As the thermal forcing is unchanged, we have to associate the modifications of the flow with the increase of the rotation rate. It is well known (Lorenz 1963) that the increase of the rotation rate destabilizes the second mode variables. On the other hand, the relative increase of $\psi_{2,0}$ [(5.2)] leads to zonal wind profiles which are barotropically unstable *with respect to small perturbations*. Hence, at sufficiently high rotation rates, the zonal wind profile may exhibit a point of inflection. The question is now to see how the concept of *linear* barotropic instability can account for the flow behavior.

Consider, at the mean level, a meridional profile of the zonal wind which is symmetric with respect to the axis of the channel ($\psi_{2,0} = 0$). At the rotation rate used here, $k = 0.275$, the value $\psi_{2,0} = 0$ is unstable and may become sufficiently large as to lead to linearly unstable zonal wind profiles. The time of maximum value of $\psi_{2,0}$ is also the one of maximum tilt of the ψ-wave (see Table III and maps 1 and 4 of Fig. 16) and corresponds to extreme values of meridional momentum transport and of momentum transport convergence (see Figs. 18a and 18b). After this time a rapid reduction of all these barotropic processes, as well as of the transfer of kinetic energy from the eddies toward the zonal flow, is observed (curve C_K of Fig. 21). Hence, the concomitant decrease of $\psi_{2,0}$ can be considered as resulting from the release of the (linear) barotropic instability of the zonal flow.

If an examination of the zonal wind profile can provide some criterion for the appearance of barotropic instability in the model, the subsequent process of redistribution of the kinetic energy between the zonal flow and the eddies is governed by (4.96)–(4.98). Note that the influence of this mechanism on the waves of the stream field is hidden by the dominant baroclinic processes.

Indeed, the increase of $K_{1,3}$ and the decrease of $K_{2,3}$, associated with the barotropic reduction of $K_{2,0}$, is not reflected in the amplitudes of the waves (1, 3) and (2, 3) of the ψ-field following extreme values of $\psi_{2,0}$.

It should be emphasized that neither $\psi_{1,3}$ nor $\psi_{2,3}$ are small when $\psi_{2,0}$ reaches its extreme values. In fact, the unstable zonal wind profile results precisely from the existence of the eddies which concentrate zonal momentum at some particular latitude. The release of barotropic instability is then operated through a reduction of energy transfer from the eddies toward the zonal flow.

Obviously, the increase of the rotation rate influences also the baroclinic processes. The most important qualitative effect is observed on $\tau_{2,0}$. When $\tau_{2,0}$ is large and positive (negative) a strong vertical wind shear prevails to the south (north) of the center of the channel. The extreme values of $\tau_{2,0}$ occur at the time when the flow pattern at the midlevel exhibits closed circulation cells. The *northern* cell is associated with a maximum *southern* vertical shear and vice versa (see Table III and Fig. 16). On the other hand, the first maximum of C_E (Fig. 21) corresponds approximately to map 3 of Fig. 16 (high latitude low), where the high latitude eddy heat transport is maximum (Fig. 18c); the second one corresponds to map 6 (low latitude high), where the low latitude eddy heat transport is maximum. Accordingly, the intense eddy transport of heat allows an intense conversion of eddy available potential energy into eddy kinetic energy. This conversion reduces the vertical wind shear at the latitude where the process is taking place but, at the same time, increases the shear in the other half of the channel, where conditions of strong baroclinic instability are then created. As a result, the associated closed circulation cells alternately disappear and reappear north and south of the center of the channel. At the same time, the zonally averaged S–N eddy heat transport (Fig. 18c) varies considerably along the S–N direction. Actually the heat transport curves at latitudes 54° and 126° (Fig. 18c) are out of phase, the small low latitude values being associated with large high-latitude values, and vice versa.

5.4. The Various Vacillation Cycles

When numerical experiments are performed at various rotation rates and thermal forcings, different kinds of vacillation can be generated. In order to easily distinguish between these circulations, we represent the flow in the subspace of the $(^{\circ}\psi_{1,3}, \psi_{2,0})$ variables, the superscript 0 to the left indicating that the coordinate system is moving with the wave (1.3) of the ψ-field.

At high thermal forcing and relatively low rotation rates, the model exhibits *unsymmetric vacillation*. This kind of flow has been discussed in Section 5.2 and is represented in Fig. 24a corresponding to $\theta_{1,0}^* = 0.25$, $k = 0.2925$. The period of vacillation is $56t_0$. Recalling that the equations of

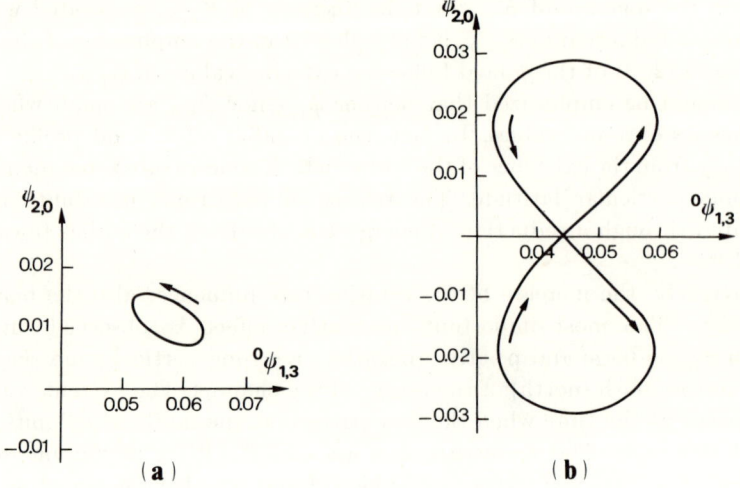

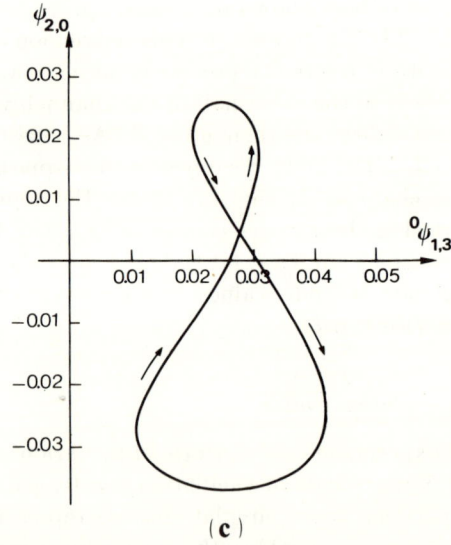

Fig. 24. Representation in the $(^{\circ}\psi_{1,3}, \psi_{2,0})$-plane of: (a) an unsymmetric vacillation $(k = 0.2925, \theta^*_{1,0} = 0.25)$; (b) a symmetric vacillation $(k = 0.275, \theta^*_{1,0} = 0.25)$; (c) an asymmetric vacillation $(k = 0.225, \theta^*_{1,0} = 0.25)$; (d) a double asymmetric vacillation $(k = 0.2, \theta^*_{1,0} = 0.25)$; (e) a double symmetric vacillation $(k = 0.25, \theta^*_{1,0} = 0.175)$ generated in the model.

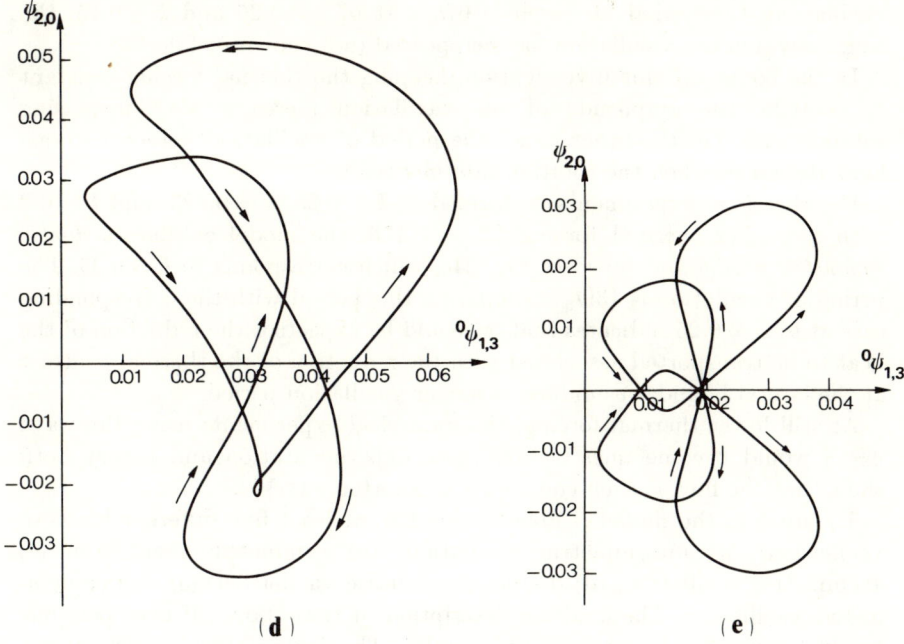

Fig 24 d, e. See facing page for legend.

the model are invariant for a change of sign of the second mode variables, the variables of the first mode being left unchanged, there also exists an other unsymmetric vacillation cycle whose representation would be the image of Fig. 24a with respect to the $°\psi_{1,3}$-axis. The initial conditions of integration decide which one of the possible circulations is ultimately realized. The same, of course, applies to each curve of Fig. 24.

When the rotation rate is increased, the *symmetric vacillation*, discussed in Section 5.3, develops. The flow represented in Fig. 24b corresponds to $\theta_{1,0}^* = 0.25$ and $k = 0.275$ and has a period of $60t_0$.

At still higher rotation rates, the model performs an *asymmetric vacillation* shown in Fig. 24c, corresponding to $\theta_{1,0}^* = 0.25$, $k = 0.225$, and a period of $40t_0$. In this flow, $\psi_{2,0}$ changes sign during the cycle but reaches extreme values of different moduli. The characteristics of the curves associated with the energy processes of the asymmetric vacillation are similar to those of the symmetric vacillation except that there are now two different maxima (and two different minima) on each curve.

For $\theta_{1,0}^* = 0.25$ and $k = 0.2$ the model exhibits still another type of vacillation, shown in Fig. 24d, which may be called a *double asymmetric* vacillation, the period of which is $67t_0$. At $\theta_{1,0}^* = 0.25$ and $k = 0.15$, the single asymmetric vacillation has reappeared (not shown in Fig. 24).

In the course of this investigation, keeping the thermal forcing constant $\theta_{1,0} = 0.25$, the amplitude of the vacillation increases with increasing rotation rate. On the other hand, the period of vacillation, referred to one loop, decreases when the rotation rate increases.

For the three experiments performed at $k = 0.25$, $k = 0.225$, and $k = 0.2$ with a weaker thermal forcing $\theta_{1,0}^* = 0.175$, the model exhibits a *double symmetric vacillation* shown in Fig. 24e, which corresponds to $k = 0.25$. The period of vacillation is $139t_0$. Comparing this period with the corresponding case at $\theta_{1,0}^* = 0.25$ indicates that, as could be expected, the reduction of the heat to be transported associated with the reduction of the thermal contrast at the lateral boundaries induces a longer vacillation period.

At still lower thermal forcing, the numerical experiments using the wave $n = 3$ would become questionable since experimentation and theory both show that the flow is then controlled by shorter waves.

To sum up, the model is able to simulate at least five different kinds of vacillation: (a) unsymmetric vacillation, (b) symmetric vacillation, (c) asymmetric vacillation, (d) double asymmetric vacillation, (e) double symmetric vacillation. The analytic description of these flows, if ever possible, would presumably be increasingly complex. The driving energy mechanisms, however, can be reasonably believed to be of the same kind. It should nevertheless be kept in mind that the relative as well as the absolute importance of the various terms of the energy budgets (4.79)–(4.82) is modified when varying the thermal forcing and the rotation rate. In this respect, it is interesting to note that the mean value $\bar{C}_Z$ of C_Z during one vacillation period drops considerably with increasing rotation rates. For instance, at the given thermal forcing $\theta_{1,0}^* = 0.25$, $\bar{C}_Z$ at $k = 0.15$ is only one half of $\bar{C}_Z$ at $k = 0.275$. This confirms the qualitative argument mentioned in Section 2, that the higher the rotation rate, the more geostrophic will be the flow. Moreover, it is observed that the influence of the nonlinear interactions between the scales (2.0), (1, 3), and (2, 3) increases with the rotation rate. Accordingly, at high rotation rates, the curves of A_Z, A_E, K_E, and K_Z become less correlated with each other, and the respective times of occurrence of their extreme values exhibit more or less important shifts.

6. Vacillation in the Atmosphere

6.1. Potential Energy Vacillation in the Atmosphere

Pfeffer *et al.* (1969) have published striking analogs between atmospheric and laboratory flows, especially regarding the index circulation changes. Figure 25, borrowed from the Pfeffer *et al.* (1969) paper, reproduces the

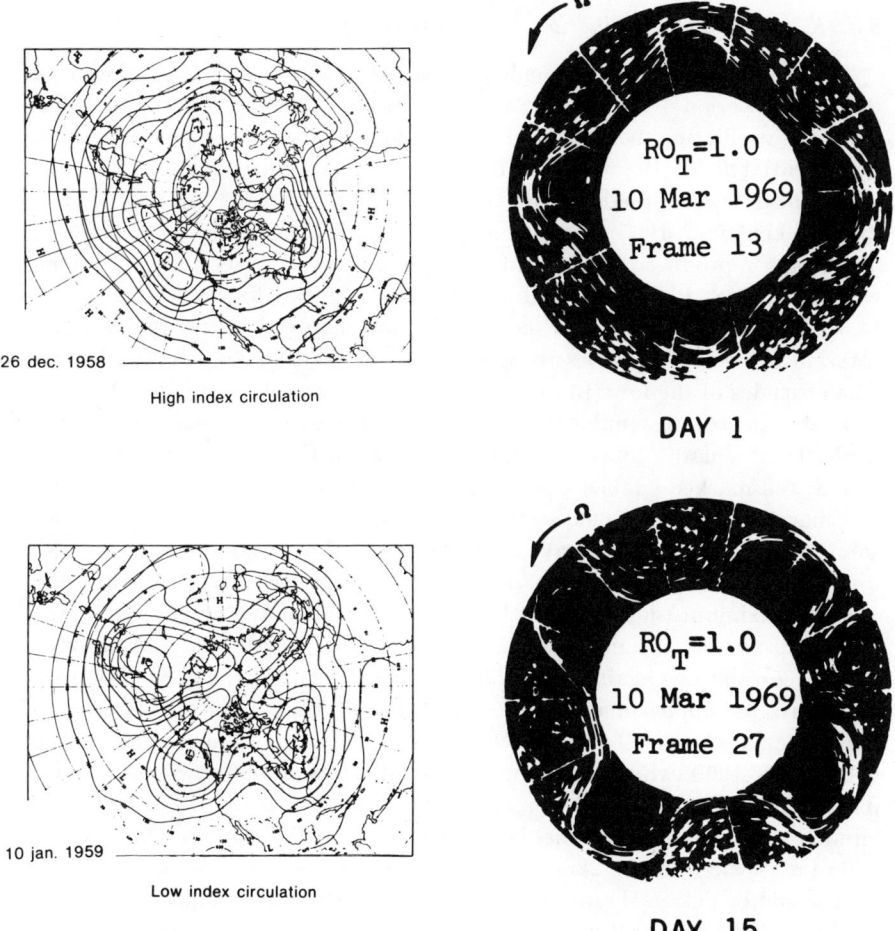

FIG. 25. Comparison between an atmospheric index circulation change (after Winston and Krueger, 1961) and the extreme stages of an amplitude vacillation generated in a rotating annulus experiment (courtesy of Pfeffer and colleagues, 1969).

500 mbar charts of 26 December 1959 and 10 January 1960, characteristic of high and low index circulation, respectively. The extreme stages of an amplitude vacillation in a rotating annulus experiment are shown to the right side of these charts. The interval between the two situations, either in the atmosphere or in the annulus, is 14 revolutions. The complete modification of the flow, from nearly symmetrical to strongly lobed waves, appears distinctly both in the atmosphere and in the laboratory experiment.

6.2. Kinetic Energy Vacillation in the Atmosphere

The routine northern hemisphere weather maps at the 500 mbar level during the period 4 February 1969–22 February 1969 suggest the existence of some vacillating behavior on a well-defined four-wave pattern of the geopotential field. In order to further investigate this situation the geopotential data were analyzed by the least squares spectral method (Eliasen and Mackenhauer, 1965) using even spherical harmonics up to ($m = 9$, $n = 9$). The maps reproduced in Fig. 26 were then drawn by retaining the zonal components (2, 0), (4, 0), and (6, 0) and only the four-wave harmonics (4, 4), (6, 4), and (8, 4) from the set of scales involved in the analysis. The interval between two successive isolines in Fig. 26 is 80 geopotential meters. The altitudes of the low (L) and high (H) pressure centers are also shown.

There is a striking similarity between these maps and those of a vacillation cycle. The "square" four-wave of the first map (4 February 1969) transforms into a well marked "lobed" pattern on the 12th and is reestablished on the last map (22 February 1969) so that the period of the cycle is 18 days. At the same time the axis of the wave, initially oriented approximately in the S–N direction (no tilt north of 50° latitude) becomes SE–NW (westward tilt) with a maximum tilt on the 10th, then changes to SW–NE (eastward tilt) with a maximum tilt on the 16th and, finally, the S–N direction of the wave axis is reestablished again on the 22nd. During this time, the wave progresses westward by approximately three quarters of the wavelength ($\simeq 70°$ of longitude). It can also be easily recognized that the high index circulation of 4 February 1969 exhibits more intense midlatitude zonal flow than the low index circulation prevailing on 12 February 1969. On the other hand, the circulation is dominantly zonal both on 4 February 1969 and on 22 February 1969 but weaker on the 22nd, after which the cycle does not reappear.

It would be extremely useful to pursue the large-scale synoptic, aerological analysis of the flow patterns of Fig. 26 and, in particular, to identify significant characteristic features on the first map (4 February 1969) in order to define initial condition criteria leading to such an evolution in the atmosphere. This would have very interesting practical implications. Indeed, although it is not possible to derive precise synoptic information from such very large-scale patterns, it should then be possible to forecast, in a very economic way, the main features of the large-scale synoptic evolution for a rather extended period of time. This, for instance, would allow one to know the sequence of weather types that would prevail during two weeks or more (18 days in the above given example).

The difference between the flow patterns on 4 February 1969 and 22 February 1969 and between their respective further evolution seems to indicate that the appearance of vacillation in the atmosphere depends

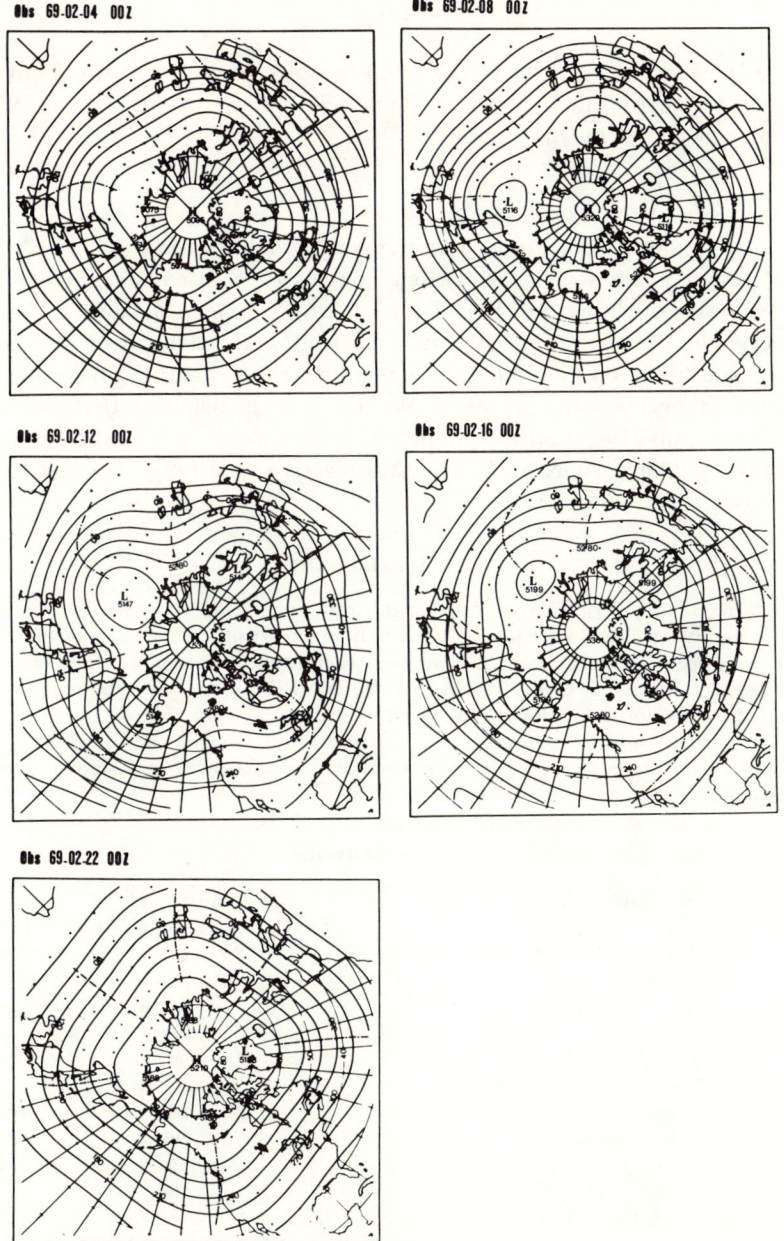

Fig. 26. Maps drawn in retaining the zonal harmonic components (2, 0), (4, 0), (6, 0), and the four-harmonics (4, 4), (6, 4), and (8, 4), of the 500 mbar geopotential field during the period 4 February–22 February 1969 and showing the existence of an atmospheric kinetic energy vacillation cycle.

critically upon the intensity of the zonal flow. Returning to Section 5 of this monograph, there is no doubt that the meridional profile of the zonal wind is also of importance in this respect.

6.3. Conclusion

The two examples of Figs. 25 and 26 amply demonstrate that vacillation is not only a characteristic laboratory flow closely related to the large-scale atmospheric circulation but has also an intrinsic pragmatic aspect. A clear synoptic identification of initial conditions leading to atmospheric vacillation would be a first step in making economic long range weather predictions, at least in favorable circumstances. Moreover, it is believed that a highly truncated numerical weather prediction model would provide, in a rather short computation time, a meaningful objective long range forecast of the large-scale atmospheric flow.

Acknowledgments

I am indebted to Professor J. Van Mieghem for suggesting writing this monograph and for careful reading and commenting on the manuscript. Let me also acknowledge the benefit received from many helpful discussions with Professors P. Defrise and J. Van Isacker. The maps of Fig. 26 have been drawn by using a subroutine program provided by the author's colleague, W. Struylaert.

I am most grateful to Professors D. Fultz and R. Pfeffer and to Dr. W. Fowlis, as well as to their publishers, who have granted me permission to reproduce previously published illustrations.

List of Symbols

A	Available potential energy of the system	C_K	Transfer rate of K_Z into K_E
A_E	Eddy available potential energy	C_Z	Conversion rate of A_Z into K_Z
		c	$2\pi^2 p_0 g^{-1} L^4 f^2$; is used as energy unit
A_Z	Zonal available potential energy	c_p	Specific heat of dry air at constant pressure
a	$[(3/4)^\kappa + (1/4)^\kappa]/2 \simeq 0.797$	c_{ijk}	Interaction coefficient [see Eq. (4.20)]
$-a_i^2$	Eigenvalue of the Laplacian operator in Cartesian coordinates	D_E	Dissipation rate of K_E by friction
B	$b^{-1}c^{-1}f$	D_Z	Dissipation rate of K_Z by friction
b	$[(3/4)^\kappa - (1/4)^\kappa]/2 \simeq 0.124$		
C_A	Transfer rate of A_Z into A_E resulting from sensible heat advection	E^α, E_α	Contravariant and covariant components of the velocity of the reference body with respect to an absolute reference body (entrainment velocity)
C_{AR}	Transfer rate of A_Z into A_E resulting from changes of static stability in time		
C_E	Conversion rate of A_E into K_E	e^{ij}, e_{ij}	e-symbol [see Eq. (3.36)]

$2F$	Coefficient of friction at surface p_0		of the reference body with respect to the coordinate system
F'	Coefficient of friction at surface p_2	T	Temperature or nondimensional contribution of scale-interaction effects to barotropic energy transfer rates [see Eq. (4.100)]
F_i	Basic function of the spectral representation		
f	Coriolis parameter (vertical component of the vorticity of the absolute motion of the reference body)		
		T_1, T_2, T_3	Nondimensional contributions of scale-interaction effects to baroclinic energy conversion and transfer rates [see Eqs. (4.106)–(4.108)]
G	Generation rate of total potential energy		
G_E	Generation rate of A_E by external heating		
G_Z	Generation rate of A_Z by external heating	T_a	Temperature at the inner wall of an annulus
G_{ER}	Generation rate of A_E due to the change of the reference state in time	T_b	Temperature at the outer wall of an annulus
		t	Time
G_{ZR}	Generation rate of A_Z due to the change of the reference state in time	t_0	Nondimensional time unit $t_0 = ft$
		U	Zonal wind
g	Acceleration of gravity	u	x-component of the isobaric wind
g_{ij}, g^{ij}	Metric tensor and its reciprocal		
		$\mathbf{V}$	Relative velocity vector of a material point with respect to a reference body
$2H$	Coefficient of heating at the lower layer		
H'	Coefficient of heating at the upper layer	V^α, V_α	Contravariant and covariant components of $\mathbf{V}$
I	Internal energy of the system	v	y-component of the isobaric wind
K	Relative kinetic energy, per unit mass in Section 3.2, for the whole system elsewhere		
		v^α, v_α	Contravariant and covariant components of the velocity of a material point with respect to the coordinate system
K_E	Eddy kinetic energy		
K_Z	Zonal kinetic energy		
k	Ff^{-1}, and is used as a measure of rotation rate	x^α	Space coordinate ($\alpha = 1, 2, 3$)
		x, y	Cartesian coordinates in a plane surface
L	Length scale		
m	Meridional wavenumber or mode	γ	Determinant of g_{ij}
		δ	Isobaric divergence of the wind
n	Longitudinal wavenumber		
P	Potential energy of the system	$\delta_\alpha{}^\beta$	Kronecker tensor [see Eq. (3.11)]
p	Pressure	ζ	Vertical component of the relative vorticity
R	Specific gas constant for dry air or the reference body		
		η	Vertical component of the absolute vorticity ($\eta = \zeta + f$)
Ro	Rossby number		
r^α, r_α	Contravariant and covariant components of the velocity		
		θ	Potential temperature

θ^*	Temperature of the underlying surface	OPERATORS	
κ	R/c_p	$\partial/\partial x^\alpha \equiv \partial_\alpha$	Partial derivative with respect to the space coordinate x^α
ρ	Density of the fluid		
σ	A measure of the static stability [see Eq. (3.77)]	$\partial/\partial t \equiv \partial_t$	Partial derivative with respect to time
τ	Stream function of the vertical wind shear	d/dt	Derivative with respect to time along the flow
		$\int$	Integration symbol
ϕ	Potential function of the body forces	dm	Mass element of the fluid
		$d\Sigma$	Element of horizontal area
$\varphi_{m,n}$	Basic function of the spectral representation	$\nabla^2 X$	Laplacian of X [see Eq. (3.42)]
χ	Velocity potential of the isobaric wind	$J(X, Y)$	Jacobian of X and Y [see Eq. (3.43)]
Ω	Angular velocity of rotation of the reference body	$Q(X, Y)$	Differential Q-operator [see Eq. (3.44)]
ω	dp/dt	$S(X, Y)$	Differential S-operator [see Eq. (3.45)]
ω_i	Nondimensional spectral component of $\nabla^2 \chi$ [see Eq. (4.12')]	$\dot{X}$	Local derivative of X with respect to t_0 ($\dot{X} \equiv \partial X/\partial t_0$)
		$\overline{X}$	Horizontal average of X
$2\omega_{\alpha\beta}$	Vorticity tensor of the absolute motion of the reference body	$[X]$	Zonal average of X
		X'	Fluctuation of X with respect to $[X]$ ($X' = X - [X]$)

INDICES

In the tensorial expressions, the Roman indices take the values 1 and 2, the Greek indices take the values 1 to 3. They are summation indices when repeated in the same term.

In Section 3, subscripts 0, 2, and 4 refer respectively to levels $p_0 = 100$ cbar, $p_2 = p_0/2$, and $p_4 = 0$ cbar; subscripts 1 and 3 refer to the intermediate levels $p_1 = 3p_0/4$ and $p_3 = p_0/4$, respectively.

Subscripts to the scalar dependent variables ψ, τ, θ, σ denote spectral components [see Eqs. (4.7)–(4.10)].

REFERENCES

Bolin, B. (1952). Studies of the general circulation of the atmosphere. *Advan. Geophys.* **1**, 87–118.

Charney, J. (1947). The dynamics of long waves in a baroclinic westerly current. *J. Meteorol.* **4**, 135–163.

Davies, T. (1953). The forced flow of a rotating viscous liquid which is heated from below. *Phil. Trans. Roy. Soc. London, Ser. A* **246**, 81–112.

Davies, T. (1959). On the forced motion due to heating of a deep rotating liquid in an annulus. *J. Fluid Mech.* **5**, 593–621.

Defrise, P. (1964). Tensor calculus in atmospheric mechanics. *Advan. Geophys.* **10**, 261–314.

Döös, B. (1969). The influence of the large-scale heat sources on the dynamics of the ultra-long waves. *Tellus* **21**, 25–39

Eady, E. (1949). Long waves and cyclone waves. *Tellus* 1, 35–52.
Eliasen, E., and Mackenhauer, B. (1965). A study of the fluctuations of the atmospheric flow patterns represented by spherical harmonics. *Tellus* 17, 220–238.
Fjörtoft, R. (1953). On the changes in the spectral distribution of kinetic energy of two-dimensional non-divergent flow. *Tellus* 5, 225–230.
Fowlis, W., and Hide, R. (1965). Thermal convection in a rotating annulus of liquid: Effect of viscosity on the transition between axisymmetric and non-axisymmetric flow regimes. *J. Atmos. Sci.* 22, 542–558.
Fultz, D. (1949). A preliminary report on experiments with thermally produced lateral mixing in a rotating hemispherical shell of liquid. *J. Meteorol.* 6, 17–33.
Fultz, D., Long, R., Owens, G., Bohan, W., Kaylor, R., and Weil, J. (1959). Studies of thermal convection in a rotating cylinder with some implications for large scale atmospheric motion. *Meteorol. Monogr.* 4, 1–104.
Fultz, D., Kaiser, J., Fain, M., Kaylor, R. E., and Weil, J. (1964). Experimental investigations of the spectrum of thermal convective motions in a rotating annulus. *In* "Research on Hydrodynamic Analogues of Large-Scale Meteorological Phenomena," pp. 2B1–2B89. Final report, Dept. Geophys. Sci., Univ. of Chicago, Chicago, Illinois.
Hide, R. (1953). Some experiments on thermal convection in a rotating liquid. *Quart. J. Roy. Meteorol. Soc.* 79, 161.
Hide, R. (1958). An experimental study of thermal convection in a rotating fluid. *Phil. Trans. Roy. Soc. London, Ser. A* 250, 441–478.
Holton, J. (1972). "An Introduction to Dynamic Meteorology," Int. Geophys. Ser. Academic Press, New York.
Krueger, A., Winston, J., and Haines, D. (1965). Computation of atmospheric energy and its transformation for the northern hemisphere for a recent five-year period. *Mon. Weather Rev.* 93, 227–238.
Kuo, H. L. (1949). Dynamic instability of two-dimensional non-divergent flow in a barotropic atmosphere. *J. Meteorol.* 6, 105–122.
Kuo, H. L. (1954). Symmetrical disturbances in a thin layer of fluid subject to a horizontal temperature gradient and rotation. *J. Meteorol.* 11, 399–411.
Kuo, H. L. (1957). Further studies of thermally driven motion in a rotating fluid. *J. Meteorol.* 14, 553–558.
Langlois, W., and Kwok, H. (1969). "Description of the Mintz-Arakawa Numerical General Circulation Model," Tech. Rep. No. 3. Dept. Meteorol., University of California, Los Angeles.
Lorenz, E., (1953). A proposed explanation for the existence of two regimes of flow in a rotating symmetrically heated cylindrical vessel. *In* "Fluid Models in Geophysics," pp. 73–80. Johns Hopkins Press, Baltimore, Maryland.
Lorenz, E. (1955). Available potential energy and the maintenance of the general circulation. *Tellus* 7, 157–167.
Lorenz, E. (1960). Energy and numerical weather prediction. *Tellus* 12, 364–373.
Lorenz, E. (1962). Simplified dynamic equations applied to the rotating-basin experiments. *J. Atmos. Sci.* 19, 39–51.
Lorenz, E. (1963). The mechanics of vacillation. *J. Atmos. Sci.* 20, 448–464.
Margules, M. (1903). Uber die Energie der Stürme. "Jahrb. kais.—kön. Zent. Meteorol." [transl. by C. Abbe in *Smithson. Misc. Collect.* 51, 533–595 (1910)].
Merilees, P. (1968). On the transition from axisymmetric and non-axisymmetric flow in a rotating annulus. *J. Atmos. Sci.* 25, 1003–1014.
Merilees, P. (1972). On the periods of amplitude vacillation. *J. Meteorol. Soc. Jap.* [2] 50, 214–225.

Namias, J. (1954). Quasi-periodic cyclogenesis in relation to the general circulation. *Tellus* **6**, 8–22.

Pfeffer, R., and Chiang, Y. (1967). Two kinds of vacillation in rotating laboratory experiments. *Mon. Weather Rev.* **95**, 75–82.

Pfeffer, R., and Fowlis, W. (1968). Wave dispersion in a rotating, differentially heated cylindrical annulus of fluid. *J. Atmos. Sci.* **25**, 361–371.

Pfeffer, R., Mardon, D., Serbenz, P., and Fowlis, W. (1965). "A new Concept of Available Potential Energy," Rep. No. 66-1. Dept. Meteorol., University of Florida, Tallahassee.

Pfeffer, R., Fowlis, W., Buzyna, G., Fein, J., and Buckley, J. (1969). Laboratory studies of the global atmospheric circulation, Proc. Stanstead Semin. 8th. *Meteorol.* **89**, 89–91.

Phillips, N. (1965). The equations of motion for a shallow rotating atmosphere and the "traditional approximation." *J. Atmos. Sci.* **23**, 626–628.

Quinet, A. (1973a). The structure of non-linear processes. *Tellus* **25**, 536–544.

Quinet, A. (1973b). Non-Linear mechanisms in a nonconservative quasi-geostrophic flow which possesses 30 degrees of freedom. *Tellus* **25**, 545–559.

Rossby, C. (1926). On the solution of problems of atmospheric motion by means of model experiments. *Mon. Weather Rev.* **54**, 237–240.

Rossby, C. (1947). On the distribution of angular velocity in gaseous envelopes under the influence of large-scale horizontal mixing process. *Bull. Amer. Meteorol. Soc.* **28**, 53–68.

Starr, V. (1948). An essay on the general circulation of the earth's atmosphere. *J. Meteorol.* **5**, 39–48.

Statistiques Quinquennales. (1971). "Observations aérologiques—station d'Uccle." Inst. Roy. Meteorol. Belg. 1971. Uccle.

Stessel, J. P. (1969). Le freinage de l'atmosphère par l'océan. Ph. D. Thesis, Univ. libre de Bruxelles, Bruxelles.

Thompson, P.. (1961). "Numerical Weather Analysis and Forecasting," Macmillan, New York.

Van Isacker, J. (1963). Conservation de la rotationnelle absolue et de l'énergie dans les modèles atmosphériques. *Contrib. Int. Symp. Dyn. Large Scale Process, 1963* (unpublished).

Van Mieghem, J. (1952). Energy conversion in the atmosphere on the scale of the general circulation. *Tellus* **4**, 334–351.

Van Mieghem, J. (1957). Energies potentielle et interne convertibles en energie cinétique dans l'atmosphére. *Beitr. Phys. Frei. Atmos.* **30**, 5–17.

Van Mieghem, J. (1973). "Atmospheric Energetics," Oxford Monographs on Meteorology. Oxford Univ. Press (Clarendon), London and New York.

Van Mieghem, J., and Vandenplas, A. (1950). Les équations de la dynamique atmosphérique en coordonnées généralisqes. Application au cas des coordonnées sphériques. *Inst. Roy. Meteorol. Belg., Mem.* **41**, 1–55.

Winston, J., and Krueger, A. (1961). Some aspects of a cycle of available potential energy. *Mon. Weather Rev.* **89**, 307–318.

Young, J. (1968). Comparative properties of some time differencing schemes for linear and non linear oscillations. *Mon. Weather Rev.* **96**, 357–364.

FILTER TECHNIQUES IN GRAVITY INTERPRETATION

Frans De Meyer

Royal Belgian Meteorological Institute, Uccle, Belgium

1. Introduction 187
2. Convolution Filtering 189
   2.1. Definition of Convolution Filtering 189
   2.2. The Frequency Response 191
   2.3. Linear, Discrete Filters 192
   2.4. Construction of Linear, Discrete Filters in the (x, y)-Plane ... 195
   2.5. Construction of Linear, Discrete Filters in the (u, v)-Plane ... 200
3. Upward and Downward Continuation of the Surface Gravity Effect 203
   3.1. Definition of the Surface Gravity Effect 203
   3.2. The Upward Continuation as a Filtering Operation 205
   3.3. Construction of a Numerical Upward Continuation Filter 210
   3.4. The Downward Continuation as a Filtering Operation 213
   3.5. Construction of a Numerical Downward Continuation Filter 219
4. Frequency Filtering 226
   4.1. Procedure for Filtering in the Frequency Plane 226
   4.2. Optimal Wiener Filtering 228
   4.3. Strakhov's Method for Extraction of Potential Field Signal 233
   4.4. Digitization of a Continuous Field—Aliasing 237
   4.5. Estimation of the Power Spectrum 240
   4.6. Convolution Filtering versus Frequency Filtering 244
5. Calculation of Derivatives of Higher Order 248
   Appendices 250
   List of Symbols 254
   References 256

1. Introduction

One of the most important problems in the interpretation of gravity measurements is that of separating a surface field, derivable from a potential, into independent components and of ascribing separate geological structures to these parts. The geological interpretation of gravity anomalies therefore consists largely of estimating the positions and shapes of the disturbing masses in the upper part of the earth's crust and of determining the density structure of the area in which they are imbedded.

In this respect gravity interpretation uses the whole of the following procedures: (1) A gravity survey yields values of the gravitational field over a limited part of the earth's surface. (2) Bouguer corrections are applied to allow for the topography of the survey area and for the attraction of the earth's spheroid; the residue, the Bouguer anomaly, reflects the heterogeneous structure of the upper part of the crust. (3) The irregularly spaced gravity data are interpolated onto a regular (rectangular or square) grid, for ready entry to a computer, and are automatically contoured. (4) A regional-residual analysis is applied, which separates the field into components of larger and smaller extent. (5) Several interpretation techniques are used for obtaining a better insight into the composition of the field, such as filtering the surface field, calculation of the vertical derivatives, upward and downward continuation. (6) A model-fitting technique is applied in order to obtain a geological model that could produce the observed part of the surface field.

It is a fact that a gravimeter, measuring the variations in the vertical component of a potential field, is sensitive enough to register the local effects of many types of geological configurations. Frequently the data are subjected to elaborate reductions and inaccurate numerical approximation methods; consequently they are not exact and are seldom precise. Finally we have a map and a distribution of data on a regular grid, giving a more or less complete picture of the behavior of the field at the earth's surface and of the internal mass distribution in the earth's crust.

Even if the surface field were known very accurately, there remains an inherent lack of determinacy of the source. Indeed, any gravitational field possesses two characteristics, preventing a unique interpretation. The first difficulty arises from the fact that the reduced anomaly at each observation point undergoes the influence of a very complex mass distribution in the upper part of the crust, and the attraction of local, shallow structures is often seen as a small deviation in the broad picture of the regional features of the map. The effects caused by these disturbing masses are of course indistinguishable, but usually we can assume that they are independent of one another, stating implicitly that they are considered as being random, affecting individually only a single observation point.

The second difficulty results from the property that any gravity field, derivable from a potential, implies the inherent ambiguity in defining the source of the potential field (Skeels, 1947). The distribution of the field in free space can be calculated uniquely, since the upward continuated field satisfies Laplace's equation, with known Dirichlet's conditions in the earth's surface. However, if knowledge of the field at a given depth is desired, then we require the solution of Poisson's equation with unknown density function. It is well known that the gravitational field is not in itself sufficient to define the mass distribution which produces it: for a given gravity field, measured at the earth's surface, an infinite number of mass distributions can be found, all accounting for that field. Even within the limits of some known physical parameters an infinite variety of solutions is possible and no degree of precision or amount of data will remove this fundamental ambiguity.

Therefore it is evident that a less ambiguous interpretation of a measured gravity field must rest on data of another kind, which are mostly geological and seismological. However, this does not prevent us from trying to perfect the interpretation techniques and to give the computational methods a more rational basis.

2. Convolution Filtering

2.1. Definition of Convolution Filtering

We consider the class $\mathscr{L}$ of the continuous, indefinitely derivable functions $f(x, y)$ of the coordinates x and y, with the property that the following norm is finite

$$(2.1) \quad \|f\|_{\mathscr{L}} = \lim_{T_x, T_y \to \infty} (1/4 T_x T_y) \int_{-T_y}^{T_y} \int_{-T_x}^{T_x} |f(x, y)|^2 \, dx \, dy < \infty \quad f \in \mathscr{L}$$

Note that the functions of $\mathscr{L}$ are not necessarily quadratically integrable, as we see from the trivial case $f(x, y) \equiv 1$.

A linear filter is, by definition, an operator which associates a function $g(x, y)$ of $\mathscr{L}$, called output or response, with any function $f(x, y)$, input or signal, of $\mathscr{L}$. Formally we write

$$(2.2) \quad g(x, y) = \mathscr{F}\{f(x, y)\}$$

where $\mathscr{F}$ is a specified operation on the functions of $\mathscr{L}$.

Two properties are required for the operator $\mathscr{F}$:

(1) Linearity or superposition principle: if for each $f_k(x, y)$ the output of the filter is

$$(2.3) \quad g_k(x, y) = \mathscr{F}\{f_k(x, y)\}$$

then we must have for any sequence of constants a_k, $1 \leq k \leq n$,

$$(2.4) \quad \mathscr{F}\left\{\sum_k a_k f_k(x, y)\right\} = \sum_k a_k \mathscr{F}\{f_k(x, y)\} = \sum_k a_k g_k(x, y)$$

This simply states the fact that the output of the linear filter $\mathscr{F}$ from a sum of inputs is equal to the sum of the outputs of the filter, with each input applied separately.

(2) The operator $\mathscr{F}$ commutes with the translation operator $\tau_{\xi, \eta}$ defined by its action on a function $f(x, y)$ of $\mathscr{L}$

$$(2.5) \quad \tau_{\xi, \eta} f(x, y) = f(x - \xi, y - \eta)$$

which means that

$$(2.6) \quad \mathscr{F} \tau_{\xi, \eta} = \tau_{\xi, \eta} \mathscr{F}$$

This property implies that the filter behavior is independent of the origin of the (x, y)-plane: if $g(x, y)$ is the response of the filter to an input $f(x, y)$, then $g(x-\xi, y-\eta)$ will be the response to the input $f(x-\xi, y-\eta)$, for any ξ and η. Indeed, applying both sides of Eq. (2.6) to the input $f(x,y)$ gives

$$\mathscr{F}\tau_{\xi,\eta}\{f(x,y)\} = \mathscr{F}\{f(x-\xi, y-\eta)\}$$

and

$$\tau_{\xi,\eta}\mathscr{F}\{f(x,y)\} = \tau_{\xi,\eta}g(x,y) = g(x-\xi, y-\eta)$$

Equality yields

$$\mathscr{F}\{f(x-\xi, y-\eta)\} = g(x-\xi, y-\eta)$$

In the theory of gravitational interpretation the filter $\mathscr{F}$ is usually defined by a two-dimensional convolution integral of the form

$$(2.7) \quad g(x,y) = h(x,y) * f(x,y) = \int_{-\infty}^{\infty}\int_{-\infty}^{\infty} h(\xi,\eta) f(x-\xi, y-\eta) \, d\xi \, d\eta$$

where the weighting function $h(x,y)$ is called the impulse response or filter function and it describes the behavior of the filter in the coordinate plane (x,y).

If we let the Dirac delta distribution $\delta(x,y)$ be a member of the class $\mathscr{L}$ and if we apply $\delta(x,y)$ as an input to the filter (2.7), then it immediately follows that

$$g(x,y) = \int_{-\infty}^{\infty}\int_{-\infty}^{\infty} h(\xi,\eta)\,\delta(x-\xi, y-\eta)\, d\xi\, d\eta = h(x,y)$$

Therefore we conclude that the filter function $h(x,y)$ may be regarded as the response of the filter to the Dirac impulse function $\delta(x,y)$.

It often happens that the impulse response is circularly symmetrical, so that $h(x,y)$ is a function of the radius variable r

$$(2.8) \quad h(x,y) = h(r) \quad \text{where} \quad r^2 = x^2 + y^2$$

Transforming to polar coordinates $(x-\xi) + i(y-\eta) = re^{i\theta}$, Eq. (2.7) can be written in the equivalent form

$$g(x,y) = \int_0^{\infty}\int_0^{2\pi} h(r\cos\theta, r\sin\theta) f(x - r\cos\theta, y - r\sin\theta) r \, dr \, d\theta$$

since the convolution product is commutative. If the filter function is a function of the radius variable only, it follows from (2.8) that

$$(2.9) \quad g(x,y) = 2\pi \int_0^{\infty} h(r) f(x,y,r) r \, dr$$

where

$$(2.10) \quad f(x,y,r) = (1/2\pi) \int_0^{2\pi} f(x - r\cos\theta, y - r\cos\theta)\, d\theta$$

is the average of the function $f(x, y)$ in the coordinate plane (x, y) on the circle of radius r with center at the point (x, y).

2.2. The Frequency Response

An equivalent form of the input–output relation of a convolution filter results from the use of the Fourier transform: if $f(x, y)$ is a member of the class $\mathscr{L}$, then we assume that its Fourier transform exists and is given by

$$(2.11) \qquad F(u, v) = \int_{-\infty}^{\infty} \int_{-\infty}^{\infty} f(x, y) e^{-i(ux+vy)} \, dx \, dy$$

with inverse relation

$$(2.12) \qquad f(x, y) = (1/4\pi^2) \int_{-\infty}^{\infty} \int_{-\infty}^{\infty} F(u, v) e^{i(ux+vy)} \, du \, dv$$

where $u/2\pi$ and $v/2\pi$ are the frequencies in cycles per unit length in the x- and y-directions, respectively. If we assume that the Fourier transforms $F(u, v)$, $G(u, v)$ and $H(u, v)$ of the respective input, output, and filter function of the convolution filter (2.7) exist, i.e. if the functions of the class $\mathscr{L}$ satisfy Dirichlet's conditions in any finite domain and are absolutely integrable, then we obtain from the convolution theorem of Fourier transforms the following relation between the signal and response

$$(2.13) \qquad G(u, v) = H(u, v) F(u, v)$$

where

$$(2.14) \qquad H(u, v) = \int_{-\infty}^{\infty} \int_{-\infty}^{\infty} h(x, y) e^{-i(ux+vy)} \, dx \, dy$$

is known as the frequency response, characterizing the filter behavior in the frequency plane (u, v); $H(u, v)$ tells us the portions of the Fourier spectrum of the input which will be amplified or attenuated by the filtering operation.

In this connection we must remark that, in a strictly mathematical sense, the Fourier representations of the input and output are not necessarily meaningful. Indeed, in order to have a definite Fourier transform these fields need to be absolutely integrable over the (x, y)-plane; any function which goes to zero for large x and y and which does not contain singularities is absolutely integrable, but a gravity field does not vanish for large x and y. The problem is apparent: since we suppose that the input field is a realization of a two-dimensional stochastic process, we have the property that it does not vanish outside a finite region of the (x, y)-plane. To overcome this mathematical difficulty we must assume that the input field will be defined over a large, but finite area of the coordinate plane and that it becomes zero outside this domain. Otherwise it would be imperative to work with Wiener's generalized Fourier transform, thus making the mathematical expressions much more complex.

It may be readily proved that the frequency response $H(u, v)$ is a circularly symmetrical function of the radial frequency ρ if the filter function depends only on the radius variable r. Indeed, transforming to polar coordinates $x + iy = re^{i\theta}$ and $u + iv = \rho e^{i\varphi}$, it follows from Eq. (2.14) that

$$H(\rho, \varphi) = \int_0^\infty \int_0^{2\pi} h(r) e^{-ir\rho \cos(\theta - \varphi)} r \, dr \, d\theta$$

Using the relation

(2.15) $$\int_0^{2\pi} e^{ip \cos \phi} \, d\phi = 2\pi J_0(p)$$

(Sneddon, 1951, p. 515), where $J_0(p)$ is the zero-order Bessel function, we obtain

(2.16) $$H(\rho) = 2\pi \int_0^\infty h(r) J_0(r\rho) r \, dr$$

with inverse relation

(2.17) $$h(r) = (1/2\pi) \int_0^\infty H(\rho) J_0(r\rho) \rho \, d\rho$$

These relations are known as the zero-order Hankel transforms.

2.3. Linear, Discrete Filters

In the practical case, the data are sampled on a rectangular grid, with sampling intervals Δx and Δy in the x and y directions, respectively. For regularly spaced observations, the convolution integral (2.7) is approximated by the following sum

(2.18) $$\tilde{g}(x, y) = \Delta x \, \Delta y \sum_{k=-M}^{M} \sum_{l=-N}^{N} h_{kl} f(x - k \Delta x, y - l \Delta y)$$

and the continuous filter function $h(x, y)$ is now replaced by a discrete set of filter weights h_{kl}. In this case the frequency response of the discrete filter (2.18) is given by the analogous form of (2.14)

(2.19) $$H_d(u, v) = \Delta x \, \Delta y \sum_{k=-M}^{M} \sum_{l=-N}^{N} h_{kl} e^{-i(ku\Delta x + lv\Delta y)}$$

Another point of view of discrete filtering results from the application of an adequate numerical integration method to Eq. (2.9); the output of the linear filter is then approximated by a weighted sum of the form

(2.20) $$\tilde{g}(x, y) = \sum_{k=0}^{N} h_k f(x, y, R_k)$$

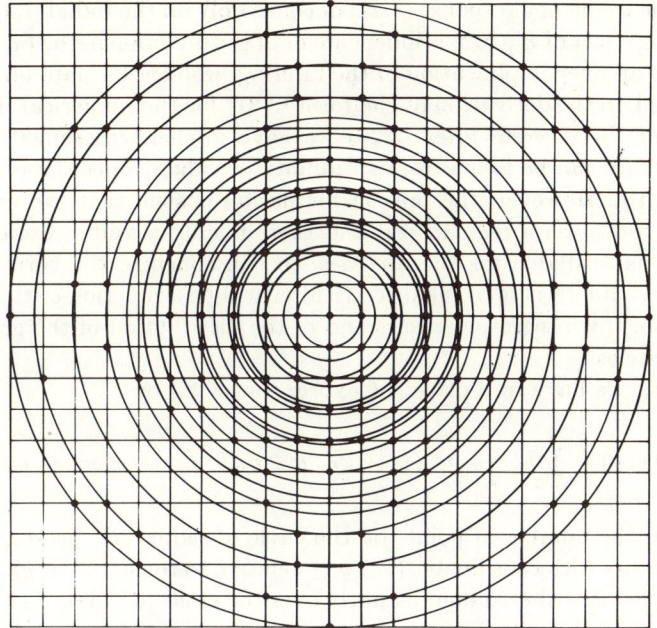

FIG. 1. Square integration grid in the (x, y)-plane.

with weights h_k, and R_k the radii of circles passing through the points of intersection of a square integration grid (see Fig. 1).

We now derive the explicit form of the frequency response of the discrete filter, as defined by Eq. (2.20). Substituting (2.10) into (2.20) we obtain

$$\tilde{g}(x, y) = (1/2\pi) \sum_{k=0}^{N} h_k \int_0^{2\pi} f(x - R_k \cos \theta, y - R_k \sin \theta) \, d\theta$$

Taking Fourier transforms of both sides of this expression we have

$$\tilde{G}(u, v) = \left\{ (1/2\pi) \sum_{k=0}^{N} h_k \int_0^{2\pi} \exp[-iR_k(u \cos \theta + v \sin \theta)] \, d\theta \right\} F(u, v)$$

Transforming to polar coordinates $u + iv = \rho e^{i\varphi}$ and using Eq. (2.15) we obtain for the frequency response of the discrete filter (2.20)

(2.21) $$H_d(u, v) = \sum_{k=0}^{N} h_k J_0[R_k(u^2 + v^2)^{1/2}]$$

Consequently, the averaging process over the circle of radius r defines a linear filter with frequency response

(2.22) $$H_r(u, v) = J_0[r(u^2 + v^2)^{1/2}]$$

Since the frequency response (2.21) depends only on the radial frequency ρ, the filtering effect of $H_d(u, v)$ is independent of the direction in the (u, v)-plane. But this is an oversimplification of the filtering problem for data on a square grid; indeed, in the derivation of the form (2.21) for the numerical frequency response of the discrete filter (2.20) we have implicitly assumed that the circle averages can be taken from an infinity of values, since these are given by Eq. (2.10). However, the practical situation is such that the input field $f(x, y)$ is known on a regular grid and there is only a finite, discrete set of observations available for the analysis; in consequence, the corresponding frequency characteristic of the discrete filter will be a function of the number of points on the integration circles and of the distribution of the grid points on these circles.

Therefore we have to replace the expression (2.20) by

$$(2.23) \quad \tilde{g}(x, y) = \sum_{k=0}^{N} h_k \frac{1}{n_k} \sum_{j_k=1}^{n_k} f(x + x_{j_k}, y + y_{j_k}), \qquad x_{j_k}^2 + y_{j_k}^2 = R_k^2$$

where n_k is the number of points on the circle of radius R_k, passing through the grid points with coordinates (x_{j_k}, y_{j_k}), relative to the point (x, y). Hence, if the points are distributed equidistantly over the circle of radius R_k, we have

$$(2.24) \qquad x_{j_k} = R_k \cos(2\pi j_k/n_k), \qquad y_{j_k} = R_k \sin(2\pi j_k/n_k)$$

Taking Fourier transforms of both sides of Eq. (2.23) we find

$$(2.25) \qquad \tilde{G}(u, v) = H_d(u, v) \, F(u, v)$$

where now $H_d(u, v)$ is defined as the frequency response of the discrete filter with weights h_k, for a square integration grid with a finite number of grid points on the integration circles,

$$(2.26) \qquad H_d(u, v) = \sum_{k=0}^{N} h_k (1/n_k) \sum_{j_k=1}^{n_k} \exp[i(ux_{j_k} + vy_{j_k})]$$

For a square grid, n_k is an even number; every grid point on the circle of radius R_k defines another point, having coordinates of opposite sign. Therefore the sum in Eq. (2.26) can be written as

$$H_d(u, v) = \sum_{k=0}^{N} h_k(1/n_k) \sum_{j_k=1}^{n_k/2} \{\exp[i(ux_{j_k} + vy_{j_k})] + \exp[-i(ux_{j_k} + vy_{j_k})]\}$$

or,

$$(2.27) \qquad H_d(u, v) = \sum_{k=0}^{N} h_k(2/n_k) \sum_{j_k=1}^{n_k/2} \cos(ux_{j_k} + vy_{j_k})$$

In conclusion, one can say that the frequency response of the filter, that consists of averaging $f(x, y)$ over a finite number n_r of points on a circle of radius r, is given by

$$(2.28) \qquad H_r(u, v) = (2/n_r) \sum_{j=1}^{n_r/2} \cos(ux_j + vy_j), \qquad x_j^2 + y_j^2 = r^2$$

The practical presentation of the problem already produces an intrinsic smoothing: the fact that the continuous field $f(x, y)$ is sampled on a square grid ($\Delta x = \Delta y = s_0$) eliminates the high frequency response of the input field. Indeed, from the sampling theorem (Bendat, 1958) it follows that the highest frequencies, left in the data after the digitalization of the continuous field, are given by

$$(2.29) \qquad |u_N| = |v_N| = \pi/s_0$$

known as the Nyquist frequencies, associated with a sampling on a square grid with spacing s_0, and the maximum radial frequency is then

$$(2.30) \qquad \rho_N = \sqrt{2}\pi/s_0$$

These are the cutoff frequencies, induced by the digital profile and they are controlled by the grid spacing.

Figure 2 shows the differences between the forms (2.22) and (2.28) in the Nyquist domain $0 \leq u \leq u_N$, $0 \leq v \leq v_N$ of the frequency plane for $n_r = 4$ and $n_r = 8$ points, distributed equidistantly over the circle of radius r.

It was found that the differences are less than 0.01 % over the Nyquist domain for $n_r = 12$ and $n_r = 16$ points on the circles and this result is nearly independent of the distribution of the points on the circles. From Fig. 2 it follows that the largest deviations for $n_r = 8$ points is 1.7 % in the point (π, π) and significant deviations occur at four-point averaging (about 30 %). In consequence one may expect that averaging over eight and more points yields a representative estimation of the true circle average; a significant deviation of the exact circle average occurs at four-point averaging and these deviations are mainly situated in the high frequency region of the Nyquist domain.

2.4. Construction of Linear, Discrete Filters in the (x, y)-Plane

In this section the construction of a discrete filter of the form (2.20) with weights h_k by an algorithm that works entirely in the coordinate plane will be considered. Suppose that the frequency response of an ideal low-pass filter

$$\begin{aligned} H(\rho) &\neq 0 & \text{for} & \quad 0 \leq \rho \leq \rho_C \\ &= 0 & \text{for} & \quad \rho > \rho_C \end{aligned}$$

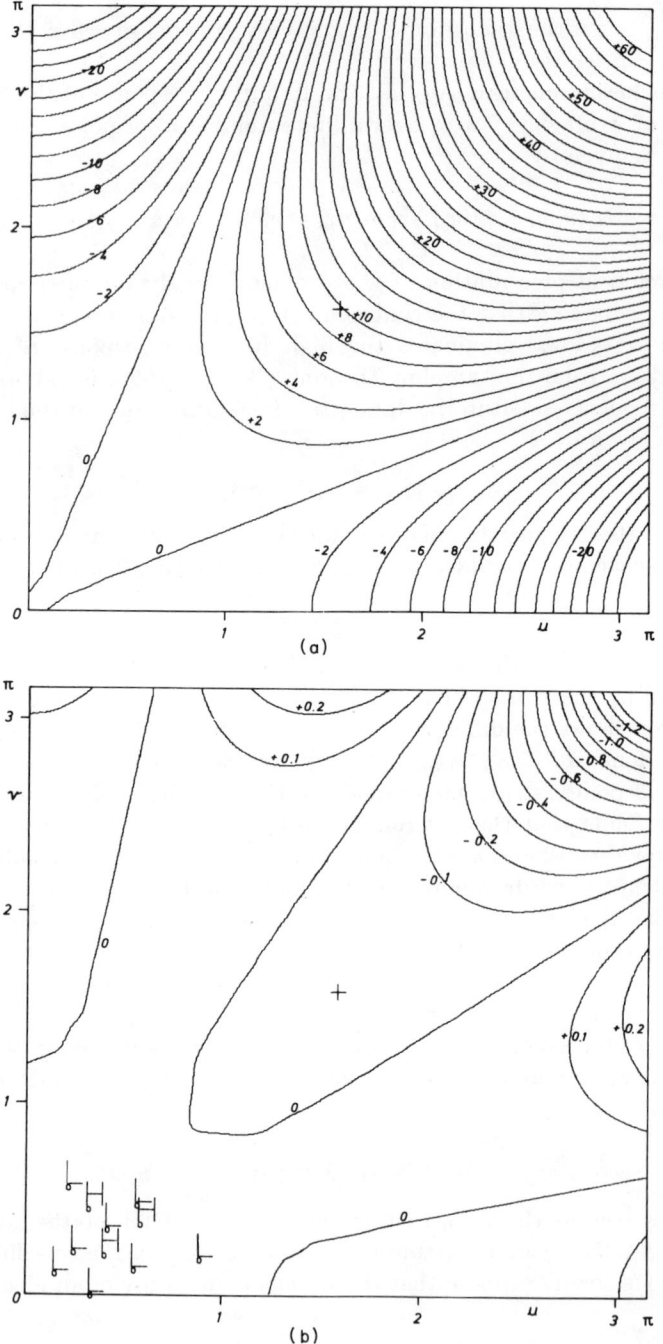

Fig. 2. Deviations for (a) $n_r = 4$ points, contour interval 2 %; (b) $n_r = 8$ points, contour interval 0.1 %.

is approximated by

(2.31) $$\bar{H}(\rho) = H(\rho) \quad \text{for} \quad 0 \leq \rho \leq \rho_C,$$
$$= H(\rho)k_\alpha(\rho) \quad \text{for} \quad \rho_C < \rho \leq \rho_k,$$
$$= 0 \quad \text{for} \quad \rho_k < \rho,$$

for some cutoff frequency ρ_C; ρ_k is the radial frequency where we want $\bar{H}(\rho)$ to become zero. The introduction of the kernel $k_\alpha(\rho)$ enables us to avoid the discontinuity at the cutoff frequency ρ_C (see Fig. 3).

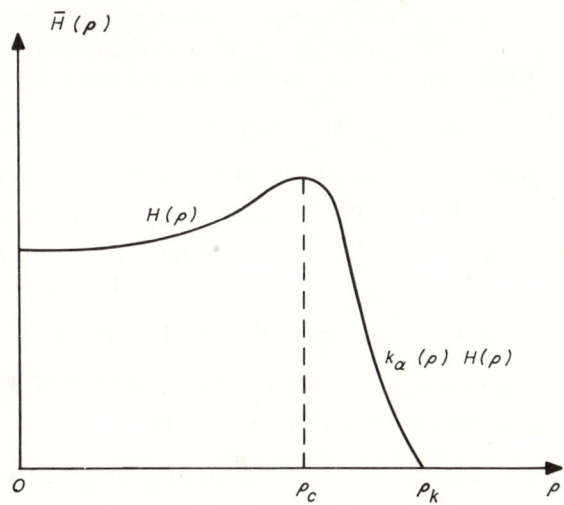

FIG. 3. Hypothetical low-pass filter.

For the function $k_\alpha(\rho)$ one can use one of the following kernels:

(2.32)
Féjer kernel $\quad k_\alpha(\rho) = [\sin \alpha(\rho - \rho_C)/\alpha \sin(\rho - \rho_C)]^2$

Jackson kernel $\quad k_\alpha(\rho) = [\sin \alpha(\rho - \rho_C)/\alpha \sin(\rho - \rho_C)]^4$

Abel–Poisson kernel $\quad k_\alpha(\rho) = 1/\{1 + [\alpha(\rho - \rho_C)]^2\}$

Weierstrass kernel $\quad k_\alpha(\rho) = \exp[-\alpha(\rho - \rho_C)^2]$

The corresponding filter function is the Fourier transform of the frequency response; using Eq. (2.17) we obtain

(2.33) $$\bar{h}(x, y) = (1/2\pi)\left[\int_0^{\rho_C} H(\rho)J_0(r\rho)\rho \, d\rho + \int_{\rho_C}^{\rho_k} H(\rho)k_\alpha(\rho)J_0(r\rho)\rho \, d\rho\right]$$

where $r^2 = x^2 + y^2$. This can also be written as

(2.34) $$\bar{h}(x, y) = (1/2\pi)\left\{\int_0^{\rho_c} H(\rho)J_0(r\rho)\rho\, d\rho + \int_0^{\rho_k-\rho_c} H(\rho+\rho_c)k_\alpha(\rho+\rho_c)J_0[r(\rho+\rho_c)](\rho+\rho_c)d\rho\right\}$$

It follows immediately that $\bar{h}(x, y)$ is only a function of the radial coordinate r in the (x, y)-plane. This filter function is seen to be infinite in length; for practical use it must be shortened to a finite length. Therefore we introduce an even function $d_i(r)$, frequently called a data kernel or covariance kernel in the theory of the spectral analysis of time series, and subject to the restrictions that $d_i(0) = 1$ and $d_i(r) = 0$ for $r > R$. The choice of the parameter R will be discussed later. Hence the modified impulse response is defined

(2.35) $$\bar{h}_f(x, y) = \bar{h}(x, y)\, d_i[(x^2 + y^2)^{1/2}]$$

having the property that $\bar{h}_f(x, y) = 0$ for $(x^2 + y^2)^{1/2} > R$. A summary of some frequently used data kernels is:

(2.36)

	Box car kernel	$d_0(r) = 1, \quad r < R$
		$= 0, \quad r > R$
	Bartlett kernel	$d_1(r) = [1 - (|r|/R)]\, d_0(r)$
	Hanning kernel	$d_2(r) = \frac{1}{2}[1 + \cos(\pi r/R)]\, d_0(r)$
	Hamming kernel	$d_3(r) = [0.54 + 0.46 \cos(\pi r/R)]\, d_0(r)$
	Arsac kernel	$d_4(r) = [1 - (r^2/R^2)]\, d_0(r)$

In order to modify the filter function $\bar{h}(x, y)$ as little as possible in the interval $(0, R)$ one can use the kernel

(2.37) $$d_5(r) = 1, \qquad\qquad\qquad\qquad 0 \leq r \leq 0.9R$$
$$= \tfrac{1}{2}\{1 - \cos[\pi(R-r)/0.1R]\}, \quad 0.9R \leq r \leq R$$

The modified frequency response corresponding to this filter function $\bar{h}_f(r)$ of finite length is then given by Eq. (2.16)

(2.38) $$\bar{H}_f(\rho) = \int_0^{\rho_c} H(\rho')D_i(\rho;\rho')\rho'\, d\rho'$$
$$+ \int_0^{\rho_k-\rho_c} H(\rho'+\rho_c)k_\alpha(\rho'+\rho_c)D_i(\rho;\rho'+\rho_c)(\rho'+\rho_c)\, d\rho',$$

where the function $D_i(\rho; \rho')$ is defined by

(2.39) $$D_i(\rho; \rho_0) = \int_0^R d_i(r)J_0(r\rho)J_0(r\rho')r\, dr$$

For the kernel $d_0(r)$ one can apply the following relations for the zero-order Bessel function (Watson, 1958, p. 134)

(2.40) $\quad (\rho' \neq \rho) \quad \int J_0(\rho'z)J_0{}^*(\rho z)z\,dz$
$$= z[\rho' J_1(\rho'z)J_0{}^*(\rho z) - \rho J_0(\rho'z)J_1{}^*(\rho z)]/(\rho'^2 - \rho^2)$$

(2.41) $\quad (\rho' = \rho) \quad \int J_0(\rho z)J_0{}^*(\rho z)z\,dz$
$$= \tfrac{1}{4}z^2[2J_0(\rho z)J_0{}^*(\rho z) - J_{-1}(\rho z)J_1{}^*(\rho z) - J^*_{-1}(\rho z)J_1(\rho z)]$$

for any complex number z, in order to obtain

(2.42)
$$(\rho' \neq \rho) \quad D_0(\rho;\rho') = R[\rho' J_1(R\rho')J_0(R\rho) - \rho J_0(R\rho')J_1(R\rho)]/(\rho'^2 - \rho^2)$$
$$(\rho' = \rho) \quad D_0(\rho;\rho) = \tfrac{1}{2}R^2[J_0{}^2(R\rho) + J_1{}^2(R\rho)]$$

The expression (2.9) for the output field is now approximated by

(2.43)
$$g(x,y) \simeq 2\pi \int_0^R \bar{h}_f(r) f(x,y,r) r\,dr$$

In order to meet the property that the filter coefficients should add up to unity, let R be the solution of the equation

(2.44)
$$I(R) \equiv 2\pi \int_0^R \bar{h}_f(r) r\,dr - 1 = 0$$

which lies nearest to the outer integration circle of radius R_N of the chosen integration grid. Hence, substituting (2.34) and (2.35) into (2.44) one obtains

(2.45) $\quad I(R) \equiv \int_0^{\rho_C} H(\rho)D_i(\rho;0)\rho\,d\rho$
$$+ \int_0^{\rho_k - \rho_C} H(\rho + \rho_C)k_\alpha(\rho + \rho_C)D_i(\rho + \rho_C;0)(\rho + \rho_C)\,d\rho - 1 = 0$$

The box car kernel $d_0(r)$ gives

(2.46)
$$D_0(\rho;0) = \int_0^R J_0(r\rho)r\,dr = (R/\rho)J_1(R\rho)$$

(Sneddon, 1951, p. 513) and R is then the solution of the equation

(2.47) $\quad I(R) \equiv R\Big\{\int_0^{\rho_C} H(\rho)J_1(R\rho)\,d\rho$
$$+ \int_0^{\rho_k - \rho_C} H(\rho + \rho_C)k_\alpha(\rho + \rho_C)J_1[R(\rho + \rho_C)]\,d\rho\Big\} - 1 = 0$$

Applying the trapezoidal rule with interval Δ to the finite integral (2.43) yields the weighted sum

(2.48) $$g(x, y) \simeq \sum_{l=0}^{L} w(r_l) f(x, y, r_l),$$

where

(2.49) $$\begin{aligned} L &= R/\Delta \\ r_l &= \Delta(l-1) \\ w(r_l) &= 2\pi\Delta \bar{h}_{\mathrm{f}}(r_l) r_l \quad \text{for} \quad 0 \leq l \leq L-1 \\ w(r_L) &= \pi\Delta \bar{h}_{\mathrm{f}}(r_L) r_L \end{aligned}$$

The form (2.48) can then be converted into the equivalent expression

(2.50) $$g(x, y) \simeq \sum_{k=0}^{N} h_k f(x, y, R_k)$$

in which the radii R_k have values so that the corresponding circles will pass through certain grid points. The conversion from (2.48) to (2.50) is made by applying an n-point interpolation formula to yield an expression for each $f(x, y, r_l)$ in terms of values $f(x, y, R_k)$ on the n nearest circles. By combining the interpolation coefficients with the values of $w(r_l)$ one obtains the coefficients h_k in Eq. (2.50).

An application of this method will be given in Section 3.5 for the process of the downward continuation of a surface gravity field.

2.5. Construction of Linear, Discrete Filters in the (u, v)-Plane

In order to construct a suitable numerical filter for the continuous filter (2.9), we want it to give the property that the corresponding frequency response $H_{\mathrm{d}}(u, v)$ must have a "good" approximation to the theoretical frequency response $H(u, v)$ in the Nyquist domain of the frequency plane.

We want the difference between the ideal output $g(x, y)$ of the theoretical filter and the numerical output $\tilde{g}(x, y)$ of the discrete filter to be a minimum, according to a given norm. Therefore let us consider the following metrics:

(1) the C-metric on the space of the continuous functions, bounded in the (x, y)-plane

(2.51) $$\|f\|_C = \max_{x, y} |f(x, y)|$$

(2) The L_1-metric on the space of the absolutely integrable functions

(2.52) $$\|f\|_{L_1} = \int_{-\infty}^{\infty} \int_{-\infty}^{\infty} |f(x, y)| \, dx \, dy,$$

(3) the L_2-metric on the space of the quadratically integrable functions

$$\|f\|_{L2} = \left[\int_{-\infty}^{\infty} \int_{-\infty}^{\infty} f^2(x, y) \, dx \, dy \right]^{1/2} \tag{2.53}$$

The form $a(x, y)$ represents the difference between the actual and numerical outputs

$$a(x, y) = g(x, y) - \tilde{g}(x, y) \tag{2.54}$$

From (2.13) and (2.25) it immediately follows that the Fourier transform of $a(x, y)$ is given by

$$A(u, v) = G(u, v) - \tilde{G}(u, v) = F(u, v)[H(u, v) - H_d(u, v)] \tag{2.55}$$

where $H_d(u, v)$ is one of the forms (2.21) or (2.27). Applying the inverse Fourier transform to both sides of (2.55) yields

$$a(x, y) = (1/4\pi^2) \int_{-\infty}^{\infty} \int_{-\infty}^{\infty} F(u, v)[H(u, v) - H_d(u, v)] e^{i(ux+vy)} \, du \, dv \tag{2.56}$$

Majoring the absolute value of the integral in (2.56) we find that

$$\|a\|_C \leq (1/4\pi^2) \int_{-\infty}^{\infty} \int_{-\infty}^{\infty} |F(u, v)| |H(u, v) - H_d(u, v)| \, du \, dv \tag{2.57}$$

Similarly, using Parseval's theorem for double Fourier transforms

$$\|a\|_{L2}^2 = \|A\|_{L2}^2 = \int_{-\infty}^{\infty} \int_{-\infty}^{\infty} |F(u, v)|^2 [H(u, v) - H_d(u, v)]^2 \, du \, dv \tag{2.58}$$

If the Fourier transform of the input field is bounded, this leads to the inequalities

$$\|a\|_C \leq c_1 \int_{-\infty}^{\infty} \int_{-\infty}^{\infty} |H(u, v) - H_d(u, v)| \, du \, dv \tag{2.59}$$

and

$$\|a\|_{L2}^2 \leq c_2 \int_{-\infty}^{\infty} \int_{-\infty}^{\infty} [H(u, v) - H_d(u, v)]^2 \, du \, dv \tag{2.60}$$

where c_1 and c_2 are constants. According to the basic idea of minimizing the difference between the true and numerical outputs, the filter coefficients h_k may be determined from the condition

$$\int_{-\infty}^{\infty} \int_{-\infty}^{\infty} |H(u, v) - H_d(u, v)| \, du \, dv = \text{minimum} \tag{2.61}$$

or from the L_2-metric

$$\int_{-\infty}^{\infty} \int_{-\infty}^{\infty} [H(u, v) - H_d(u, v)]^2 \, du \, dv = \text{minimum} \tag{2.62}$$

Furthermore, it is known that the study of the frequency behavior of the filters in question may be restricted to the Nyquist domain; hence, one has the following criteria for determining the filterweights h_k

$$(2.63) \qquad \int_{-v_N}^{v_N} \int_{-u_N}^{u_N} |H(u, v) - H_d(u, v)| \, du \, dv = \text{minimum}$$

and

$$(2.64) \qquad \int_{-v_N}^{v_N} \int_{-u_N}^{u_N} [H(u, v) - H_d(u, v)]^2 \, du \, dv = \text{minimum}$$

The minimalization according to Eq. (2.64) leads naturally to the method of least-squares. Indeed, writing $I(h)$ for the left-hand side of Eq. (2.64), the coefficients h_j are obtained from the conditions

$$(2.65) \qquad \partial I(h)/\partial h_j = 0, \qquad 0 \leq j \leq N$$

Writing $H_{Rk}(u, v)$ for the frequency response of the averaging process over the circle R_k, as given by Eq. (2.22) or Eq. (2.28), this leads to the equations

$$(\partial/\partial h_j) \int_{-v_N}^{v_N} \int_{-u_N}^{u_N} \left[H(u, v) - \sum_{k=0}^{N} h_k H_{Rk}(u, v) \right]^2 du \, dv = 0$$

and to the following linear system

$$(2.66) \qquad \sum_{k=0}^{N} h_k \int_{-v_N}^{v_N} \int_{-u_N}^{u_N} H_{Rk}(u, v) H_{Rj}(u, v) \, du \, dv$$

$$= \int_{-v_N}^{v_N} \int_{-u_N}^{u_N} H(u, v) H_{Rj}(u, v) \, du \, dv, \qquad 0 \leq j \leq N$$

from which the weights h_k may be computed.

The minimalization of the integral on the left-hand side of Eq. (2.63) leads us to the method of linear programming which is an extremely useful numerical aid to this kind of approximation problem. The algebraic pattern of linear programming problems has the following general formulation

$$(2.67) \qquad \sum_{k=1}^{n} a_{jk} x_k \leq b_j, \qquad 1 \leq j \leq m$$

where $m \geq n$ or $m \leq n$ and there are an infinite number of solutions. In order to choose an optimal solution from all possible solutions one also introduces the condition

$$(2.68) \qquad z = \sum_{k=1}^{n} c_k x_k = \text{minimum}$$

where all the coefficients a_{jk}, b_j, and c_k are known and the x_k have to be determined. One often deals with the additional condition

(2.69) $$x_k \leq 0, \quad 1 \leq k \leq n$$

Among several numerical methods for solving this system of linear inequalities, the simplex algorithm of Dantzig (1963) is perhaps the most used. This technique can be applied to the problem of the approximation of a function by a finite series of the form

(2.70) $$g(\rho) \simeq \sum_{k=1}^{n} h_k\, f_k(\rho)$$

with given functions $f_k(\rho)$. For m values ρ_j of the variable ρ, let the general formulation of the approximation be

(2.71) $$-\varepsilon \leq g(\rho_j) - \sum_{k=1}^{n} h_k\, f_k(\rho_j) \leq \varepsilon, \quad 1 \leq j \leq m$$

and the task is to find the coefficients h_k such that ε should be minimum, without the restriction (2.69). The weighted sum in (2.70) will then provide a numerical function which approaches $g(\rho)$ in a band of width 2ε. This technique will be used in Section 3.3 for the construction of upward continuation filters.

3. Upward and Downward Continuation of the Surface Gravity Effect

3.1. Definition of the Surface Gravity Effect

The potential of the gravitational field of the earth at a point $P(\mathbf{r})$ on or outside the surface is given by the well-known Newton's expression

(3.1) $$V(\mathbf{r}) = -G \int_v \frac{\rho(\mathbf{r}')\, d\mathbf{r}'}{|\mathbf{r} - \mathbf{r}'|}$$

with $d\mathbf{r}'$ the volume element of the volume v of the earth, $\mathbf{r}'$ the position vector of an arbitrary point of the volume v, $\rho(\mathbf{r}')$ the density, G the gravitational constant, and

$$|\mathbf{r} - \mathbf{r}'| = r^2 + r'^2 - 2rr' \cos \gamma$$

The origin of the orthogonal coordinate system (x, y, z) is at the mass center of the earth, with the z-axis coinciding with the rotation axis; γ is the angle between the directions of $\mathbf{r}$ and $\mathbf{r}'$.

At points outside the earth, the integral in Eq. (3.1) is nonsingular and the gravity potential satisfies Laplace's equation

$$(3.2) \qquad \nabla^2 V(\mathbf{r}) \equiv \frac{\partial^2 V}{\partial x^2} + \frac{\partial^2 V}{\partial y^2} + \frac{\partial^2 V}{\partial z^2} = 0$$

whereas for the interior points the integrand in (3.1) becomes singular at $\mathbf{r}' = \mathbf{r}$; therefore, inside the earth the potential is the solution of Poisson's equation

$$(3.3) \qquad \nabla^2 V(\mathbf{r}') = 4\pi G \rho(\mathbf{r}')$$

The total potential W of the earth consists of two parts, the principal one, V, caused by the Newton's attraction, and the other by the earth's rotation. If the point P is on the earth's surface, the gravity vector is defined as the gradient of the total gravity potential on the surface

$$(3.4) \qquad \mathbf{g}(P) = \nabla W(P)$$

whose direction is called the vertical in P; the modulus g is known as the gravitational acceleration or the gravitational field and the unit of g is the gal = 1 cm/sec^2.

Explicitly, the tidal variations in the vector $\mathbf{g}$, in amplitude as well as in direction, due to the changing positions of the sun and the moon with respect to the earth are not considered and, therefore, it is assumed that the contributions of the tidal effects to the gravity vector are eliminated from the measurements, when performing an analysis of an observed surface field.

Indeed, the aim is to study the local deviations of the gravity field in the earth's surface, caused by subsurface masses. Consequently it will be assumed that the earth is divided into a regular part, bounded by the earth spheroid, and a deviation part. The regular part is supposed to have the same mass as the earth as a whole and the earth spheroid to have the volume and flattening of the geoid. Clearly it is implicitly assumed that the deviation part of the earth's field consists partly of negative masses, so that the total mass of the deviation part is zero. The potential caused by these deviation masses of the earth is denoted by U and it is the difference between the total gravity potential and the potential of the spheroidal earth.

Since the contributions of these local anomalies are inherent in the gravitational field of the earth, a gravimeter can only measure the superimposed influence of the gravity field of the earth spheroid and of the deviation masses. The attraction of local anomalous bodies is often seen only as a minor distortion of the pattern due to some major structure. The gravimeter, being leveled in the total gravitational field, can respond only to the vertical component of the gravity field of the disturbing masses; this is called the "gravity effect" by Grant and West (1965) and it describes in fact the heterogeneous structure of the upper part of the earth's crust. For a point P on the surface we define the gravity effect by

(3.5) $$\Delta g(P) = \partial U(P)/\partial n$$

where n is the internal normal to the surface.

In this connection, the assumption has been implicitly made that the direction of the gravity vector is not disturbed by the presence of these subsurface masses, as we have supposed that Δg is measured in the direction of normal gravity. The assumption that $\Delta g \ll g$ is physically justified by the fact that changes in the densities of the crust within the anomalous region will produce variations in g which will generally not exceed 10 mgal, so that one can say that the ratio $\Delta g/g$ is of the order of 10^{-5}.

3.2. The Upward Continuation as a Filtering Operation

The first boundary value problem of potential theory, also known as Dirichlet's problem, is stated as follows: it is assumed that the earth's surface is replaced by a plane and that a local coordinate system (x, y, z) has been chosen, with the z-axis in the direction of the internal normal and the x- and y-axis defining the horizontal plane $z = 0$. Let the boundary values be the values of a function $f(x, y, 0) = f_0(x, y)$ in the plane $z = 0$, which has continuous third derivatives on and above the horizontal plane. The problem is to find the solution of Laplace's equation $\nabla^2 f = 0$ which is continuous on and above the surface and regular in the half-space $z \leq 0$, and coincides with $f_0(x, y)$ on the plane $z = 0$.

It can be proved that this solution, under the above given Dirichlet's conditions, exists and that it is unique. The rigorous proof of this statement and the derivation of the form of the solution are based on the theory of the Green function, corresponding to the given boundary value problem (Morse and Feshbach, 1953; Courant and Hilbert, 1962). It is certainly not the intention here to explain the general outline of the theory of the Laplace equation; reference is therefore made to Kellog (1960) and Grant-West (1965). From these works one obtains the solution of the first boundary value problem for the gravity effect

(3.6) $$\Delta g(x, y, z) = \frac{|z|}{2\pi} \int_{-\infty}^{\infty} \int_{-\infty}^{\infty} \frac{\Delta g(\xi, \eta, 0)}{[(x-\xi)^2 + (y-\eta)^2 + z^2]^{3/2}} d\xi\, d\eta, \quad z \leq 0$$

showing that the distribution of the gravity effect above the earth's surface is completely determined by the integration of the surface data. Hence, the problem of the upward continuation of the gravity effect is reduced to the numerical integration of Eq. (3.6).

Equation (3.6) is clearly a convolution integral of the form (2.7) and thus defines a linear filter, with the surface effect $\Delta g(x, y, 0)$ as input and as output the desired field $\Delta g(x, y, z)$ at height z, using a filter function

(3.7) $\qquad h^{(-)}(x, y, z) = (|z|/2\pi) \cdot 1/(x^2 + y^2 + z^2)^{3/2}$

This intrinsic relation between the problem of upward continuation and the general linear filter theory was first clearly pointed out by Dean (1958).

In the frequency plane the equivalent form of Eq. (3.6) becomes

(3.8) $\qquad G(u, v, z) = H^{(-)}(u, v, z) G(u, v, 0)$

in view of the convolution theorem of Fourier transforms. $G(u, v, z)$, $G(u, v, 0)$, and $H^{(-)}(u, v, z)$ are the Fourier transforms of $\Delta g(x, y, z)$, $\Delta g(x, y, 0)$ and $h^{(-)}(x, y, z)$, respectively. The explicit form of the frequency response of the upward continuation filter follows from Eq. (2.16); $H^{(-)}(u, v, z)$ is then the zero-order Hankel transform of the function $|z|(r^2 + z^2)^{-3/2}$. Hence (Sneddon, 1951, p. 528)

(3.9) $\qquad H^{(-)}(u, v, z) = e^{-|z|\rho} = \exp[-|z|(u^2 + v^2)^{1/2}], \qquad \rho^2 = u^2 + v^2$

This result can be easily interpreted if one notes that small masses mainly contribute to the Fourier spectrum of the surface effect in the high frequency region and that the broader structures induce waves of small frequency. Combining (3.9) and (3.8) one obtains the fundamental relation

(3.10) $\qquad G(u, v, z) = \exp[-|z|(u^2 + v^2)^{1/2}] G(u, v, 0)$

The appearance of the decreasing exponential in Eq. (3.10) implies that the field will be damped out with increasing height and that the high frequency part of the Fourier spectrum will be smoothed out faster than the low frequency region. The filter (3.10) clearly has the properties of a low-pass filter, yet without a sharp cutoff: when the field is represented at different heights, first the fine details of local extent, corresponding to small shallow masses, will disappear, leaving at last a picture of the regional features of the field.

Transforming to polar coordinates $(x - \xi) + i(y - \eta) = re^{i\theta}$, Eq. (3.6) can be written as

(3.11) $\qquad \Delta g(x, y, z) = |z| \int_0^\infty [\Delta g(x, y, r)/(r^2 + z^2)^{3/2}] r \, dr,$

where

(3.12) $\qquad \Delta g(x, y, r) = (1/2\pi) \int_0^{2\pi} \Delta g(x - r\cos\theta, y - r\sin\theta, 0) \, d\theta$

is the average of the surface gravity effect over the circle of radius r, with center at the point (x, y). From Eq. (2.20) it follows that the integral (3.11) can be reduced to a weighted sum of the form

$$(3.13) \qquad \widetilde{\Delta g}(x, y, z) = \sum_{k=0}^{N} h_k \, \Delta g(x, y, R_k)$$

by using an adequate numerical integration method.

Without going into too much detail one can essentially state that the methods of Peters (1949), reviewed by Choudhury (1972) and Henderson (1960), mainly differ in the way the transition from (3.11) to (3.13) is made, as well as in the coefficients h_k of the weighted sum $\widetilde{\Delta g}(x, y, z)$.

Peters replaces the integral (3.11) by the sum

$$(3.14) \qquad \widetilde{\Delta g}(x, y, z) = |z| \sum_{k=0}^{N} \int_{R_k}^{R_{k+1}} [\Delta g(x, y, r)/(r^2 + z^2)^{3/2}] r \, dr$$

and assumes that the value of $\Delta g(x, y, r)$ in the interval (R_k, R_{k+1}) may be approximated by the mean of the values at the end points

$$\widetilde{\Delta g}(x, y, r) \simeq \tfrac{1}{2}[\Delta g(x, y, R_k) + \Delta g(x, y, R_{k+1})]$$

for r in (R_k, R_{k+1}). Evaluation of the integrals in Eq. (3.14) leads to an expression of the form (3.13).

Henderson uses the following approximation for the field at the height $z = -ms_0$

$$(3.15) \qquad \widetilde{\Delta g}(x, y, -ms_0) = ms_0 \sum_{k=0}^{N} \frac{1}{R_{k+1} - R_k} \int_{R_k}^{R_{k+1}} \Delta g(x, y, r) r \, dr$$
$$\times \left[\frac{1}{(R_k^2 + m^2 s_0^2)^{1/2}} - \frac{1}{(R_{k+1}^2 + m^2 s_0^2)^{1/2}} \right],$$

where s_0 is the mesh size of the square grid and m an integer. By expressing $\Delta g(x, y, r)$ in terms of polynomials of the second degree over the intervals (R_k, R_{k+1}) and using the formula for the gravity anomaly of the sphere, Henderson arrives at a weighted sum of the form (3.13) with coefficients shown in Appendix 1.

Although it must be clearly kept in mind that a suitable numerical integration method (3.13) has to be a compromise between working with a large amount of surface data (we ask for a small number of integration coefficients, in order to reduce the calculation time and the edge effects, imposed by the square grid of finite extent) and obtaining a sufficient accuracy for the calculated field $\widetilde{\Delta g}(x, y, z)$ at a given height, this leads to the following conclusions:

(1) From the point of view of the numerical treatment of the surface data, both approximations are of the same type as they amount to the computation of a weighted sum of the form (3.13).

(2) Peters assumes linearity of the surface field over the integration intervals (R_k, R_{k+1}), while Henderson constructs his coefficients using a physical model, thus allowing for the nonlinear variation of the field over most integration intervals. This seems to be a reasonable improvement of Peters' method.

(3) Peters arbitrarily overweights his last coefficient so that the sum of the coefficients is unity; Henderson's method does not take this restriction into account.

(4) Both authors use almost the same number of integration circles, but the extent of the integration grid is different: $R_N = \sqrt{125} \cong 11$ units for Peters and $R_N = \sqrt{625} = 25$ units for Henderson.

(5) The surface extent of both integration arrays is the same for the continuations at different heights and only the values of the coefficients change, whereas, in fact, the extent of the integration grid should be proportional to the height were one wants to compute the field.

(6) As a final conclusion, the construction of these weighting sequences is based upon empirical, more or less physically justified assumptions about the observed surface field; therefore the quantitative success of a given set of coefficients will directly depend on the degree to which a measured field meets these assumptions. Hence it would be preferable to construct the weights in a more "objective" way, that is on the basis of the mathematical mechanism of the process of the upward continuation of a surface field.

From the above it is evident that the question of the "best" set of coefficients for the numerical computation of the integral (3.11) is difficult to answer by the absence of a mathematical norm for the construction of the weights. Indeed, the accuracy in the upward calculated field will depend more on the frequency structure of the surface field in question than on the use of any particular weighting sequence. The grid spacing, the number of integration circles, and the weights, all empirically chosen, have an important bearing on the configuration of the final map.

The differences between the theoretical upward frequency response $H^{(-)}(u, v, z)$ and the discrete frequency characteristic $H_d(u, v)$, given by Eq. (2.27), are shown in Fig. 4 for Peters' and Henderson's filters $z = 1$, $s_0 = 1$, $h = 1$.

Note the direction dependence of $H_d(u, v)$ in the frequency plane, which implies that the asymmetries in the calculated field at height h are not only properties of the surface gravity data, but are also dependent upon the kind of filter one uses. In the low frequency region the deviations are of the order of 10 % and we conclude that Peters' and Henderson's filters deform significantly the features that may be ascribed to the larger anomalies of the gravity

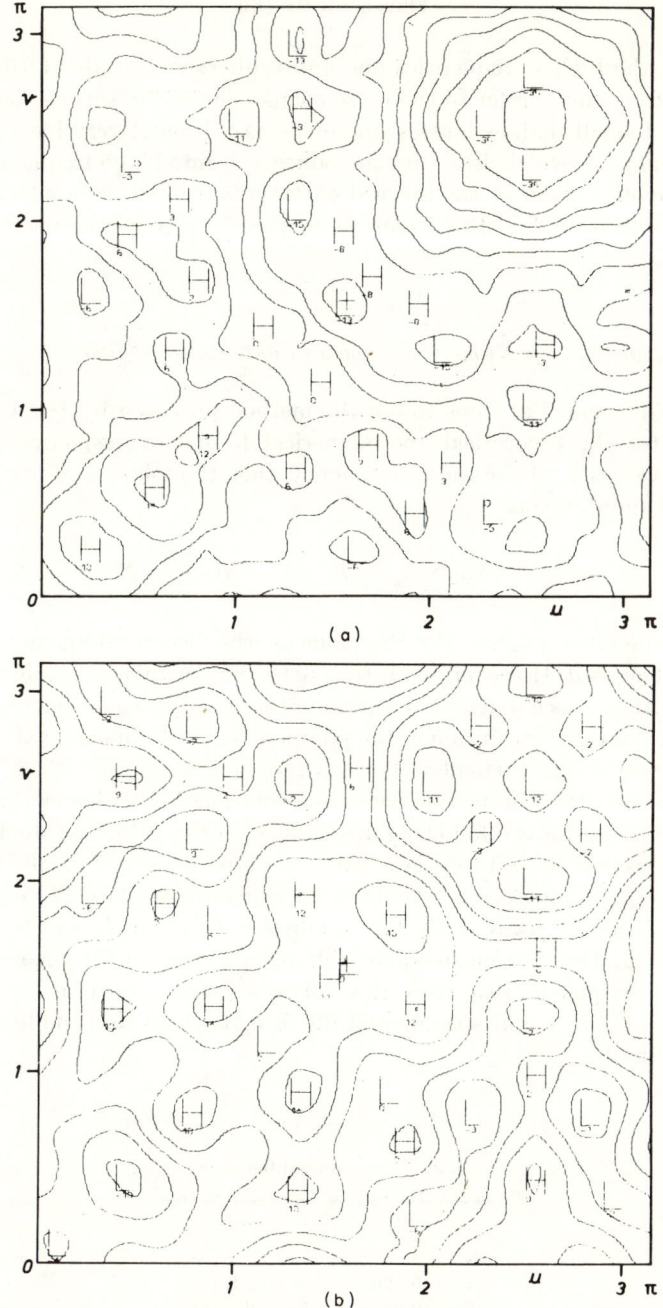

FIG. 4. Deviations for (a) Peters' filter $z = 1$, $s_0 = 1$, $h = 1$ contour interval 4%: (b) Henderson's filter $z = 1$, $s_0 = 1$, $h = 1$, contour interval 4%.

field. For the higher frequencies one notes differences of about 15 %; since the high frequencies refer to the noise component of the surface field and to the field of small shallow masses, these effects will be of very local extent in the final maps. Nevertheless, they introduce unwanted high frequency ripples in the smooth output of the upward continuation filter. As a conclusion we can say that Peters' and Henderson's filters are frequency-wise nearly equivalent.

3.3. Construction of a Numerical Upward Continuation Filter

At first it seemed obvious to use the method proposed by Dean (1958): if one equates the theoretical and numerical frequency responses at $N+1$ different values ρ_j of the radial frequency, then the solution of the resulting system of simultaneous equations

$$(3.16) \qquad e^{-|z|\rho_j} = \sum_{k=0}^{N} h_k J_0(R_k \rho_j), \qquad 0 \leq j \leq N$$

will give the filter weights. For the moment, the discrete frequency response (2.21) will be used. However, while testing the numerical solution of the linear system (3.16) it was noted that the accuracy of the approximation was strongly dependent on the distribution of the points ρ_j on the frequency axis and, to a lesser degree, on the particular choice of the integration grid R_k.

In order to show the instability of the solution of this linear system, the results are given for the following distribution of the points ρ_j: we divide the Nyquist interval $(0, \rho_N)$ into two parts of equal length $I_1 = (0, \tfrac{1}{2}\rho_N)$ and $I_2 = (\tfrac{1}{2}\rho_N, \rho_N)$. For $N+1 = 20$ filter coefficients we place N_1 equidistant points in I_1 and $N_2 = N - N_1 + 1$ equidistant points in I_2 and let N_1 vary from 3 to 12. The root-mean-square differences between the theoretical and numerical frequency responses thus obtained are shown in Table I for $z = 1$, $s_0 = 1$, $h = 1$ unit spacing and the integration grid in the first column of Appendix 2.

TABLE I

Root-mean-square differences

N_1		N_2	
3	0.026	8	1.015
4	0.032	9	11.944
5	0.023	10	38.817
6	0.107	11	25.390
7	0.264	12	11.660

It is noted here that there is a considerable increase in the root-mean-square differences when passing from $N_1 = 5$ to $N_1 = 6$ points. For $N_1 \leq 5$ we obtain coefficient sets yielding a reasonable approximation to the theoretical frequency characteristic $H^{(-)}(u, v, z)$; for the higher frequencies $\rho \geq 1$ the curves of the theoretical and numerical frequency responses match very closely, but the largest deviations occur in the low frequency range $\rho \leq 1$. For $N_1 \geq 8$ points the numerical frequency response is completely different from the theoretical curve. Because the larger masses mainly contribute to the low frequency part of the Fourier spectrum of the surface field, it is required that the numerical frequency response $H_d(u, v)$ closely match the decreasing exponential $H^{(-)}(u, v, z)$ for the lower frequencies in order to be protected against an undesirable deformation of the larger anomalies of the field when projecting it upwards. Therefore, it is concluded that Dean's method does not give reliable results.

We now try to construct the filter coefficients from the approximation of the discrete frequency response $H_d(u, v)$, given by Eq. (2.27), to the theoretical function $H^{(-)}(u, v, z)$, by the methods of Section 2.5. As the ideal frequency characteristic $H^{(-)}(u, v, z)$ is already a function of the radial frequency, we choose the points (u_i, v_i) at the intersections of a bundle of radials through the origin of the frequency plane with a bundle of concentric circles around the origin. Also note that the function $H_d(u, v)$ in Eq. (2.27) is symmetric with respect to the bisector of the first quadrant of the (u, v)-plane, so that one may restrict the approximation to the triangle $v \leq u$, $0 \leq u \leq u_N$ of the Nyquist domain.

When using the method of least-squares and the method of linear programming, one has to compute a vector

$$b_i = \exp[-|z|(u_i^2 + v_i^2)^{1/2}], \quad 1 \leq i \leq m$$

and a matrix

$$a_{ik} = (2/n_k) \sum_{j_k=1}^{n_k/2} \cos(u_i x_{j_k} + v_i y_{j_k}), \quad 1 \leq i \leq m, 1 \leq k \leq n$$

The coefficients obtained with these methods can be found in Appendix 2.

It is evident that the accuracy of this approximation will be better for larger values of n. In our case this implies that the calculated field at a desired height will be quantitatively better for a finer or larger integration grid R_k, or both. But a large number of integration circles entails a considerable increase in the computer time for the numerical integration (3.13); a broader integration grid implies knowledge of an extensive part of the surface field and a considerable loss of accuracy at the edges of the observed part of the field. There has to be a compromise between obtaining sufficient accuracy with a relative small number of integration coefficients by limiting the extent of the integration grid and reducing the calculation time.

As the work is done with a finite weighted sum in Eq. (3.13), the question of the truncation error resulting from disregarding the field beyond the last integration circle must be examined. As

$$(3.17) \qquad (|z|/2\pi) \int_{-\infty}^{\infty} \int_{-\infty}^{\infty} d\xi\, d\eta/[(x-\xi)^2 + (y-\eta)^2 + z^2]^{3/2} = 1$$

one can write for the difference between the true field value at height z and its value at the surface

$$(3.18)$$

$$\Delta g(x, y, z) - \Delta g(x, y, 0) = |z| \int_0^{\infty} \{[\Delta g(x, y, r) - \Delta g(x, y, 0)]/(r^2 + z^2)^{3/2}\} r\, dr$$

using Eqs. (3.11) and (3.12). Now, applying the numerical integration method as in Eq. (3.13) to the right-hand side of (3.18) one obtains

$$(3.19) \qquad \Delta g(x, y, z) - \Delta g(x, y, 0) \simeq \sum_{k=0}^{N} h_k [\Delta g(x, y, R_k) - \Delta g(x, y, 0)]$$

or

$$\Delta g(x, y, z) \simeq \widetilde{\Delta g}(x, y, z) + \Delta g(x, y, 0)\left[1 - \sum_{k=0}^{N} h_k\right]$$

Let the circle of radius R represent the area taken into account in the numerical integration of the right-hand side of Eq. (3.18). Then it follows that the true field at height z may be written as

$$(3.20) \qquad \Delta g(x, y, z) \simeq \widetilde{\Delta g}(x, y, z) + |z| \int_R^{\infty} [\Delta g(x, y, r)/(r^2 + z^2)^{3/2}] r\, dr$$

$$\simeq \widetilde{\Delta g}(x, y, z) + \Delta g(x, y, 0)|z|/(R^2 + z^2)^{1/2}$$

using $\Delta g(x, y, 0)$ as a measure of the average of the surface field outside the circle of radius R.

Combining Eqs. (3.19) and (3.20) leads to the conclusion that the contribution of the surface field within the circle of radius R is represented by $\widetilde{\Delta g}(x, y, z)$ and the difference between the sum of the coefficients and unity, multiplied by the center point value $\Delta g(x, y, 0)$, may be attributed to the contribution from the part of the surface field outside the circle of radius R. Hence, it is noted that one possibility of reducing the truncation error is to make the sum of the coefficients equal to unity, as in Peters' method. Otherwise, if the filter weights do not add up to unity, the surface value $\Delta g(x, y, 0)$ must be multiplied by the difference between the sum of the coefficients and unity and then the result added to the computed value $\widetilde{\Delta g}(x, y, z)$.

Whether or not the sum of the filter coefficients should be made equal to unity depends entirely upon personal judgment. However, if one requires that the weights add up to unity, it is found to result in relatively large values of ε in the method of linear programming and it was difficult to reduce the value of this ε. On the other hand, not taking this restriction into account one obtains a faster convergence to an acceptable solution, with a minor loss of accuracy (rounding-off errors) and in addition small values of ε. In consequence preference is given to the second alternative.

The significance of this ε must be kept clearly in mind: in the method of least-squares, ε means the root-mean-square difference between the theoretical and numerical frequency responses, while in the method of linear programming ε stands for the maximum difference between both curves. Therefore, one can compare the following results:

$z=1, s_0=1, h=1, N+1=20, R_N=\sqrt{100}$, least-squares: $\varepsilon = 1.7\%$, maximum difference 5.5%,

$z=1, s_0=1, h=1, N+1=20, R_N=\sqrt{100}$, linear programming: $\varepsilon = 3.5\%$, maximum difference 3.5%.

Hence, we have constructed a discrete filter, simulating the process of upward continuation, giving an approximation to the theoretical filter $H^{(-)}(u, v, z)$ for $z=1$, with an accuracy of 3.5% over the Nyquist domain of the frequency plane.

The reduction of the integration grid with $R_N = \sqrt{274} \simeq 16.5$ units to a grid with $R_N = \sqrt{100} = 10$ units results in a small increase in the value of ε: from 1.3% to 1.7% in the method of least-squares and from 3.2% to 3.5% for the method of linear programming. This is an interesting result since a limited integration grid enables us to reduce considerably the deformation of the field near the edges of the surface grid on which the field is known.

The considerable gain of this method must be found in the generality of its principles: the coefficients are derived from the knowledge of the frequency response function, describing the mathematical mechanism of the upward continuation of a surface field in the frequency plane. The method does not use any empirical assumption about the field, as was the case with Peters' and Henderson's filters. Also the flexibility of the method of linear programming in these approximation techniques is remarkable: the construction of the weights h_k is practically independent of the choice of the distribution of the frequency points (u_i, v_i), because one can take m several times larger than n.

3.4. The Downward Continuation as a Filtering Operation

The following problem will now be considered: given the normal component of a gravitational field in the earth's surface, how may one determine the value of the field at a given depth. This calls for the solution of Poisson's equation with unknown density function. The problem is obviously indeterminate because of the inherent ambiguity of inverse boundary value problems, common to all potential fields.

Skeels (1947) points out that, for a given surface field, an infinite number of mass distributions can be found, which all account for that field. It is true that a measured field generally introduces several physical parameters, such as maximum and minimum depths of the anomalous masses, largest areal extent, maximum density changes, etc., but even within these limits an infinite variety of solutions is possible. This ambiguity does not depend upon the spacing or precision of the observations in the earth's surface and no degree of precision or amount of data will remove these fundamental uncertainties.

To this are added the facts that one may expect the observations to show errors, that the measured field is the superposition of the anomalous fields of a very complicated ensemble of structures, that the measurements undergo rough reductions (such as Bouguer corrections), and that the number of observations is limited and that they are known on a regular grid of finite extent. Hence it is evident that a quantitative interpretation of gravitational anomalies can only give partial information about the form, location, depth, and density of the subsurface masses.

It is essential to avoid dealing with the Poisson equation, inasmuch as the mass distribution is unknown; consequently one has to rely upon more or less physically justified assumptions about the mechanism of the process of downward continuation of a surface field.

The distribution of the field in a plane at depth z is to be determined; Grant and West (1965) suppose that no disturbing masses intervene between the surface and the level z, assuming implicitly the validity of Laplace's equation between these two levels. The distribution of the field at the surface is now defined as the result of the upward continuation of the field at depth z; using Eq. (3.6) we postulate the following relationship between these fields

(3.21) $\quad \Delta g(x, y, 0) = (z/2\pi)$

$$\times \int_{-\infty}^{\infty} \int_{-\infty}^{\infty} \Delta g(\xi, \eta, z)/[(x - \xi)^2 + (y - \eta)^2 + z^2]^{3/2} \, d\xi \, d\eta, \qquad z > 0$$

This hypothesis leads to an inevitable divergence in the mathematical expressions, describing the downward continuation process, because the earth's surface is a real discontinuity plane and anomalous masses do occur between the two levels.

The expression (3.21) is an integral equation for the field at depth z; with the same notations as in Eq. (3.8) and applying the convolution theorem of Fourier transforms we obtain

(3.22) $\qquad\qquad G(u, v, 0) = H^{(-)}(u, v, z) G(u, v, z)$

Defining the frequency response of the downward continuation filter as

(3.23) $$H^{(+)}(u, v, z) = H^{(-)}(u, v, z)^{-1}$$
$$= \exp[z(u^2 + v^2)^{1/2}], \qquad z > 0$$

it follows that

(3.24) $$G(u, v, z) = G(u, v, 0)\exp[z(u^2 + v^2)^{1/2}]$$

Consequently, the process of downward continuation defines a linear filter with as input the surface effect $\Delta g(x, y, 0)$ and as output the field $\Delta g(x, y, z)$ at depth z, using the frequency characteristic $H^{(+)}(u, v, z)$. In the divergence of this frequency response lies the explanation of the fact that downward continuation of a field, derivable from a potential, is so intractable. This divergent nature of the theoretical frequency response for increasing depth and high frequencies prevents us from obtaining a finite filter function in the coordinate plane: only in the case where the surface field has a Fourier spectrum that attenuates with the larger frequencies more rapidly than the exponential rises, will the downward continuation integral (3.21) converge. The divergence of $H^{(+)}(u, v, z)$ finds its cause in the intrinsic amplification of the field toward its source, in the existence of small, shallow disturbing masses, and in the presence of the field of random errors, which is inherent to the observed field, whether they come from measurement, digitalization, or reduction of the data. Therefore it is imperative to eliminate the high frequency contributions from the Fourier spectrum of the surface field and to amplify the low frequency waves, building up the larger masses of the field in a way that is consistent with Eq. (3.23).

Peters (1949) and Henderson (1960) give coefficient sets, simulating the downward continuation filter. Both methods can be compared on the basis of their approximation to the theoretical frequency response $H^{(+)}(u, v, z)$. Although both algorithms are equivalent from the point of view of practical application, they differ mainly in the principles on which they depend: Peters sees the downward projected field in the form of a Maclaurin series; Henderson, however, first computes the field at different heights by the expression (3.15) and extrapolates the distributions at these levels, together with the surface field, into depth. Using the analogous form of (2.21) for the frequency response of the numerical downward projection filter $(h_0, h_1, \ldots, h_N)$

(3.25) $$H_{\mathrm{d}}^{(+)}(u, v) = \sum_{k=0}^{N} h_k J_0[R_k(u^2 + v^2)^{1/2}]$$

the curves illustrated in Fig. 5 are obtained. The corresponding coefficients are given in Appendix 3.

If we take for a norm the best approximation of the numerical frequency response $H_{\mathrm{d}}^{(+)}(u, v)$ to $H^{(+)}(u, v, z)$, then Henderson's method is more satis-

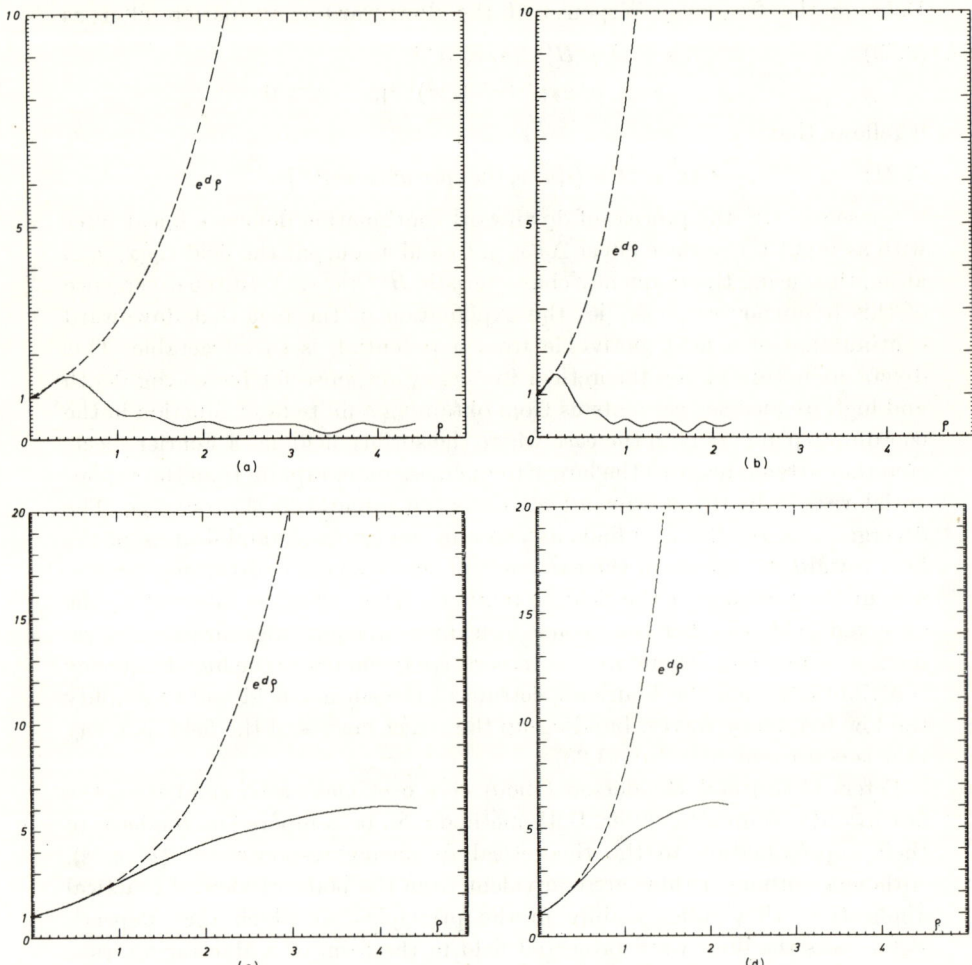

Fig. 5. Frequency response for (a) Peters' filter $z = 1$, $s_0 = 1$, $d = 1$; (b) Peters' filter $z = 1$, $s_0 = 1$, $d = 2$; (c) Henderson's filter $z = 1$, $s_0 = 1$, $d = 1$; (d) Henderson's filter $z = 1$, $s_0 = 2$, $d = 2$.

factory than Peters' algorithm. On the other hand, the frequency decay is sharper in Peters' method and Henderson's coefficients more greatly amplify the high frequency waves in the input field, so that we may expect larger fluctuations in the projected field than with Peters' weights. The conclusion is obvious: we need to construct a filter somewhere between these two methods, that is, a filter matching $H^{(+)}(u, v, z)$ very closely in the low frequency range and cutting off rather sharply.

Although the divergence of the theoretical frequency response is restricted to the Nyquist domain of the frequency plane, we still have the problem that the high frequency waves, corresponding to uninteresting features of the surface effect, are more greatly amplified than the low frequencies, so that the frequencies in the neighborhood of the Nyquist frequencies, that is waves with periods of the order of $2s_0$, greatly deform the pattern of the field of the broader anomalies in the output maps.

The expression (3.21) has the form of the general integral equation of the type

(3.26)
$$\varphi(x_1, \ldots, x_n) = \int_{-\infty}^{\infty} \cdots \int_{-\infty}^{\infty} \psi(\xi_1, \ldots, \xi_n) k(x_1 - \xi_1, \ldots, x_n - \xi_n) \, d\xi_1 \cdots d\xi_n$$

where φ is a measured field, k an analytic function, and ψ is to be determined. We further assume that φ and ψ are bounded and that the kernel k belongs to the class L_1. By definition, a function $f(x_1, \ldots, x_n)$ is L_p if its p-norm is bounded

(3.27) $\quad \|f\|_p = \left[\int_{-\infty}^{\infty} \cdots \int_{-\infty}^{\infty} |f(x_1, \ldots, x_n)|^p \, dx_1 \cdots dx_n \right]^{1/p} < \infty$

Kreisel (1948) showed that ψ cannot be determined uniquely from measurements of φ alone: if $\psi_0(x_1, \ldots, x_n)$ is a solution of the integral equation (3.26), then

$$\psi_1(x_1, \ldots, x_n) = \psi_0(x_1, \ldots, x_n) + A(2\pi)^{-n/2} \exp\left(i \sum_{j=1}^{n} p_j x_j \right)$$

also satisfies (3.26), where A may be taken arbitrarily large provided that $p_1^2 + \cdots + p_n^2$ is also sufficiently large. Therefore ψ is not determined continuously from measurements of φ, because rapidly oscillating ψ produce a small φ. In our case of gravitational interpretation this means that it is always possible to add to any solution $\Delta g(x, y, z)$ a harmonic mass distribution of arbitrarily great amplitude provided that the frequencies, induced by the added distribution, are sufficiently large. Therefore the observations at the earth's surface can give no information about changes in density occurring in a horizontal distance small compared with the depth, and the solution required is one that does not contain these rapid oscillations.

Accordingly one has to look for partial information about ψ which is insensitive to rapid oscillations in ψ. For this reason Kreisel defines the smoothed version of ψ

(3.28)
$$\bar{\psi}(x_1, \ldots, x_n) = \int_{-\infty}^{\infty} \cdots \int_{-\infty}^{\infty} \psi(\xi_1, \ldots, \xi_n) w(x_1 - \xi_1, \ldots, x_n - \xi_n) \, d\xi_1 \cdots d\xi_n$$

where w is a normalized filter function of a suitable low-pass filter.

If K and W are the Fourier transforms of the analytic functions k and w, respectively, and if (1) k and w belong to the class L_1, (2) $\Lambda = W/K$ is defined to be zero if $W=0$, and Λ is L_1, (3) the inverse Fourier transform λ of Λ is L_1, and (4) ψ is bounded and integrable in any finite domain, then Kreisel shows that (1) φ is bounded and integrable in any finite domain, and (2) the smoothed version of ψ is given by the expression

(3.29)
$$\bar{\psi}(x_1, \ldots, x_n) = \int_{-\infty}^{\infty} \cdots \int_{-\infty}^{\infty} \varphi(\xi_1, \ldots, \xi_n) \lambda(x_1 - \xi_1, \ldots, x_n - \xi_n) \, d\xi_1 \cdots d\xi_n$$

Applying this theorem to the integral equation (3.21) we define the smoothed version of the field at depth z by

(3.30)
$$\overline{\Delta g}(x, y, z) = \int_{-\infty}^{\infty} \int_{-\infty}^{\infty} \Delta g(\xi, \eta, z) w(x - \xi, y - \eta) \, d\xi \, d\eta$$

and from (3.29) it then follows that

(3.31)
$$\overline{\Delta g}(x, y, z) = \int_{-\infty}^{\infty} \int_{-\infty}^{\infty} \Delta g(\xi, \eta, 0) \bar{h}^{(+)}(x - \xi, y - \eta, z) \, d\xi \, d\eta$$

with

(3.32)
$$\bar{h}^{(+)}(x, y, z) = (1/4\pi^2) \int_{-\infty}^{\infty} \int_{-\infty}^{\infty} [W(u, v)/K(u, v)] e^{i(ux+vy)} \, du \, dv$$

Using (3.23) we obtain

(3.33)
$$\bar{h}^{(+)}(x, y, z) = (1/4\pi^2) \int_{-\infty}^{\infty} \int_{-\infty}^{\infty} H^{(+)}(u, v, z) W(u, v) e^{i(ux+vy)} \, du \, dv$$

Substitution of (3.33) into (3.31) yields the following frequency relation between the Fourier spectrum of the surface field and the Fourier spectrum of the smoothed field at depth z

(3.34)
$$\bar{G}(u, v, z) = \bar{H}^{(+)}(u, v, z) G(u, v, 0)$$

where $\bar{G}(u, v, z)$ is the Fourier transform of $\overline{\Delta g}(x, y, z)$. We conclude that Eq. (3.34) defines a modified downward continuation process by specifying a linear filter with impulse response $\bar{h}^{(+)}(x, y, z)$ and frequency response $\bar{H}^{(+)}(u, v, z)$. The smoothed field at depth z has a bounded Fourier spectrum for all frequencies provided that

(3.35)
$$\bar{H}^{(+)}(u, v, z) = W(u, v) \exp[z(u^2 + v^2)^{1/2}], \quad z > 0$$

belongs to the class L_1. Only in the case where the smoothing function $w(x, y)$ fulfils this condition can we apply Kreisel's theorem, in order to find a solution of the inverse potential problem, not for the field at depth z itself, but for the components of the field which are not eliminated from the Fourier spectrum of the surface field by the frequency response $W(u, v)$.

3.5. Construction of a Numerical Downward Continuation Filter

Bullard-Cooper (1949) and Grant-West (1965) considered the mathematical filter whose weights are proportional to the ordinates of the normal distribution

$$w(x, y) = (1/4\pi\gamma)e^{-(x^2+y^2)/4\gamma} \tag{3.36}$$

with corresponding frequency response

$$W(u, v) = e^{-\gamma(u^2+v^2)} \tag{3.37}$$

thus defining the modified downward continuation filter

$$\bar{H}^{(+)}(u, v, z) = \exp[z(u^2 + v^2)^{1/2} - \gamma(u^2 + v^2)] \tag{3.38}$$

γ is a constant and determines the degree of smoothing.

This analytic filter can be simulated by a discrete filter of the form (3.25); the differences between the theoretical and numerical frequency responses can be seen in Fig. 6 for the filters of Grant-West (Appendix 4). As a criticism of this algorithm we could say that the cutoff is not sharp, that the deviation in the low frequency range is considerable, and that the high frequency response is relatively high.

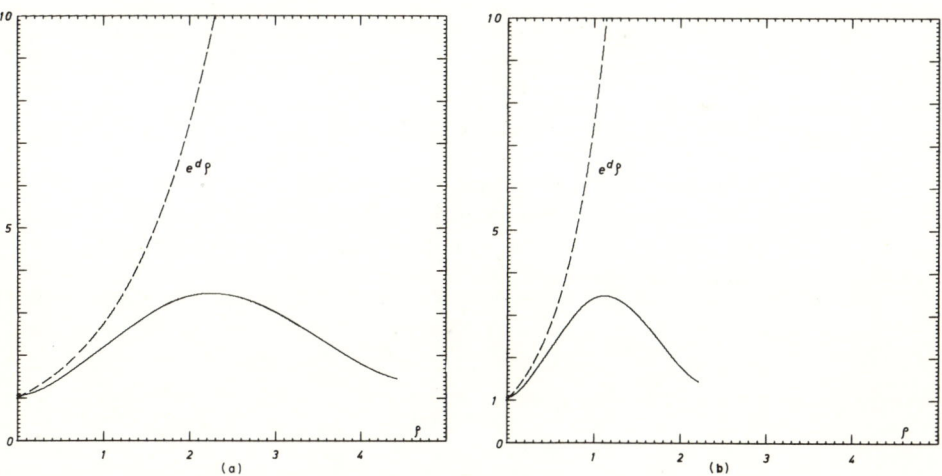

Fig. 6. Frequency response for (a) Grant's filter $z = 1$, $s_0 = 1$, $d = 1$, $\gamma = 1/6$; (b) Grant's filter $z = 1$, $s_0 = 2$, $d = 2$, $\gamma = 1/6$.

In order to avoid the amplification of the high frequency waves by the factor $H^{(+)}(u, v, z)$ we introduce the low-pass filter

$$(3.39) \quad \begin{aligned} W(\rho) &= 1 & \text{for} \quad & 0 \leq \rho \leq \rho_C, \rho^2 = u^2 + v^2 \\ &= k_\alpha(\rho) & & \rho_C < \rho \leq \rho_k \\ &= 0 & & \rho_k < \rho \end{aligned}$$

for some cutoff frequency ρ_C; ρ_k is the radial frequency when we want $W(\rho)$ to become zero. The kernel $k_\alpha(\rho)$ is one of the forms in Eq. (2.32).

From Eq. (3.35) it follows that we now work with the modified downward continuation filter

$$(3.40) \quad \begin{aligned} \bar{H}^{(+)}(\rho, z) &= e^{z\rho} & \text{for} \quad & 0 \leq \rho \leq \rho_C \\ &= e^{z\rho} k_\alpha(\rho) & & \rho_C < \rho \leq \rho_k \\ &= 0 & & \rho_k < \rho \end{aligned}$$

This function is illustrated schematically in Fig. 7.

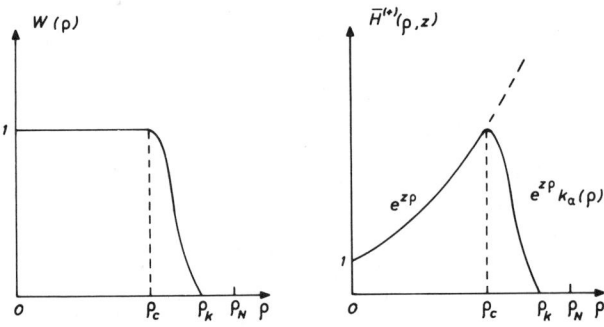

FIG. 7. The functions $W(\rho)$ and $\bar{H}^{(+)}(\rho, z)$.

The results of Section 2.4 may then be applied directly, if we substitute the function $e^{z\rho}$ for the general frequency response $H(\rho)$ in Eq. (2.31). Using Eq. (2.34) this immediately yields the filter function corresponding to the frequency response $\bar{H}^{(+)}(u, v, z)$

$$(3.41) \quad \bar{h}^{(+)}(x, y, z) = (1/2\pi) \left\{ \int_0^{\rho_C} e^{z\rho} J_0(r\rho) \rho \, d\rho \right.$$
$$\left. + \int_0^{\rho_k - \rho_C} e^{z(\rho + \rho_C)} k_\alpha(\rho + \rho_C) J_0[r(\rho + \rho_C)](\rho + \rho_C) \, d\rho \right\}$$

which must be shortened to a finite length, defining the filter function

$$(3.42) \quad \bar{h}_f^{(+)}(x, y, z) = \bar{h}^{(+)}(x, y, z) \, d_i[(x^2 + y^2)^{1/2}]$$

as in Eq. (2.35). The frequency response corresponding to this impulse response of finite length may be obtained from Eq. (2.38)

(3.43)
$$\bar{H}_{\mathrm{f}}^{(+)}(\rho, z) = \int_0^{\rho_C} e^{z\rho'} D_i(\rho; \rho')\rho' \, d\rho'$$
$$+ \int_0^{\rho_k - \rho_C} \exp[z(\rho' + \rho_C)] k_\alpha(\rho' + \rho_C) D_i(\rho; \rho' + \rho_C)(\rho' + \rho_C) d\rho'$$

where the function $D_i(\rho; \rho')$ is defined by Eq. (2.39) for the data kernel $d_i(r)$. The value R for the radial coordinate r where $d_0(r)$ becomes zero is obtained from the nonlinear equation in (2.47)

(3.44) $I(R) \equiv R$

$$\times \left\{ \int_0^{\rho_C} e^{z\rho} J_1(R\rho) \, d\rho + \int_0^{\rho_k - \rho_C} e^{z(\rho + \rho_C)} k_\alpha(\rho + \rho_C) J_1[R(\rho + \rho_C)] \, d\rho \right\} - 1 = 0$$

Using the results in Eqs. (2.48) and (2.49) one finally obtains the following weighted sum for the smoothed field at depth z

(3.45) $$\overline{\Delta g}(x, y, z) \simeq \sum_{k=0}^{N} h_k \, \Delta g(x, y, R_k)$$

with weights h_k.

There is a problem in defining the parameters ρ_C, ρ_k, and α. The choice of ρ_k and α is rather arbitrary and depends upon the kernel one decides to use. If $k_\alpha(\rho)$ is a Féjer kernel, then one can define ρ_k as the first zero of $k_\alpha(\rho)$ and we obtain the relation

(3.46) $$\alpha(\rho_C - \rho_k) = \pi$$

If $k_\alpha(\rho)$ is a Weierstrass kernel, then one defines ρ_k as the frequency where the frequency response $\bar{H}_{\mathrm{f}}^{(+)}(\rho, z)$ is equal to a small number δ

$$\exp[z\rho_k - \alpha(\rho_k - \rho_C)^2] = \delta$$

which yields the relation

(3.47) $$\alpha(\rho_k - \rho_C)^2 = z\rho_k + |\ln \delta|$$

The choice of the cutoff frequency ρ_C, however, is fundamental and will depend largely on the Fourier spectrum of the observed part of the surface field. Therefore, the choice of ρ_C will remain fairly subjective. In this connection one might argue for two alternatives:

(1) If it happens that the Fourier spectrum of the observed field falls off sharply at a certain radial frequency, then we can use this frequency as the

definition of ρ_C and we know that the components of the field, building up the low frequency structure of the Fourier spectrum, are more or less significantly projected downward.

(2) In the opposite case one might choose *a priori* a value for ρ_C, thus defining "signal" and "noise" components in the surface field, and then one knows the part of the field that is continuated downward and which part is left at the surface.

However, there is no mathematical criterion allowing a choice between these alternatives, as this can be immediately related to the impossibility of solving uniquely the inverse potential problem with unknown density function.

Using this method the following approximations are arrived at:

Test (1): $z=1$, $\rho_C=1$, $R=5.98725$, $\Delta=0.1$, $N=13$, 3-point Lagrange interpolation, Weierstrass kernel with $\alpha=6.61$ and $\rho_k=2$.

Test (2): $z=1$, $\rho_C=1$, $R=9.00739$, $\Delta=0.1$, $N=20$, 4-point Lagrange interpolation, Weierstrass kernel, with $\alpha=6.61$ and $\rho_k=2$.

Test (3): $z=1$, $\rho_C=1$, $R=6.55906$, $\Delta=0.1$, $N=13$, 3-point Lagrange interpolation, Féjer kernel with $\alpha=2\pi$ and $\rho_k=1.5$.

Test (4): $z=1$, $\rho_C=1$, $R=9.28465$, $\Delta=0.1$, $N=20$, 4-point Lagrange interpolation, Féjer kernel with $\alpha=2\pi$ and $\rho_k=1.5$.

The corresponding coefficients can be found in Appendix 5 and the frequency responses are illustrated in Fig. 8.

It immediately follows that the choice of the particular kernel $k_\alpha(\rho)$ is irrelevant. In the low frequency range $0 \leq \rho \leq \rho_C$ the approximation $N=20$ is evidently better than $N=13$ and the cutoff is sharper; in the high frequency interval $\rho > \rho_C$ all curves show oscillations of the same magnitude (of the order of 0.3) so that these filters have an almost equivalent effect in this interval. However, from the fact that the outer integration circle has a value of $\sqrt{50} \simeq 7.1$ units for $N=13$, as opposed to $\sqrt{100}=10$ units for $N=20$, it is concluded that the filters for $N=13$ are qualitatively equivalent to the filters for $N=20$. The oscillations above the cutoff frequency ρ_C are somewhat misleading: indeed, in this approximation technique these oscillations cannot be avoided and appear naturally, but the considerable gain of this form of low-pass filtering is found in the fact that the amplitudes of these deviations of the zero level are relatively small, compared to the increase of the exponential $H^{(+)}(\rho, z)$ in the Nyquist interval, attaining a value of about 77 in the point (π, π) for $z=1$. This implies that the noise part of the surface effect can be kept sufficiently under control when projecting the field downwards.

It is also possible to simulate the low-pass downward continuation filter $\bar{H}^{(+)}(u, v, z)$ by using the method of linear programming, as discussed in Section 2.5. Although it followed from the numerical frequency response that the function $\bar{H}^{(+)}(u, v, z)$ was relatively well approximated by the discrete filters we obtained with this method, the results were not significant. Indeed,

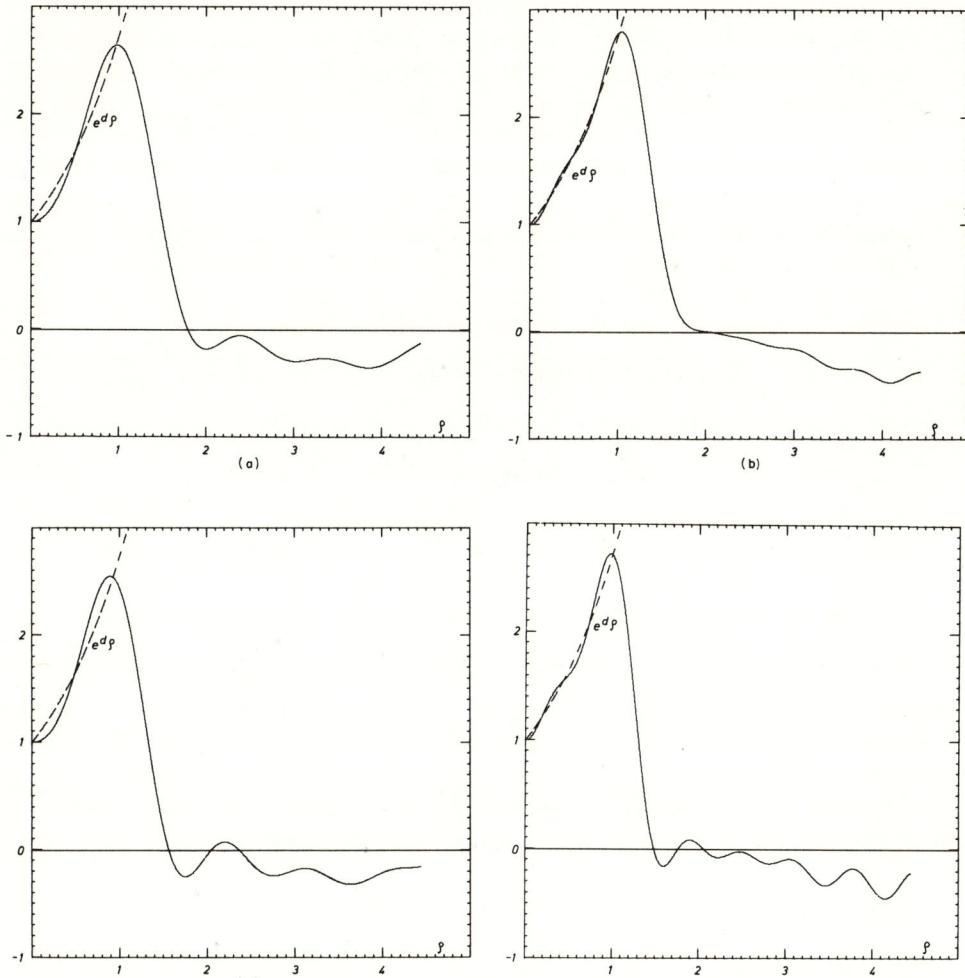

Fig. 8. Frequency response corresponding to (a) Test (1); (b) Test (2); (c) Test (3); (d) Test (4).

we found large coefficients (some of the order of 10^6) bearing alternating signs; the existence of observational errors in the measured field, constituting a realization of a random field, prevents recognition of any meaning to the output of these filters, because we compute weighted sums of circle averages multiplied by these coefficients. A test on an artificial field, consisting of the superposition of the gravitational anomalies of a series of spheres and a random field, confirmed this presumption.

Consequently it was necessary to limit the filter weights in amplitude. This was possible by the generality of the method of linear programming. Up to this point we tried to find the optimal solution of a system of linear inequalities of the form

$$(3.48) \qquad -\varepsilon \le \sum_{k=1}^{n} a_{jk} x_k - b_j \le \varepsilon, \qquad 1 \le j \le m$$

Several extensions of this system are possible.

If one requires that

$$(3.49) \qquad |x_k| \le \gamma \qquad \text{for} \qquad 1 \le k \le n$$

for a specific γ, then this requirement can be written as

$$(3.50) \qquad |x_k|/\gamma_1 \le \varepsilon \qquad \text{with} \qquad \gamma_1 = \gamma/\varepsilon > 0$$

adding the following system of linear inequalities to (3.48)

$$(3.51) \qquad -\varepsilon \le \sum_{k=1}^{n} a_{m+j, k} x_k - b_{m+j} \le \varepsilon, \qquad 1 \le j \le n$$

with

$$(3.52) \qquad a_{m+j, k} = (1/\gamma_1)\, \delta_{jk}, \qquad 1 \le j \le n,\, 1 \le k \le n$$

and

$$(3.53) \qquad b_{m+j} = 0, \qquad 1 \le j \le n$$

Further one can demand that the sum of the coefficients should differ slightly from unity, adding to the system (3.48) the condition

$$(3.54) \qquad \left| \sum_{k=1}^{n} x_k - 1 \right| \le \varepsilon/\beta$$

or

$$(3.55) \qquad -\varepsilon \le \sum_{k=1}^{n} a_{m+n+1, k} x_k - b_{m+n+1} \le \varepsilon$$

with

$$(3.56) \qquad \begin{aligned} a_{m+n+1, k} &= \beta, \qquad 1 \le k \le n \\ b_{m+n+1} &= \beta \end{aligned}$$

Finally it is possible to introduce approximation degrees over the Nyquist interval. Indeed, suppose that we divide the interval $J = \{1 \le j \le m\}$ into subintervals $J_i = \{m_i \le j \le m_{i+1}\}$, $1 \le i \le s$, $m_1 = 1$, and $m_{s+1} = m$. If we now want to obtain the accuracies ε/α_i in the intervals J_i then we can write the system

(3.57) $$-\varepsilon \leq \sum_{k=1}^{n} a'_{jk} x_k - b'_j \leq \varepsilon, \qquad 1 \leq j \leq m$$

if we define

(3.58) $$a'_{jk} = \alpha_i a_{jk}, \, b'_j = \alpha_i b_j, \qquad 1 \leq i \leq s, 1 \leq k \leq n$$
$$m_i \leq j \leq m_{i+1}$$

Among the several filters obtained with this method, are the weights given in Appendix 6, for the following choice of parameters:

Test (1): $z = 1$, $\rho_C = 1$, $N = 13$ filter weights, Weierstrass kernel with $\rho_k = 2$ and $\alpha = 6.61$, $\gamma_1 = 14$, $\beta = 4$.

Test (2): $z = 1$, $\rho_C = 1$, $N = 13$ filter weights, Weierstrass kernel with $\rho_k = 2$ and $\alpha = 6.61$, $\gamma_1 = 14$, $\beta = 4$, and approximation degrees

$$\tfrac{1}{2}\varepsilon \text{ in the interval } 0 \leq \rho_C \leq \rho_k$$
$$5\varepsilon \text{ in the interval } \rho_C < \rho \leq \rho_k$$
$$\varepsilon \text{ in the interval } \rho_k < \rho$$

The corresponding frequency responses are illustrated in Fig. 9.

The curve corresponding to Test (1) shows that the maximum deviation from the modified downward continuation frequency response is about 14.4% in the low frequency range of the Nyquist interval. The Test (2) yields a value for ε of about 10.7%, which must be interpreted as follows: in the low frequency range $0 \leq \rho \leq \rho_C$ we have an approximation of about $\tfrac{1}{2}\varepsilon = 5.3\%$

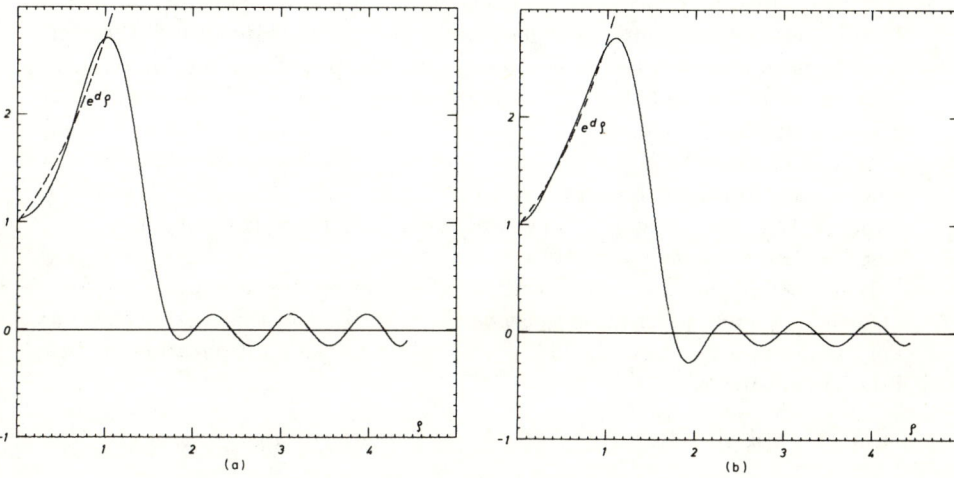

Fig. 9. Frequency response corresponding to (a) Test (1); (b) Test (2).

and in the high frequency range $\rho > \rho_k$ the oscillations of the frequency response have amplitudes of $\varepsilon = 10.7\%$. Since the introduction of the kernel $k_\alpha(\rho)$ enables us to avoid the sharp cutoff at the point ρ_C, it is immaterial what the accuracy of the approximation will be in the interval $\rho_C < \rho \leq \rho_k$.

As a final conclusion we can state that the method of linear programming yields coefficients, simulating sufficiently the modified downward continuation process, and constructs discrete filters which are defined as a function of the observed surface field. The only parameter to be estimated is the cutoff frequency ρ_C, that is the point where we want the frequency response to become essentially zero. This technique enables us to derive filters as functions of the Fourier spectrum of the surface effect, in opposition to the methods of Peters and Henderson. In contrast with the upward continuation, which defines a stable filter, it is now desirable to separate the input field into signal and noise components. For this reason we remark that the coefficients obtained are only indicative, but the method is general enough to compute readily numerical filters, once the cutoff frequency is estimated.

4. Frequency Filtering

4.1. Procedure for Filtering in the Frequency Plane

As a norm for the construction of a suitable weighting sequence h_k, describing the effect of a numerical filter by the discrete convolution (2.20), the approximation of the numerical frequency response of the filter h_k to the theoretical frequency characteristic has been defined. Section 2.2 discussed the fact that the representation of the input field in the frequency plane (u, v) is not at all evident, because the fields are almost never periodic, so that a Fourier series description is not necessarily meaningful. In spite of the non-periodic nature of most two-dimensional data, the practical situation requires that the observed fields be known only on a regular grid of finite extent, and it seems intuitively plausible to suppose the measured part of the input field to be a realization of a periodic process. Indeed, when one works with measured fields it is natural to think in terms of "signal" and "noise" components, implying that the possibility exists of separating these components on the basis of their frequency structure.

It seems that the most interesting point in treating the data under the form of their frequency description must be found in the fact that taking the convolution of two functions f and h is equivalent to the multiplication of their Fourier transforms

$$(4.1) \quad g(x, y) = \int_{-\infty}^{\infty} \int_{-\infty}^{\infty} f(\xi, \eta) h(x - \xi, y - \eta) \, d\xi \, d\eta \Leftrightarrow G(u, v)$$
$$= F(u, v) H(u, v)$$

As a consequence, no calculation of a discrete filter coefficient set has to be performed, as the filtering can be done entirely in the frequency plane by specifying the theoretical frequency response.

In this connection the following procedure for the calculation of the upward continuated field is plausible: (1) Calculate the Fourier transform $G(u, v, 0)$ of the measured part of the surface field $\Delta g(x, y, 0)$. (2) Multiply $G(u, v, 0)$ with the theoretical frequency response $H^{(-)}(u, v, z)$ over the Nyquist domain of the frequency plane. (3) Inverse Fourier transformation of the product $H^{(-)}(u, v, z)G(u, v, 0)$ yields the field at height z.

Filtering by multiplication in the frequency plane is often preferred to conventional convolution because no filter coefficients need to be calculated, very complicated filters are readily applied by specifying the frequency response, no data are lost at the edges of the data grid, and a fast Fourier transforming algorithm is available (Cooley and Tukey, 1965). In spite of the advantages of this procedure it is dangerous to rely upon it unconditionally. It is a fact that only a finite part of the surface field is known or can be treated with a computer; in addition, working with a Fourier series means that this part is considered to be extended periodically over the (x, y)-plane, but the true frequency structure of the field as a whole is generally much more complicated than this finite portion would indicate. This provides the key to the structural problems we deal with later on when we perform the filtering in the frequency domain.

In view of the fast divergence of the theoretical frequency response $H^{(+)}(u, v, z)$ it is obvious that filtering in the frequency plane is meaningless in the process of downward continuation, since the inverse Fourier transform of the product $H^{(+)}(u, v, z)G(u, v, 0)$ does not exist, in general. In addition, we remark that the signal component will be superposed by a random noise field, finding its origin in observational errors, digitalization and numerical treatment of the data, and in the composition of the upper part of the earth's crust itself (small, shallow masses). As the modern measuring instruments are almost stable, one may assume the instrumental noise to be small, implying that the larger part of the noise is caused by the medium and digitalization of the surface field.

For the problem of the separation of an observed field into signal and noise components, the origin of this noise is immaterial, though the knowledge of the statistical description of the noise is necessary. Therefore one assumes that the random noise is homogeneous (stationary) and Gaussian; this hypothesis is only acceptable when the space of the noise generating masses is large compared to the space of the signal producing structures. The presence of this noise leads to divergences in the mathematical expressions, governing the downward continuation process. In this respect we shall apply the Wiener filter technique for the optimalization of the downward continuation filter.

4.2. *Optimal Wiener Filtering*

We assume that the input field $f_i(x, y)$ consists of a deterministic signal $s(x, y)$, superposed by an additive noise component $n(x, y)$

(4.2) $$f_i(x, y) = s(x, y) + n(x, y)$$

The field $n(x, y)$ is thought to be a realization of a stationary and ergodic random process.

Our aim is to extract the signal $s(x, y)$ from the total message $f_i(x, y)$ by means of a linear filter, specified by the impulse response $h(x, y)$, with minimum error in accordance with a chosen norm of error. Because of the presence of this noise whose statistical properties generally are unknown, it is expected that complete recovery of the signal by the linear filter transformation will be impossible, unless the situation is an exceptional one. Nevertheless, we define a desired output $f_d(x, y)$, which may be any operation on the signal component of the input process.

The difference between the actual output, given by the convolution of the input field with the filter function $h(x, y)$

(4.3) $$f_o(x, y) = \int_{-\infty}^{\infty} \int_{-\infty}^{\infty} h(\xi, \eta) f_i(x - \xi, y - \eta) \, d\xi \, d\eta$$

and the desired output, is the system error

(4.4) $$e_h(x, y) = f_o(x, y) - f_d(x, y)$$

These fields are usually interpreted as being realizations of stochastic processes in two dimensions, obeying a probability law, so that the mathematical expectation operator E is defined for the processes in question. Wiener (1950) determines the optimal filter function $h_{\text{opt}}(x, y)$ by minimizing the mean-square system error, defined by

(4.5) $$E\{e_h^2\} = E\{[f_o(x, y) - f_d(x, y)]^2\}$$

In the case of stationary and ergodic random processes, this expected value is often interpreted as being defined by

(4.6) $$E\{g(x, y)\} = \lim_{T_x, T_y \to \infty} (1/4T_x T_y) \int_{-T_y}^{T_y} \int_{-T_x}^{T_x} g(x, y) \, dx \, dy$$

for any field $g(x, y)$.

Substitution of Eq. (4.3) into (4.5) yields

(4.7) $$E\{e_h^2\} = \int_{-\infty}^{\infty} \int_{-\infty}^{\infty} h(\xi, \eta) \, d\xi \, d\eta \int_{-\infty}^{\infty} \int_{-\infty}^{\infty} h(\lambda, \mu) \gamma_{f_i}(\xi - \lambda, \eta - \mu) \, d\lambda \, d\mu$$
$$- 2 \int_{-\infty}^{\infty} \int_{-\infty}^{\infty} h(\xi, \eta) \gamma_{f_i f_d}(\xi, \eta) \, d\xi \, d\eta + \gamma_{f_d}(0, 0),$$

where $\gamma_{f_i}(\xi, \eta)$ is the autocovariance of the input field $f_i(x, y)$

(4.8) $$\gamma_{f_i}(\xi, \eta) = E\{f_i(x, y)f_i(x+\xi, y+\eta)\}$$

and $\gamma_{f_i f_d}(\xi, \eta)$ is the covariance function of $f_i(x, y)$ and $f_d(x, y)$

(4.9) $$\gamma_{f_i f_d}(\xi, \eta) = E\{f_i(x, y)f_d(x+\xi, y+\eta)\}$$

Here it is assumed that all the fields considered have zero mean values; if this is not true for an observed field, we suppose that the arithmetical mean is substracted from the field values.

Note how naturally correlation functions arise in this type of problem. It must be said, however, that the validity of the expression (4.6) as the definition of the expected value of a field $g(x, y)$ rests on two assumptions: the first is the hypothesis of ergodicity, stating that the required statistical information about the stochastic process, of which $g(x, y)$ is a specific sample function, may be extracted from one typical realization of the random process. The second assumption is the hypothesis of stationarity of the field $g(x, y)$, implying that the statistical quantities are independent of the particular choice of the origin in the coordinate plane. As a result of these assumptions we conclude that all statistical properties of the processes in question may be evaluated by appropriate integrals over the (x, y)-plane, thus avoiding the necessity of defining them by ensemble averages over the realizations of the stochastic process.

The form (4.7) gives the mean-square system error for a particular filter, as specified by its weighting function $h(x, y)$. The quantity $E\{e_h^2\}$ is now minimized as a function of the filter function $h(x, y)$, in order to find that linear filter $h_{opt}(x, y)$, giving the least possible mean-square system error. The existence and uniqueness of this optimum filter are tacitly assumed, but will be usually guaranteed from the physical presentation of the problem.

It may be shown (Bendat, 1958) that a necessary and sufficient condition that $h_{opt}(x, y)$ be the filter function of the optimal filter, given by the minimalization of the mean-square system error (4.7), is that it is the solution of the Wiener-Hopf integral equation

(4.10) $$\gamma_{f_i f_d}(\xi, \eta) = \int_{-\infty}^{\infty} \int_{-\infty}^{\infty} h_{opt}(\lambda, \mu)\gamma_{f_i}(\xi-\lambda, \eta-\mu) \, d\lambda \, d\mu$$

By substituting (4.10) into (4.7) we find that the minimum mean-square system error resulting from this optimal filter function is given by

(4.11) $$E\{e_{h_{opt}}^2\} = \gamma_d(0, 0) - \int_{-\infty}^{\infty} \int_{-\infty}^{\infty} h_{opt}(\xi, \eta)\gamma_{f_i f_d}(\xi, \eta) \, d\xi \, d\eta$$

In the case of gravity interpretation let us define the filter input to be the surface field $\Delta g(x, y, 0)$ and then assume that it consists of a signal component $s(x, y)$ and an additive noise $n(x, y)$

(4.12) $$\Delta g(x, y, 0) = s(x, y) + n(x, y)$$

The desired field now becomes the signal component of the input field

(4.13) $$f_d(x, y) = s(x, y)$$

According to Eq. (4.10) the optimal impulse response $h_{opt}(x, y)$ of the linear filter that passes $s(x, y)$ and rejects $n(x, y)$ satisfies the Wiener-Hopf equation

(4.14) $$\gamma_{\Delta g_0, s}(\xi, \eta) = \int_{-\infty}^{\infty} \int_{-\infty}^{\infty} h_{opt}(x, y) \gamma_{\Delta g_0}(\xi - x, \eta - y)\, dx\, dy$$

where Δg_0 is a shorthand notation for the surface gravity effect, $\gamma_{\Delta g_0}(x, y)$ is the autocovariance of the surface field, and $\gamma_{\Delta g_0, s}(x, y)$ is the covariance between the input field and the signal component.

Applying the convolution theorem of Fourier transforms, the optimal frequency response $H_{opt}(u, v)$ follows directly from Eq. (4.14)

(4.15) $$H_{opt}(u, v) = P_{\Delta g_0, s}(u, v) / P_{\Delta g_0}(u, v)$$

where $P_{\Delta g_0}(u, v)$ is the power spectrum of the surface field, defined as the Fourier transform of the autocovariance $\gamma_{\Delta g_0}(x, y)$

(4.16) $$P_{\Delta g_0}(u, v) = \int_{-\infty}^{\infty} \int_{-\infty}^{\infty} \gamma_{\Delta g_0}(x, y) e^{-i(ux + vy)}\, dx\, dy$$

and $P_{\Delta g_0, s}(u, v)$ is the cross-power spectrum between $\Delta g(x, y, 0)$ and $s(x, y)$, that is the Fourier transform of $\gamma_{\Delta g_0, s}(x, y)$.

The minimum mean-square system error may be expressed in terms of power spectra. With the definition of the function

(4.17) $$\chi(x, y) = \gamma_s(x, y) - \int_{-\infty}^{\infty} \int_{-\infty}^{\infty} h_{opt}(\xi, \eta) \gamma_{\Delta g_0, s}(x - \xi, y - \eta)\, d\xi\, d\eta$$

it is easy to see that

(4.18) $$E\{e_{h_{opt}}^2\} = \chi(0, 0)$$

But the Fourier transform of $\chi(x, y)$ is given by the convolution theorem of Fourier transforms

$$P_s(u, v) - H_{opt}(u, v) P_{\Delta g_0, s}(u, v) = P_s(u, v) - \frac{P_{\Delta g_0, s}^2(u, v)}{P_{\Delta g_0}(u, v)}$$

Hence (inversion formula)

$$(4.19) \quad E\{e_{h_{opt}}^2\} = \frac{1}{4\pi^2} \int_{-\infty}^{\infty} \int_{-\infty}^{\infty} \frac{P_{\Delta g_0}(u, v) P_s(u, v) - P_{\Delta g_0, s}^2(u, v)}{P_{\Delta g_0}(u, v)} \, du \, dv$$

From (4.12) and (4.8) it follows immediately that

$$(4.20) \quad \gamma_{\Delta g_0}(x, y) = \gamma_s(x, y) + \gamma_{sn}(x, y) + \gamma_{ns}(x, y) + \gamma_n(x, y)$$

If it is assumed that the signal and noise are uncorrelated

$$\gamma_{sn}(x, y) = 0,$$

that is,

$$P_{sn}(u, v) = 0 \quad \text{for any } x, y, u, v,$$

we then have that

$$(4.21) \quad \gamma_{\Delta g_0}(x, y) = \gamma_s(u, v) + \gamma_n(u, v)$$

or in terms of frequency

$$(4.22) \quad P_{\Delta g_0}(u, v) = P_s(u, v) + P_n(u, v)$$

We also find that

$$(4.23) \quad \gamma_{\Delta g_0, s}(x, y) = \gamma_s(x, y) \quad \text{or} \quad P_{\Delta g_0, s}(u, v) = P_s(u, v)$$

Combining (4.22), (4.23), and (4.15) one finally obtains for the optimal frequency response

$$(4.24) \quad H_{opt}(u, v) = P_s(u, v)/[P_s(u, v) + P_n(u, v)]$$

and the minimum mean-square system error is given by (4.19)

$$(4.25)$$

$$E\{e_{h_{opt}}^2\} = (1/4\pi^2) \int_{-\infty}^{\infty} \int_{-\infty}^{\infty} P_s(u, v) P_n(u, v)/[P_s(u, v) + P_n(u, v)] \cdot du \, dv$$

Now we define the modified downward continuation process by calculating the downward projected field of the signal component of the surface field. Hence, the field at a desired depth will be the output of two linear filters in series: the first filter extracts the signal component optimally from the input field, and the second filter amplifies the signal part toward its source, according to the theoretical filter $H^{(+)}(u, v, z)$. This is just the procedure already mentioned in Eq. (3.35), where now the frequency response $W(u, v)$ is replaced by the optimum frequency characteristic $H_{opt}(u, v)$ of the Wiener filter.

Suppose that $f_1(x, y)$ is the input to a first filter with filter function $h_1(x, y)$; the output $f_1'(x, y)$ of this filter serves as input to a second filter with weighting function $h_2(x, y)$. Then it is easy to see that the filter function of the compound filter is the convolution of both weighting functions

$$(4.26) \qquad h(x, y) = \int_{-\infty}^{\infty} \int_{-\infty}^{\infty} h_1(\xi, \eta) h_2(x - \xi, y - \eta) \, d\xi \, d\eta$$

whence

$$(4.27) \qquad H(u, v) = H_1(u, v) H_2(u, v)$$

Indeed, according to Eq. (2.7), the output of the compound filter is

$$f_o(x, y) = \int_{-\infty}^{\infty} \int_{-\infty}^{\infty} h_2(\lambda, \mu) \, d\lambda \, d\mu$$
$$\times \int_{-\infty}^{\infty} \int_{-\infty}^{\infty} h_1(\xi, \eta) f_1(x - \xi - \lambda, y - \eta - \mu) \, d\xi \, d\eta$$

When we substitute ξ for $\xi + \lambda$ and η for $\eta + \mu$ we obtain

$$(4.28) \qquad f_o(x, y) = \int_{-\infty}^{\infty} \int_{-\infty}^{\infty} h(\xi, \eta) f_1(x - \xi, y - \eta) \, d\xi \, d\eta$$

if the filter function of the compound filter is defined by Eq. (4.26).

Combining (4.26), (4.24), and (3.23) one sees that the frequency response of the modified downward continuation filter is given by

$$(4.29) \qquad H_{\text{opt}}^{(+)}(u, v, z) = P_s(u, v) / [P_s(u, v) + P_n(u, v)] \cdot \exp[z(u^2 + v^2)^{1/2}]$$

and we conclude that the optimal frequency response of the modified downward continuation is the product of the theoretical frequency response $H^{(+)}(u, v, z)$ of the downward continuation filter with the frequency characteristic of the optimum Wiener filter to pass the signal $s(x, y)$ and reject the noise $n(x, y)$.

In conclusion, the following procedure is arrived at, based on the optimal Wiener filtering technique: (1) Calculate the Fourier transform $G(u, v, 0)$ of the surface field $\Delta g(x, y, 0)$. (2) Compute the power spectrum $P_{\Delta g_0}(u, v)$ of the surface field. (3) Separate $P_{\Delta g_0}(u, v)$ into signal and noise components $P_s(u, v)$ and $P_n(u, v)$. (4) Multiply $G(u, v, 0)$ by $H_{\text{opt}}^{(+)}(u, v, z)$ over the Nyquist domain. (5) Inverse Fourier transformation of the product $H_{\text{opt}}^{(+)}(u, v, z) G(u, v, 0)$ yields the downward continuated field at depth z.

When stating Eq. (4.29) it was implicitly assumed that the output of the filter (4.24) defines a band-limited process, that is, that the Fourier transform of the residue left in the surface field after Wiener filtering vanishes outside a finite region of the frequency plane.

In order to apply the Wiener filtering method, several restrictions need to be considered. The optimal downward continuation filter is constructed as a function of the observed surface field and this is in conformity with our conclusion at the end of Section 3.5. But the fundamental problem in applying Eq. (4.29) is to obtain the necessary statistical description of signal and noise, that is, the separation of the estimated power spectrum of the surface field into spectra corresponding to the signal and noise components. This leads to the adoption of a statistical model for the random noise. In the absence of any statistical information, the easiest (and perhaps the best physically motivated) assumption is that of white noise, and then estimate the white noise level from the power spectrum of the surface field. However, such a separation will remain subjective, as it is *a priori* impossible to estimate the noise term correctly without knowing explicitly all of its statistical properties. This can be immediately related to the ambiguity of the inverse potential problem, excluding an unambiguous separation of the input field.

Another weak point in the Wiener filtering method is the hypothesis that the noise and the signal are uncorrelated. This is a strong requirement. For gravity fields, derivable from a potential, this assumption could be made acceptable by defining the noise as the part of the surface field that finds its cause in random observational errors, digitalization, and the distribution of small, shallow disturbing masses.

The Wiener filter theory is justified in the case where the fields in question are realizations of Gaussian stochastic processes. Only in this case will a linear filter be able to minimize the mean-square system error, as defined by Eq. (4.5) (Clarke, 1969; Van Trees, 1968). However, in the absence of any statistical information the Wiener filter theory offers the best method for the available information.

We also have assumed that all statistical quantities, such as covariance functions, may be calculated by averaging a single sample function of the stochastic process over the (x, y)-plane. This corresponds to the ergodicity hypothesis of stationary random processes. We think that more weight ought to be given to the question whether or not as measured field belongs to a stationary and ergodic random process, although admittedly this problem is an extremely difficult one.

4.3. Strakhov's Method for Extraction of Potential Field Signal

Let $\Delta g(x, y, 0)$ be continuously defined over the horizontal plane $z = 0$ and let it be expressible in the form

(4.30) $$\Delta g(x, y, 0) = s(x, y) + n(x, y)$$

where $s(x, y)$ is some deterministic (nonrandom) field, called the signal, and $n(x, y)$ a two-dimensional random function. We shall make the following assumptions regarding these components:

(1) $s(x, y)$ is absolutely and square integrable

$$(4.31) \qquad \|s\|_{L_1} = \int_{-\infty}^{\infty} \int_{-\infty}^{\infty} |s(x, y)| \, dx \, dy < \infty$$

and

$$(4.32) \qquad \|s\|_{L_2} = \left[\int_{-\infty}^{\infty} \int_{-\infty}^{\infty} s^2(x, y) \, dx \, dy \right]^{1/2} < \infty$$

(2) $n(x, y)$ is a realization of a stationary random process with zero mean

$$(4.33) \qquad E\{n\} = 0$$

and autocovariance

$$(4.34) \qquad \gamma_n(\xi, \eta) = E\{n(x, y) n(x + \xi, y + \eta)\}$$

The physical origin of the noise is immaterial; unfortunately, the statistical characteristics of the noise are not known beforehand and this is a serious disadvantage of Wiener's method. This difficulty is partially overcome by Strakhov's method (1964, a, b), reviewed by Naidu (1966, 1967).

Suppose that the observed values in (4.30) are smoothed by a linear filter with weights h_{kl}, $-M \leq k \leq M$, $-N \leq l \leq N$; then one can write for the output of this filter

$$(4.35) \qquad \overline{\Delta g}(x, y, 0) = \sum_{k=-M}^{M} \sum_{l=-N}^{N} h_{kl} \Delta g(x - k \Delta x, y - l \Delta y, 0)$$
$$= \bar{s}(x, y) + \bar{n}(x, y),$$

with

$$(4.36) \qquad \bar{s}(x, y) = \sum_{k=-M}^{M} \sum_{l=-N}^{N} h_{kl} s(x - k \Delta x, y - l \Delta y)$$

and

$$(4.37) \qquad \bar{n}(x, y) = \sum_{k=-M}^{M} \sum_{l=-N}^{N} h_{kl} n(x - k \Delta x, y - l \Delta y)$$

the smoothed versions of the signal and the noise; Δx and Δy stand for the grid spacings in the x- and y-directions, respectively.

The mean of the filtered noise is zero, since

$$(4.38) \qquad E\{\bar{n}\} = E\{n\} \sum_{k=-M}^{M} \sum_{l=-N}^{N} h_{kl} = 0$$

in view of the assumption (4.33), and the autocovariance is given by

$$(4.39) \quad \gamma_{\tilde{n}}(\xi, \eta) = \sum_{k=-M}^{M} \sum_{l=-N}^{N} \sum_{k'=-M}^{M} \sum_{l'=-N}^{N} \gamma_n[\xi + (k-k')\Delta x, \eta + (l-l')\Delta y]$$

We attempt to determine the filter coefficients h_{kl} such that (1) the distorted signal $\tilde{s}(x, y)$ is as close to $s(x, y)$ as possible (in some metric), and (2) the variance $\sigma_{\tilde{n}}$ of the random process, of which $\tilde{n}(x, y)$ is a realization, is as small as possible.

The distortion of the signal caused by the linear operator h_{kl} may be estimated by the L_2-metric. Using Parseval's theorem for double Fourier transforms, we obtain

$$(4.40) \quad \|s - \tilde{s}\|_{L_2}^2 = \|S - \tilde{S}\|_{L_2}^2 = \int_{-\infty}^{\infty} \int_{-\infty}^{\infty} [S(u, v) - \tilde{S}(u, v)]^2 \, du \, dv$$

where $S(u, v)$ and $\tilde{S}(u, v)$ are the Fourier transforms of $s(x, y)$ and $\tilde{s}(x, y)$, respectively. From (4.36) it immediately follows that

$$(4.41) \quad \tilde{S}(u, v) = H(u, v) S(u, v)$$

where $H(u, v)$ is the frequency response of the discrete filter h_{kl},

$$(4.42) \quad H(u, v) = \sum_{k=-M}^{M} \sum_{l=-N}^{N} h_{kl} e^{-i(ku\Delta x + lv\Delta y)}$$

Combining (4.41) and (4.40) yields

$$(4.43) \quad \|s - \tilde{s}\|_{L_2}^2 = \int_{-\infty}^{\infty} \int_{-\infty}^{\infty} S^2(u, v)[1 - H(u, v)]^2 \, du \, dv$$

Polya (1929) shows that every potential field signal has a unique smallest convex region containing all singularities of the potential field. If we let d stand for the depth of the uppermost singularity we have the following inequality

$$(4.44) \quad |G(u, v, 0)| \leq (1/2\pi) \|\Delta g(x, y, d - \varepsilon)\|_{L_1} \exp[-(d - \varepsilon)(u^2 + v^2)^{1/2}]$$

where ε is a small, positive number. We also note that $\|\Delta g(x, y, d - \varepsilon)\|_{L_1}$ is a constant and may be interpreted as the total excess mass within the anomalous region. Introducing Polya's inequality into (4.43) results in

$$(4.45) \quad \|s - \tilde{s}\|_{L_2}^2 \leq [\|\Delta g(x, y, d)\|_{L_1}/2\pi]$$

$$\int_{-\infty}^{\infty} \int_{-\infty}^{\infty} \exp[-2d(u^2 + v^2)^{1/2}][1 - H(u, v)]^2 \, du \, dv$$

The following condition minimizes the signal distortion on the L_2-metric

$$(4.46) \quad \int_{-\infty}^{\infty} \int_{-\infty}^{\infty} \exp[-2d(u^2 + v^2)^{1/2}][1 - H(u, v)]^2 \, du \, dv = \text{minimum}$$

The simple minimization of the form (4.46) is not sufficient, as the filter coefficients could then be easily determined from the condition that $H(u, v) \equiv 1$ over the frequency plane. Indeed, the condition that the variance of the reduced noise be small must still be satisfied. Equation (4.39) readily gives for the variance of the reduced noise

$$\sigma_{\bar{n}} \equiv \gamma_{\bar{n}}(0, 0) = \sum_{k=-M}^{M} \sum_{l=-N}^{N} h_{kl}^2 \gamma_n(0, 0)$$

$$+ 2 \sum_{k=1}^{2M} \sum_{l=1}^{2N} \sum_{k'=-M}^{M} \sum_{l'=-N}^{N} h_{k'l'} h_{k+k', l+l'} \gamma_n(k\Delta x, l\Delta y)$$

Dividing both sides of this expression by the variance $\gamma_n(0, 0)$ of the noise $n(x, y)$ it follows that

(4.47) $\quad \sigma_{\bar{n}}/\sigma_n = \sum_{k=-M}^{M} \sum_{l=-N}^{N} h_{kl}^2$

$$+ \sum_{k=1}^{2M} \sum_{l=1}^{2N} \sum_{k'=-M}^{M} \sum_{l'=-N}^{N} h_{k'l'} h_{k+k', l+l'} \gamma_n(k\Delta x, l\Delta y)/\gamma_n(0, 0)$$

If the noise is approximately random we obtain the simple form

(4.48) $\quad\quad\quad \sigma_{\bar{n}}/\sigma_n \simeq \sum_{k=-M}^{M} \sum_{l=-N}^{N} h_{kl}^2 = h^2$

and this has been called the "prescribed noise reduction factor" by Naidu (1967).

In practice it is impossible to estimate the autocovariance of the noise with a sufficient accuracy; therefore, all the terms in Eq. (4.47), except the first, are indefinite. Strakhov's method consists of minimizing the form (4.46), with the condition that the noise should be reduced by a prescribed factor h^2. The remaining terms in Eq. (4.47) may then be interpreted as correction factors to be applied to this prescribed noise reduction factor for a given statistical structure of the noise $n(x, y)$. The filter coefficients h_{kl} can be found from the conditions (4.46) and (4.48) by a modification of the method of steepest descent; a computer program in FORTRAN II language can be found in the paper by Naidu (1967). The single parameter to be estimated from the surface field is the depth d of the uppermost singularity of the potential field.

It is interesting that Strakhov's method does not explicitly require knowledge of the statistical properties of the signal, neither of the noise. However, Naidu's introduction of a prescribed noise reduction factor is based upon a white noise assumption, and this is very similar to what has been said about Wiener's method. In Strakhov's method the signal distortion and variance of the noise are minimized separately, while in Wiener's method the sum of signal distortion and random noise is minimized. Wiener's method has the advantage that the filtering is performed entirely in the frequency plane; the method of Strakhov first supplies the filter coefficients, which yield a smooth output of the surface field by discrete convolution, and then this filtered signal may be transformed to the frequency plane. It is difficult to say which method is superior; the choice is left to the interpreter.

4.4. Digitization of a Continuous Field—Aliasing

The analogy between Section 3.5 and Section 4.2 is quite remarkable. When the downward continuation is performed by the conventional convolution procedure we have to simulate the downward continuation frequency response by a linear, discrete filter, but the calculations are carried out in the (x, y)-plane. The main problem then consists of obtaining a sufficient approximation to the modified downward continuation filter. If, however, the filtering is performed in the frequency plane we can use the theoretical downward continuation filter $H^{(+)}(u, v, z)$, but the resulting problems are related to the calculation of the Fourier transform of the digitized surface field of finite extent. Therefore we will now concentrate on the relation between the Fourier transform of the continuous field as a whole and the Fourier spectrum of the observed part of the field.

We suppose that the surface field $f(x, y)$, distributed continuously over the (x, y)-plane, is sampled on a rectangular grid, with spacings Δx and Δy in the x- and y-directions, respectively. This yields the digitized field

$$(4.49) \qquad f(k\,\Delta x, l\,\Delta y) = [d_0(x, y)\,\delta(x, y)f(x, y)]_{\substack{x = k\,\Delta x \\ y = l\Delta y}}$$

$1 \leq k \leq m$, $1 \leq l \leq n$, where $\delta(x, y)$ is the infinite Dirac comb in two dimensions (Blackman and Tukey, 1958, p. 71)

$$(4.50) \qquad \delta(x, y) = \Delta x\,\Delta y \sum_{q=-\infty}^{\infty} \sum_{r=-\infty}^{\infty} \delta(x - q\,\Delta x)\,\delta(y - r\,\Delta y)$$

and

$$(4.51) \qquad \begin{aligned} d_0(x, y) &= 1, \qquad \Delta x \leq x \leq m\,\Delta x,\ \Delta y \leq y \leq n\,\Delta y \\ &= 0, \qquad \text{otherwise} \end{aligned}$$

The aliased, finite version of the Fourier spectrum, corresponding to the observations $f(k\,\Delta x, l\,\Delta y)$, is now defined by

$$F_a(u, v) = \sum_{k=-\infty}^{\infty} \sum_{l=-\infty}^{\infty} f(k\,\Delta x, l\,\Delta y) \exp[-i(u\,k\Delta x + vl\,\Delta y)]\,\Delta x\,\Delta y$$

which can be written as

$$F_a(u, v) = \int_{-\infty}^{\infty} \int_{-\infty}^{\infty} d_0(x, y)\delta(x, y)f(x, y)\exp[-i(ux + vy)]\, dx\, dy$$

Using the convolution theorem for Fourier transforms, this expression becomes

(4.52) $\qquad F_a(u, v) = D_0(u, v) * \nabla(u, v) * F(u, v)$

where $*$ denotes convolution and $D_0(u, v)$, $\nabla(u, v)$, and $F(u, v)$ are the Fourier transforms of $d_0(x, y)$, $\delta(x, y)$, and $f(x, y)$, respectively. Since the Fourier transform of the infinite Dirac comb $\delta(x, y)$ is an infinite Dirac comb in frequency

(4.53) $\qquad \nabla(u, v) = \sum_{q=-\infty}^{\infty} \sum_{r=-\infty}^{\infty} \delta(u - (2\pi q/\Delta x))\, \delta(v - (2\pi r/\Delta y))$

the form (4.52) becomes

(4.54) $\quad F_a(u, v) = \int_{-\infty}^{\infty} \int_{-\infty}^{\infty} D_0(u - u', v - v')$
$$\times \left[\sum_{q=-\infty}^{\infty} \sum_{r=-\infty}^{\infty} F(u' - (2\pi q/\Delta x), v' - (2\pi r/\Delta y)) \right] du'\, dv'$$

with

(4.55) $\qquad D_0(u, v) = 4mn\, \Delta x\, \Delta y (\sin nu\, \Delta x / nu\, \Delta x)(\sin mv\, \Delta y / mv\, \Delta y)$

The expression (4.54) must be interpreted as follows: suppose that the sampled version of $f(x, y)$ is available all over the (x, y)-plane. Then Eq. (4.49) becomes

$$f(k\, \Delta x, l\, \Delta y) = [\delta(x, y)f(x, y)]_{\substack{x = k\, \Delta x \\ y = l\, \Delta y}}, \qquad -\infty \leq k \leq \infty,\ -\infty \leq l \leq \infty$$

and (4.54) reduces to

(4.56) $\qquad \tilde{F}_a(u, v) = \sum_{q=-\infty}^{\infty} \sum_{r=-\infty}^{\infty} F(u - (2\pi q/\Delta x), v - (2\pi r/\Delta y))$

If the sampling rates Δx and Δy are so chosen that $f(x, y)$ contains no waves of frequencies outside the Nyquist domain

(4.57) $\qquad -\pi/\Delta x \leq u \leq \pi/\Delta x, \qquad -\pi/\Delta y \leq v \leq \pi/\Delta y$

then it is completely determined by giving its ordinates at the grid points. As no information about $f(x, y)$ is available between the grid points, there is no means of directly estimating the amplitudes of frequencies outside the Nyquist domain. In other words, Eq. (4.56) shows that the sampled Fourier spectrum $\tilde{F}_a(u, v)$ is obtained by folding the true Fourier spectrum $F(u, v)$

into the Nyquist domain and adding these contributions inside the low frequency region. This spectrum distortion has been called "aliasing" by J. W. Tukey. If $f(x, y)$ has the property that its Fourier transform $F(u, v)$ is essentially zero outside the Nyquist domain, then the terms for $q \neq 0$ and $r \neq 0$ in (4.56) do not contribute to the sampled Fourier spectrum and the digitization process loses no information. Since the Nyquist frequencies $u_N = \pi/\Delta x$ and $v_N = \pi/\Delta y$ are controlled by the grid spacings, we conclude that the practical application of the concept of the sampled Fourier spectrum consists of choosing grid spacings that are small enough to reduce the aliasing to an acceptable level.

Furthermore we note from Eq. (4.54) that if $f(x, y)$ is sampled on a rectangular grid of finite dimensions, then the aliased Fourier spectrum $F_a(u, v)$ will be the convolution of the sampled Fourier spectrum $\tilde{F}_a(u, v)$ with the Fourier transform $D_0(u, v)$ of the function $d_0(x, y)$, thus introducing a second kind of frequency distortion. Even if aliasing were absent, the true Fourier spectrum $F(u, v)$ would be blurred out by the frequency kernel $D_0(u, v)$.

The power spectrum may be treated in a completely analogous way. Indeed, the best one can obtain from the sampled field is the knowledge of the "aliased" autocovariance

$$(4.58) \qquad \gamma_a(k\,\Delta x, l\,\Delta y) = [d_0(x, y)\,\delta(x, y)\gamma(x, y)]_{\substack{x=k\,\Delta x \\ y=l\,\Delta y}}$$

where $\gamma(x, y)$ is the true autocovariance of $f(x, y)$. It immediately follows that the sampled, finite version of the power spectrum $P(u, v)$ is given by

$$(4.59) \quad P_a(u, v) = \int_{-\infty}^{\infty} \int_{-\infty}^{\infty} D_0(u - u', v - v')$$

$$\times \left[\sum_{q=-\infty}^{\infty} \sum_{r=-\infty}^{\infty} P(u' - (2\pi q/\Delta x), v' - (2\pi r/\Delta y)) \right] du'\, dv'$$

and the above results hold for $P_a(u, v)$ and $\tilde{P}_a(u, v)$.

Another relation between $P_a(u, v)$ and $P(u, v)$ may be derived if we note that Eq. (4.58) can be written as

$$(4.60) \qquad \gamma_a(k\,\Delta x, l\,\Delta y) = [\delta_{mn}(x, y)\gamma(x, y)]_{\substack{x=k\,\Delta x \\ y=l\,\Delta y}}$$

where $\delta_{mn}(x, y)$ is the finite Dirac comb in two dimensions

$$(4.61) \qquad \delta_{mn}(x, y) = \delta_m(x)\,\delta_n(y)$$

and

$$(4.62) \quad \delta_m(x) = \tfrac{1}{2}\Delta x[\delta(x + m\,\Delta x) + \delta(x - m\,\Delta x)] + \Delta x \sum_{q=-m+1}^{m-1} \delta(x - q\,\Delta x)$$

In complete analogy with Eq. (4.59) the aliased, finite power spectrum is given by the convolution

$$(4.63) \qquad P_a(u, v) = \int_{-\infty}^{\infty} \int_{-\infty}^{\infty} V_{mn}(u - u', v - v') P(u', v') \, du' \, dv'$$

Substituting (4.62) into the expression

$$V_{mn}(u, v) = \int_{-\infty}^{\infty} \int_{-\infty}^{\infty} \delta_m(x) \, \delta_n(y) e^{-i(ux+vy)} \, dx \, dy$$

and using the fact that $\delta(x - x_0)$ and e^{-iux_0} are Fourier transform pairs, together with the trigonometric relation

$$(4.64) \qquad \sum_{k=0}^{m} \cos ku = \cos \tfrac{1}{2}(m+1)u \sin \tfrac{1}{2}mu / \sin \tfrac{1}{2}u$$

we find that

$$(4.65) \qquad V_{mn}(u, v) = \Delta x \, \Delta y \, \cotg \tfrac{1}{2} u \, \Delta x \, \cotg \tfrac{1}{2} v \, \Delta y \, \sin nu \, \Delta x \, \sin mv \, \Delta y$$

Hence, the aliased power spectrum may also be regarded as the convolution of the true power spectrum with the aliased kernel $V_{mn}(u, v)$.

4.5. Estimation of the Power Spectrum

When writing Eq. (4.58) it was explicitly mentioned that the best one could obtain from the sampled field $f(k \Delta x, l \Delta y)$, $1 \leq k \leq m$, $1 \leq l \leq n$, are the values $\gamma(k \Delta x, l \Delta y)$, $-m+1 \leq k \leq m-1$, $-n+1 \leq l \leq n-1$ of the true autocovariance $\gamma(x, y)$ of the surface field as a whole. This surely is an optimistic statement. Indeed, the indirect method of power spectrum estimation starts with the computation of the following estimator of the autocovariance

$$(4.66) \qquad \bar{\gamma}_0(k, l) = \frac{1}{m-k} \frac{1}{n-l} \sum_{p=1}^{m-k} \sum_{q=1}^{n-l} f(p, q) f(p+k, q+l), \qquad \begin{array}{l} 0 \leq k \leq m-1, \\ 0 \leq l \leq n-1 \end{array}$$

where it is assumed that the mean of the surface field is zero. Parzen (1957a) advocates the use of the estimator

$$(4.67) \qquad \hat{\gamma}_0(k, l) = \frac{1}{mn} \sum_{p=1}^{m-k} \sum_{q=1}^{n-l} f(p, q) f(p+k, q+l), \qquad \begin{array}{l} 0 \leq k \leq m-1, \\ 0 \leq l \leq n-1 \end{array}$$

It is well known that $\bar{\gamma}_0(k, l)$ is an unbiased estimate of the autocovariance of $f(x, y)$, with the expected value given by

(4.68)
$$E\{\bar{\gamma}_0(k, l)\} = \frac{1}{m-k}\frac{1}{n-l}\sum_{p=1}^{m-k}\sum_{q=1}^{n-l} E\{f(p, q)f(p+k, q+l)\} = \gamma(k, l)$$

for any finite m and n, but $\hat{\gamma}_0(k, l)$ is biased for finite m and n and asymptotically unbiased, since

(4.69)
$$E\{\hat{\gamma}_0(k, l)\} = \lambda_m(k)\lambda_n(l)\gamma(k, l)$$

with

(4.70)
$$\lambda_m(k) = 1 - (|k|/m), \quad -m+1 \leq k \leq m-1$$
$$= 0, \quad |k| > m-1$$

Bias, however, is not the only, and certainly not the most important criterion for a "good" estimate and a preference for the biased estimate $\hat{\gamma}_0(k, l)$ is sometimes voiced on the grounds that it has, in general, smaller mean-square error than the unbiased estimate $\bar{\gamma}_0(k, l)$ (Parzen, 1957b).

These estimates are computable for values of k and l up to $|k| = m - 1$ and $|l| = n - 1$, but $\bar{\gamma}_0(m-1, n-1)$ is then estimated by the single product $f(1, 1)f(m, n)$. Hence we cannot estimate $\gamma(k, l)$ beyond $|k| = m - 1$, $|l| = n - 1$. It was empirically observed by M. G. Kendall that the sample autocovariances fail to damp down to zero for increasing values of k and l (m and n), although the true autocovariance does damp down to zero for $x \to \infty$ and $y \to \infty$. To avoid the discontinuity of the estimates at the end points, one introduces an even function $d(k, l)$, called a covariance kernel (or covariance window), and subject to the restrictions that $d(0, 0) = 1$ and $d(k, l) = 0$ for $|k| > m' - 1$ and $|l| > n' - 1$, where $m' \leq m$ and $n' \leq n$ are integers. The modified autocovariance estimates are then defined by

(4.71)
$$\bar{\gamma}_e(k, l) = d(k, l)\bar{\gamma}_0(k, l), \quad \hat{\gamma}_e(k, l) = d(k, l)\hat{\gamma}_0(k, l)$$
$$-m'+1 \leq k \leq m'-1, \; -n'+1 \leq l \leq n'-1$$

and it is interesting to see that $\bar{\gamma}_e(k, l)$ and $\hat{\gamma}_e(k, l)$ are defined for all k and l and vanish for $|k| > m' - 1$ and $|l| > n' - 1$, although the estimates $\bar{\gamma}_0(k, l)$ and $\hat{\gamma}_0(k, l)$ are not defined there.

In consequence, the modified power spectrum is defined by

(4.72)
$$\bar{P}_e(u, v) = \sum_{k=-m'+1}^{m'-1} \sum_{l=-n'+1}^{n'-1} \bar{\gamma}_e(k, l)e^{-i(ku+lv)}$$

or

(4.73)
$$\hat{P}_e(u, v) = \sum_{k=-m'+1}^{m'-1} \sum_{l=-n'+1}^{n'-1} \hat{\gamma}_e(k, l)e^{-i(ku+lv)}.$$

It may be shown that these estimates of the power spectrum are asymptotically unbiased since, for example, the expected value of $\bar{P}_e(u, v)$ is given by

$$(4.74) \qquad E\{\bar{P}_e(u, v)\} = D(u, v) * D_0(u, v) * \nabla(u, v) * P(u, v)$$

where $D(u, v)$ is the Fourier transform of $d(k, l)$, and $D_0(u, v)$ and $\nabla(u, v)$ are given by Eqs. (4.55) and (4.53), respectively.

Because the variance of $\hat{P}_e(u, v)$ is approximately given by

$$(4.75) \qquad \text{var}\{\hat{P}_e(u, v)\} \simeq (m'/m)(n'/n) P^2(u, v) \int_{-1}^{1} \int_{-1}^{1} h^2(x, y) \, dx \, dy$$

for large m and n, where the function h is defined by

$$d(x, y) = h(x/m', y/n'),$$

it may be expected that $\hat{P}_e(u, v)$ is a consistent spectral estimate if we choose the parameters m' and n' in such a way that $m'/m \to 0$ and $n'/n \to 0$ as $m \to \infty$ and $n \to \infty$. Otherwise one obtains a reduction in the variance by a factor that depends upon the ratios m'/m and n'/n and the covariance kernel used.

The starting point in the direct method of power spectrum estimation is the Schuster periodogram, which can be defined in the form

$$(4.76) \qquad P_{mn}(u, v) = (1/mn) \left| \sum_{k=1}^{m} \sum_{l=1}^{n} f(k, l) e^{-i(ku + lv)} \right|^2$$

Elementary algebraic manipulation immediately yields that

$$(4.77) \qquad P_{mn}(u, v) = \sum_{k=-m+1}^{m-1} \sum_{l=-n+1}^{n-1} \lambda_m(k) \lambda_n(l) \bar{\gamma}_0(k, l) e^{-i(ku + lv)}$$

from which the expected value of the periodogram estimates follows

$$E\{P_{mn}(u, v)\} = \sum_{k=-m+1}^{m-1} \sum_{l=-n+1}^{n-1} \lambda_m(k) \lambda_n(l) \gamma(k, l) e^{-i(ku + lv)}$$

or

$$(4.78) \qquad E\{P_{mn}(u, v)\} = \int_{-\pi}^{\pi} \int_{-\pi}^{\pi} \Lambda_{mn}(u - u', v - v') P(u', v') \, du' \, dv'$$

with $\Lambda_{mn}(u, v)$ the two-dimensional Féjer kernel, associated with the periodogram estimate of $P(u, v)$

$$(4.79) \qquad \Lambda_{mn}(u, v) = (1/mn)(\sin \tfrac{1}{2} mu / \sin \tfrac{1}{2} u)^2 (\sin \tfrac{1}{2} nv / \sin \tfrac{1}{2} v)^2$$

The expected value of $P_{mn}(u, v)$ is not $P(u, v)$, but a weighted integral of the power spectrum and, therefore one has to conclude that the periodogram is a biased estimate of the density spectrum of the continuous field for finite

m and n. But in the limit as $m \to \infty$ and $n \to \infty$, the Féjer kernel $\Lambda_{mn}(u, v)$ approaches the two-dimensional Dirac distribution and the periodogram becomes an asymptotically unbiased estimate of $P(u, v)$. However, it is not a consistent estimate since the variance of the periodogram estimates is given approximately by

$$(4.80) \quad \lim_{\substack{m \to \infty \\ n \to \infty}} \text{var}\{P_{mn}(u, v)\} \simeq P^2(u, v), \qquad u, v \neq 0 \text{ or } \pm \pi$$

$$\simeq 2P^2(u, v), \qquad u, v = 0 \text{ or } \pm \pi$$

Consequently, if we want to use this approximation for the power spectrum, we are forced to look for methods to reduce the variance of this periodogram. This lack of consistency is not due to any fundamental inefficiency of $P_{mn}(u, v)$, since the periodogram can, in the case of a normal stochastic process, be regarded as the maximum likely estimate of the power spectrum and should hardly be capable of improvement from the purely statistical point of view.

There are several methods of reducing the variance of the periodogram. One of them consists of smoothing the periodogram values $P_{mn}(r, s)$, $1 \leq r \leq m$, $1 \leq s \leq n$ by averaging over adjacent values; hence we define the smoothed periodogram by

$$(4.81) \quad \bar{P}_{mn}(r, s) = \sum_{\lambda = -p}^{p} \sum_{\mu = -q}^{q} W(\lambda, \mu) P_{mn}(r - \lambda, s - \mu)$$

where the coefficients $W(\lambda, \mu)$ are the weights of a suitable filter. It may be shown that the expected value of the smoothed periodogram will be given by

$$(4.82) \quad \lim_{\substack{m \to \infty \\ n \to \infty}} E\{\bar{P}_{mn}(r, s)\} \simeq P(2\pi r/m, 2\pi s/n) \sum_{\lambda = -p}^{p} \sum_{\mu = -q}^{q} W(\lambda, \mu)$$

This relation produces the criterion by which the filter coefficients should add up to unity, in order to obtain an asymptotically unbiased estimate of the density spectrum.

The variance of the smoothed periodogram is given to a good approximation by

$$(4.83) \quad \lim_{\substack{m \to \infty \\ n \to \infty}} \text{var}\{\bar{P}_{mn}(r, s)\} \simeq P^2(2\pi r/m, 2\pi s/n) \sum_{\lambda = -p}^{p} \sum_{\mu = -q}^{q} W^2(\lambda, \mu)$$

and one sees that the variance of the unsmoothed periodogram will be decreased by a factor

$$(4.84) \quad R = \sum_{\lambda = -p}^{p} \sum_{\mu = -q}^{q} W^2(\lambda, \mu)$$

The variance (4.83) is minimized, for given p and q, by the sequence

$$W(\lambda, \mu) = 1/(2p+1)(2q+1), \qquad -p \leq \lambda \leq p, -q \leq \mu \leq q$$

giving a reduction factor

$$(4.85) \quad R = \sum_{\lambda=-p}^{p} \sum_{\mu=-q}^{q} 1/(2p+1)^2(2q+1)^2 = 1/(2p+1)(2q+1)$$

Hence, we conclude that, for averaging over nine adjacent ordinates ($p=1$, $q=1$), we have that $R = 1/9$.

Another method consists of dividing the surface grid into K nonoverlapping blocks of equal size. For each block k the periodogram $P^{(k)}_{\Delta m, \Delta n}(u, v)$ is computed and averaged over the K blocks, leading to the spectral estimate

$$(4.86) \quad P^{(K)}_{\Delta m, \Delta n}(u, v) = (1/K) \sum_{k=1}^{K} P^{(k)}_{\Delta m, \Delta n}(u, v)$$

where $m = M \Delta m$, $n = N \Delta n$, $K = MN$, and Δm and Δn are the dimensions of each block. From (4.86) it immediately follows that the expected value of this spectral estimate is approximated by

$$(4.87) \quad E\{P^{(K)}_{\Delta m, \Delta n}(u, v)\} \simeq P(u, v)$$

for large m and n and that the estimate is asymptotically unbiased.

As the variance is given by

$$(4.88) \quad \text{var}\{P^{(K)}_{\Delta m, \Delta n}(u, v)\} \simeq (1/K)P^2(u, v), \quad u, v \neq 0 \quad \text{or} \quad \pm \pi$$

the variance of this spectral estimate is reduced by approximately a factor $1/K$, if the blocks do not overlap.

Still another method consists of multiplying the observations $f(k, l)$ by a data kernel, which is a function that is equal to unity for the larger part of the surface grid, but vanishes outside this region. For further computational details the reader is referred to the list of papers in the references.

4.6. Convolution Filtering versus Frequency Filtering

From the above sections it is clear that the interpreter has two alternatives whenever a filtering operation on a large map has to be performed: either he decides that the filtering ought to be done in the space domain, implying that he needs a set of convolution coefficients, simulating the theoretical frequency response, or the filtering may be carried out in the frequency plane by simple multiplication of the Fourier transform of the map with the desired frequency response. Let us now examine some advantages and disadvantages of both methods.

4.6.1. Convolution Filtering. (1) The problem in conventional filter design is that of obtaining the best fit between the frequency response of a discrete, linear filter and the desired frequency characteristic with a minimum number of coefficients. A small coefficient set will minimize the number of arithmetic operations involved in the discrete convolution.

(2) The frequency response is usually band-limited; hence the corresponding filter function is infinite in length and needs to be tailed off. However, a steep cutoff for the frequency response is needed in order to minimize the overlapping of the spectra of different anomalies in the surface field. Nevertheless it is a fact that several anomalies may contribute to the same region of the frequency domain and therefore it is *a priori* impossible to separate unambiguously a surface field into components, which may be assigned to different structures. But the finite size of the set of filter coefficients limits the steepness of the cutoff of the frequency response and unwanted components may be passed by the finite filter.

(3) Another reason for requiring small filters may be found in the fact that fewer points will be lost at the edges of the data grid in the convolution process. In order to minimize fluctuations in the frequency response the coefficients have to be attenuated toward the edges of the filtering grid, favoring again a loss of steepness at the cutoff.

(4) We have also seen that rectangular or square shapes of the data grid and of the coefficient set introduce directional effects in the frequency domain. These effects may be rather severe and ought to be treated accordingly. This difficulty may be avoided by defining coefficient sets with a circular symmetry.

(5) The main problem of conventional filter design is concerned with approximation techniques, but the data map will be left unchanged and no Fourier transformations need to be performed.

4.6.2. Frequency Filtering. (1) One advantage of frequency filtering is that no coefficient array has to be calculated and the availability of fast multidimensional Fourier transform methods indicates a routine in which filtering can be performed by multiplication in the frequency domain, thus reducing considerably the computing time. Very complicated filters are readily applied and one works with the best possible frequency response.

(2) No data are lost at the edges of the data grid. One way to reduce this loss in the convolution method is to surround the data with zeros or to use the measured data to extrapolate the known part of the surface field. This extrapolation may be done by fitting a polynomial surface to the observations. The procedure of filtering in the frequency plane reduces these edge effects because it extrapolates the data by fitting a trigonometric series to the observations, and the measured part of the field may be thought to be extended periodically over the space domain. It is also wise to extract a linear trend from the data, so that the new data have zero mean and zero regional trend. Indeed, a regional trend introduces spurious components at all frequencies and hence distorts the power spectrum in the Nyquist domain.

(3) The main problem of frequency filtering is evidently concerned with the fact that we work with a Fourier series of discrete data and the relation with the true Fourier transform is always problematic, as shown in Section 4.4. The practical computations of power spectra also impose several restrictions. In consequence it is not uncommon to observe distortion in the size and shape of the anomalies of the transformed field. These deviations evidently arise from the fundamental fact that the Fourier series is an incomplete approximation to the exact Fourier transform of the field as a whole. In this connection the importance of the problem of aliasing must be recognized: if the grid spacings are so chosen that the Fourier spectrum contains appreciable power above the Nyquist frequencies, then we may expect that the calculated spectra may differ considerably from the true spectra at low frequencies, that is in the region where the frequency contributions of the broader anomalies are situated.

Considering these deficiencies of both methods it is difficult to decide which procedure should give the most reliable results. It is true that convolution filtering is more time consuming, but frequency filtering supposes knowledge of the desired output and involves individually designed filters for each problem. A choice between both methods may be voiced on the grounds of practical considerations regarding computer time and storage. Until the advent of fast Fourier transform techniques, filtering in the frequency plane demanded excessive computing time as compared with convolution. This leads to a brief examination of the mechanics of Cooley–Tukey's method (1965).

The discrete Fourier transform of a one-dimensional sample function $f(k)$, $0 \leq k \leq m-1$, is defined by

(4.89) $$F(r) = \sum_{k=0}^{m-1} f(k) e^{-2\pi i k r/m}, \qquad 0 \leq r \leq m-1$$

To solve for $F(r)$ one would normally select a value for r and perform the summation over k. Suppose now that m is factorable and that it may be expressed as

(4.90) $$m = m_1 m_2$$

The indices k and r are now written as

(4.91) $$\begin{aligned} k &= k_1 m_2 + k_0, & 0 \leq k_0 \leq m_2 - 1, \; 0 \leq k_1 \leq m_1 - 1 \\ r &= r_1 m_1 + r_0, & 0 \leq r_0 \leq m_1 - 1, \; 0 \leq r_1 \leq m_2 - 1 \end{aligned}$$

With these notations Eq. (4.89) becomes

$$F(r) = \sum_{k_0=0}^{m_2-1} \sum_{k_1=0}^{m_1-1} f(k_0, k_1) e^{-2\pi i (k_1 m_2 r/m + k_0 r/m)}$$

From the periodicity of the complex exponential it follows that

$$e^{-2\pi i k_1 m_2 r/m} = e^{-2\pi i k_1 m_2 r_0/m}$$

and the inner sum over k_1 depends only on r_0; therefore one can define a new array

(4.92) $$f_1(k_0) = \sum_{k_1=0}^{m_1-1} f(k_0, k_1) e^{-2\pi i k_1 m_2 r_0/m}$$

and the result may be written as

(4.93) $$F(r) = \sum_{k_0=0}^{m_2-1} f_1(k_0) e^{-2\pi i k_0 r/m}$$

This algorithm can be generalized if one can express m as a product $m = m_1 \cdots m_p$ and the above reasoning may be repeated for each step.

The fast Fourier transform of two-dimensional data may be written in terms of one-dimensional transforms. Indeed, considering the two-dimensional analog of Eq. (4.89)

(4.94) $$F(r, s) = \sum_{k=0}^{m-1} \left[\sum_{l=0}^{n-1} f(k, l) e^{-2\pi i l s/n} \right] e^{-2\pi i k r/m}$$

we first calculate the transforms of the rows

$$F_1(k, s) = \sum_{l=0}^{m-1} f(k, l) e^{-2\pi i l s/n}$$

and then take the transforms of the columns

$$F(r, s) = \sum_{k=0}^{m-1} F_1(k, s) e^{-2\pi i k r/m}$$

where $F_1(k, s)$ is the Fourier transform of the kth column.

One of the interesting properties of this method is the considerable reduction in computer time. In the classical method the number of operations required to compute a discrete Fourier transform of a $N \times N$ matrix is approximately N^4, whereas by the fast Fourier transform it is about $4N^2 \log_2 N$ (Cochran, 1967), if we use a radix 2 algorithm. Thus the number of operations is reduced by a factor $4 \log_2 N/N^2$. For $N = 128 = 2^7$ points, this factor equals $28/65536 \simeq 1/2340$ and the fast Fourier transform should be 2340 times faster than the conventional method.

In connection with filtering it is known that, for two-dimensional data with N^2 points, the number of multiplications involved in filtering in the frequency plane is about $4N^2 \log_2 N + \frac{1}{2}N^2$; convolution filtering with M^2 weights requires $M^2 N^2$ multiplications. For $N = 256$ and $M = 20$ we obtain a reduction factor of the order of $1/14$ and frequency filtering should be 14 times faster than convolution filtering.

5. Calculation of Derivatives of Higher Order

In the last twenty-five years much effort has been devoted to improving the methods of calculation of derivatives of higher order of a surface field, derivable from the gravity potential. The second vertical derivative in particular is a very interesting aid in the interpretation of local features. Because the anomalous gravity effect $\Delta g(x, y, z)$ satisfies the Laplace equation, the following relation between the vertical second derivative and the two horizontal derivatives in the directions x and y exists

(5.1) $$\partial^2 \Delta g/\partial z^2 = -(\partial^2 \Delta g/\partial x^2 + \partial^2 \Delta g/\partial y^2)$$

Therefore the second vertical derivative may be interpreted as a measure of the total curvature of the anomaly surface, in two orthogonal directions. It is clear that the curvature of the anomaly surface is greater for local than for broader structures; in this context the second vertical derivative may be used to amplify the contributions of small, shallow masses to the surface gravity map, and the method is based upon the fact that nearby sources have greater influence on gravity gradients than on gravity itself.

Garland (1965) pointed out that the edges of an isolated structure are often indicated on the anomaly surface by the inflection points of the surface; as these are the points where the second vertical derivative becomes zero, the zero contour of the second vertical derivative map may be interpreted as showing roughly the edges of the anomalous masses. However, in view of the fact that an observed gravity field is the result of the superposition of a very complicated mass distribution, this zero contour will only outline approximately the positions and the areal extent of the anomalous masses. Nevertheless, one may conclude that, whenever the spacing of the surface grid is small and the precision of the measurements is high, the second derivative method is particularly interesting for resolving and sharpening anomalies of small areal extent.

We will next show how the vertical derivatives define linear filters, amplifying the frequency contributions of local effects. From Eq. (3.24) it follows that the field at depth z may be written as

(5.2) $$\Delta g(x, y, z) = (1/4\pi^2) \int_{-\infty}^{\infty} \int_{-\infty}^{\infty} G(u, v, 0) \exp[z(u^2 + v^2)^{1/2}] e^{i(ux + vy)} \, du \, dv$$

Differentiating both sides of Eq. (5.2) n times with respect to z results in

(5.3) $$(\partial^n/\partial z^n) \Delta g(x, y, z) = (1/4\pi^2) \int_{-\infty}^{\infty} \int_{-\infty}^{\infty} G(u, v, 0)(u^2 + v^2)^{n/2}$$
$$\times \exp[z(u^2 + v^2)^{1/2}] e^{i(ux + vy)} \, du \, dv$$

and in the limit as $z \to 0$

(5.4)
$$[(\partial^n/\partial z^n)\,\Delta g(x, y, z)]_{z=0} = (1/4\pi^2) \int_{-\infty}^{\infty} \int_{-\infty}^{\infty} G(u, v, 0)(u^2 + v^2)^{n/2}\, e^{i(ux+vy)}\, du\, dv$$

Writing $G^{(n)}(u, v, 0)$ for the Fourier transform of the vertical derivative of order n at the surface we have the following frequency relation

(5.5)
$$G^{(n)}(u, v, 0) = (u^2 + v^2)^{n/2} G(u, v, 0)$$

Hence, the vertical derivative of order n defines a linear filter on the surface gravity effect with frequency response

(5.6)
$$H^{(n)}(u, v) = (u^2 + v^2)^{n/2}$$

It is also obvious that the derivatives of higher order are independent of the internal mass distribution and can be computed unambiguously, for any observed surface field. The appearance of the function $H^{(n)}(u, v)$ in Eq. (5.5) is the mathematical justification of the observation that small anomalies and observational errors will be more amplified than the broader features in the surface field.

In particular, Elkins (1950) showed that the second vertical derivative may be approximated by a discrete, linear filter of the form

(5.7)
$$\partial^2\,\Delta g/\partial z^2 \simeq (1/s_0^2) \sum_{k=0}^{N} h_k\,\Delta g(x, y, R_k)$$

where s_0 is the grid spacing of the square grid. Several sets of coefficients h_k were derived by Elkins (1950), Rosenbach (1953), Henderson and Zietz (1949), Peters (1949), and Henderson (1960). Meskŏ (1966) gave the frequency responses of these discrete filters and indicated that Elkins' filters do not give good estimates of the second vertical derivative, because the discrete frequency response matches the theoretical frequency characteristic $u^2 + v^2$ only in the low frequency region. However, although Elkins' filters are low-pass filters and do not amplify the high frequency waves unnecessarily, they give easily interpretable results, but the corresponding output maps may not be interpreted as being second derivative maps. The other convolution coefficient sets yield better approximations to the second derivative frequency response and therefore estimate the second derivative with higher accuracy, but the structure of the output maps is often very complicated.

Anyhow it is important to amplify only the waves which are characteristic for the major parts of the mass distribution and to reject the noise field in the surface data. Consequently we are led to consider, once again, low-pass filters, cutting off rather sharply. Adequate coefficient arrays can be constructed with the methods of Sections 2.4 and 2.5, where now the general frequency characteristic $H(u, v)$ has to be replaced by $H^{(n)}(u, v)$. This requires a detailed inspection of the power spectrum of the surface field and the interpreter must decide which waves he wants to amplify and which parts of the Fourier spectrum he has to reject, thus gaining more control over the efficiency of his calculations. From Section 4.2 it is also evident that the Wiener filter method will be very useful, because all we have to do is to replace the frequency response $H^{(+)}(u, v, z)$ by $H^{(n)}(u, v)$.

APPENDIX 1.

PETERS' AND HENDERSON'S FILTERS FOR UPWARD CONTINUATION

R_k	Peters		Henderson				
	$h = 1$	$h = 2$	$h = 1$	$h = 2$	$h = 3$	$h = 4$	$h = 5$
$\sqrt{0}$	0.14645	0.05279	0.11193	0.04034	0.01961	0.01141	0.00742
$\sqrt{1}$	0.21132	0.09175	0.32193	0.12988	0.06592	0.03908	0.02566
$\sqrt{2}$	0.14943	0.11388	0.06062	0.07588	0.05260	0.03566	0.02509
$\sqrt{5}$	0.12645	0.12541	0.15206	0.14559	0.10563	0.07450	0.05377
$\sqrt{8}$			0.05335	0.07651	0.07146	0.05841	0.04611
$\sqrt{8.5}$	0.08627	0.11512					
$\sqrt{13}$			0.06586	0.09902	0.10226	0.09173	0.07784
$\sqrt{17}$	0.07771	0.12062					
$\sqrt{25}$			0.06650	0.11100	0.12921	0.12915	0.11986
$\sqrt{34}$	0.05276	0.09122					
$\sqrt{50}$			0.05635	0.10351	0.13635	0.15474	0.16159
$\sqrt{58}$	0.03462	0.06369					
$\sqrt{99}$	0.02055	0.03896					
$\sqrt{125}$	0.09459	0.18656					
$\sqrt{136}$			0.03855	0.07379	0.10322	0.12565	0.14106
$\sqrt{274}$			0.02273	0.04464	0.08323	0.08323	0.09897
$\sqrt{625}$			0.03015	0.05998	0.08917	0.11744	0.14458
Sum	1	1	0.98003	0.96014	0.94043	0.92100	0.90195

Appendix 2.

Coefficients for the Upward Continuation, $h = 1$

R_k	$L_S{}^a$	$L_P{}^b$	R_k	L_S	L_P
$\sqrt{0}$	0.13756	0.13353	$\sqrt{0}$	0.13767	0.13852
$\sqrt{1}$	0.23755	0.24163	$\sqrt{1}$	0.23676	0.23594
$\sqrt{2}$	0.13161	0.13578	$\sqrt{2}$	0.13159	0.13346
$\sqrt{4}$	0.04998	0.05134	$\sqrt{4}$	0.05149	0.05508
$\sqrt{5}$	0.07890	0.08909	$\sqrt{5}$	0.07934	0.07335
$\sqrt{8}$	0.02868	0.02379	$\sqrt{8}$	0.02624	0.01975
$\sqrt{9}$	0.02734	0.02922	$\sqrt{9}$	0.02354	0.02050
$\sqrt{10}$	0.03527	0.04815	$\sqrt{10}$	0.03495	0.03521
$\sqrt{13}$	0.02339	0.03711	$\sqrt{13}$	0.02829	0.03215
$\sqrt{17}$	0.02736	0.03371	$\sqrt{16}$	0.00825	0.00200
$\sqrt{20}$	0.01711	0.00840	$\sqrt{17}$	0.02091	0.02656
$\sqrt{25}$	0.03324	0.00326	$\sqrt{20}$	0.01493	0.01775
$\sqrt{34}$	0.00426	0.01371	$\sqrt{25}$	0.02984	0.00931
$\sqrt{40}$	0.02336	0.01527	$\sqrt{29}$	0.01124	0.02277
$\sqrt{50}$	0.01388	0.02293	$\sqrt{34}$	0.00193	0.01223
$\sqrt{65}$	0.02096	0.01221	$\sqrt{40}$	0.02277	0.03995
$\sqrt{85}$	0.01488	0.02686	$\sqrt{50}$	0.01665	0.01536
$\sqrt{100}$	0.01433	0.01893	$\sqrt{65}$	0.02694	0.01428
$\sqrt{125}$	0.02582	0.01045	$\sqrt{85}$	0.02417	0.03928
$\sqrt{274}$	0.01756	0.01224	$\sqrt{100}$	0.02673	0.02147
Sum	0.96304	0.96760	Sum	0.95423	0.96493
ε	0.013	0.032	ε	0.017	0.035

[a] L_S = method of least squares.
[b] L_P = method of linear programming.

Appendix 3.

Peters' and Henderson's Filters for the Downward Continuation

R_k	Peters		Henderson				
	$d=1$	$d=2$	$d=1$	$d=2$	$d=3$	$d=4$	$d=5$
$\sqrt{0}$	0.3696	0.6434	4.8948	16.1087	41.7731	92.5362	183.2600
$\sqrt{1}$	0.3025	0.5573	−3.0113	−13.2209	−38.2716	−89.7403	−183.9380
$\sqrt{2}$	0.3355	0.4889	0.0081	0.4027	1.7883	5.1388	11.8804
$\sqrt{5}$	0.2746	0.3440	−0.5604	−1.9459	−4.7820	−9.9452	−18.6049
$\sqrt{8}$			−0.0376	0.0644	0.5367	1.7478	4.2324
$\sqrt{8.5}$	0.2229	0.2085					
$\sqrt{13}$			−0.0689	−0.0596	0.1798	0.8908	2.4237
$\sqrt{17}$	0.0346	−0.1061					
$\sqrt{25}$			−0.0605	−0.0522	0.1342	0.6656	1.7777
$\sqrt{34}$	−0.2219	−0.5047					
$\sqrt{50}$			−0.0534	−0.0828	−0.0560	0.0718	0.3606
$\sqrt{58}$	−0.3464	−0.6497					
$\sqrt{99}$	0.1134	0.2498					
$\sqrt{125}$	−0.1120	−0.2315					
$\sqrt{136}$			−0.0380	−0.0703	−0.0900	−0.0890	−0.0571
$\sqrt{274}$			−0.0227	−0.0443	−0.0639	−0.0802	−0.0921
$\sqrt{625}$			−0.0302	−0.0600	−0.0891	−0.1173	−0.1444
Sum	1	1	1.0199	1.0393	1.0595	1.0790	1.0983

Appendix 4.

Grant–West Filters for Downward Continuation

R_k	$Z=0.75$		$Z=1.00$		$Z=1.25$	
	$\gamma=1/9$	$\gamma=1/6$	$\gamma=1/6$	$\gamma=1/4$	$\gamma=1/4$	$\gamma=1/3$
$\sqrt{0}$	2.1652	1.2361	2.5038	1.2024	1.3168	1.2609
$\sqrt{1}$	0.3844	1.8444	1.4621	2.6644	3.4369	3.5492
$\sqrt{2}$	−1.2647	−1.7333	−2.6018	−2.2727	−4.0842	−2.8342
$\sqrt{5}$	−0.1081	−0.1712	−0.1531	−0.3683	−0.4216	−0.7242
$\sqrt{8}$	−0.0481	−0.0356	−0.0455	−0.0298	−0.0174	−0.0134
$\sqrt{10}$	−0.0232	−0.0418	−0.0341	−0.0310	−0.0323	−0.0407
$\sqrt{13}$	−0.0267	−0.0279	−0.0330	−0.0389	−0.0443	−0.0437
$\sqrt{17}$	−0.0220	−0.0230	−0.0330	−0.0278	−0.0327	−0.0354
$\sqrt{20}$	−0.0167	−0.0135	−0.0142	−0.0198	−0.0241	−0.0253
$\sqrt{25}$	−0.0041	−0.0215	−0.0262	−0.0292	−0.0343	−0.0335
$\sqrt{34}$		−0.0145	−0.0237	−0.0216	−0.0267	−0.0297
$\sqrt{40.5}$		−0.0018	−0.0013	−0.0151	−0.0196	−0.0164
$\sqrt{50}$				−0.0126	−0.0164	−0.0199
$\sqrt{64.5}$						−0.0206

APPENDIX 5.

Coefficients for the Downward Continuation $d = 1$, Approximation in the (u, v)-Plane

R_k	Weierstrass kernel		Féjer kernel	
	$z=1, \rho_c=1, \rho_k=2, \alpha=6.61$		$z=1, \rho_c=1, \rho_k=1.5, \alpha=2\pi$	
$\sqrt{0}$	0.24497	0.19155	0.18407	0.14391
$\sqrt{1}$	1.32371	1.72299	0.99415	1.34005
$\sqrt{2}$	0.91269	0.23966	0.90935	0.34145
$\sqrt{4}$		0.62357		0.62636
$\sqrt{5}$	0.35727	−0.00880	0.59457	0.20118
$\sqrt{8}$	−0.29751	−0.41586	0.05064	0.03488
$\sqrt{9}$		0.33522		0.10951
$\sqrt{10}$	−0.57614	−0.80877	−0.37375	−0.46622
$\sqrt{13}$	−0.66704	−0.50324	−0.53409	−0.41646
$\sqrt{16}$		−0.57096		−0.57087
$\sqrt{17}$	−0.57360	−0.06802	−0.59571	−0.04116
$\sqrt{20}$	−0.28797	−0.31652	−0.56818	−0.64196
$\sqrt{25}$	0.09675	−0.02038	−0.35135	−0.30824
$\sqrt{29}$		0.18419		−0.05637
$\sqrt{34}$	0.47108	0.29053	0.11223	0.10694
$\sqrt{40}$	−0.00413	0.45039	0.53983	0.52919
$\sqrt{50}$		0.14494	0.03834	0.70833
$\sqrt{65}$		−0.29889		−0.05065
$\sqrt{85}$		−0.24677		−0.61816
$\sqrt{100}$		0.03518		0.02832
Sum	1.00006	1.00001	1.00010	1.00002

APPENDIX 6.

COEFFICIENTS FOR THE DOWNWARD CONTINUATION $d = 1$, LINEAR PROGRAMMING METHOD

R_k	Test (1)	Test (2)
$\sqrt{0}$	0.48901	0.47795
$\sqrt{1}$	0.66810	0.87574
$\sqrt{2}$	1.46550	1.32928
$\sqrt{5}$	-0.21308	0.02968
$\sqrt{8}$	1.09735	0.40451
$\sqrt{10}$	-2.00891	-1.49369
$\sqrt{13}$	-0.52898	-0.77445
$\sqrt{17}$	0.71144	0.44360
$\sqrt{20}$	-2.00891	-1.49369
$\sqrt{25}$	1.00588	1.11425
$\sqrt{34}$	-0.34729	0.10756
$\sqrt{40}$	1.21442	0.68057
$\sqrt{50}$	-0.50866	-0.67464
Sum	1.03587	1.02668
ε	0.1435	0.1067

ACKNOWLEDGMENTS

The author acknowledges his indebtedness to Prof. A. De Vuyst, whose encouragement and interest were a constant stimulation to accomplish this work; to Prof. J. van Isacker for offering helpful suggestions and for the use of the computer at the Royal Meteorological Institute; and to Prof. J. van Mieghem for presenting the manuscript.

LIST OF SYMBOLS

$d_i(r)$	Covariance kernel	$\overline{\Delta g}(x, y, z)$	Calculated field at depth z
$\delta(x, y)$	Infinite Dirac comb		
$\delta_{mn}(x, y)$	Finite Dirac comb	$\mathscr{F}$	Linear filter operator
$\nabla(u, v)$	Fourier transform of infinite Dirac comb	$f(x, y)$	General two-dimensional function
$\nabla_{mn}(u, v)$	Fourier transform of finite Dirac comb	$F(u, v)$	Fourier transform of $f(x, y)$
$\Delta g(x, y, r)$	Average of $\Delta g(x, y, 0)$ over the circle of radius r	$f_i(x, y)$	Input to linear filter
		$f_o(x, y)$	Output of linear filter
		$f_d(x, y)$	Desired output of Wiener filter
$\Delta g(x, y, z)$	Gravity effect at the level z	$F_a(u, v)$	Fourier transform of sampled field
$\widetilde{\Delta g}(x, y, z)$	Calculated field at height z	G	Gravitational constant

g	Gravity vector
$g(x, y)$	Output of linear, continuous filter
$\bar{g}(x, y)$	Output of linear, discrete filter
$G(u, v, z)$	Fourier transform of $\Delta g(x, y, z)$
$\gamma(x, y)$	Covariance function
$\gamma_a(x, y)$	Covariance function of sampled field
$\bar{\gamma}_0(k, l), \hat{\gamma}_0(k, l)$	Covariance estimates
$h(x, y)$	Filter function of linear continuous filter
$H(u, v)$	Frequency response of linear, continuous filter
$H_d(u, v)$	Frequency response of linear, discrete filter
$\bar{h}(x, y)$	Filter function of modified low-pass filter
$\bar{H}(u, v)$	Frequency response of modified low-pass filter
$\bar{h}_f(x, y)$	Finite filter function of modified low-pass filter
$\bar{H}_f(u, v)$	Frequency response corresponding to $\bar{h}_f(x, y)$
h_{kl}	Filter coefficients of two-dimensional filter
$h^{(-)}(x, y, z)$	Upward continuation filter function
$H^{(-)}(u, v, z)$	Upward continuation frequency response
$H_d^{(-)}(u, v)$	Discrete upward continuation frequency response
$H^{(+)}(u, v, z)$	Downward continuation frequency response
$H_d^{(+)}(u, v)$	Discrete downward continuation frequency response
$\bar{h}^{(+)}(x, y, z)$	Filter function of modified continuation process
$\bar{h}_f^{(+)}(x, y, z)$	Finite filter function of modified downward continuation process
$\bar{H}^{(+)}(u, v, z)$	Modified downward continuation frequency response
$\bar{H}_f^{(+)}(u, v, z)$	Frequency response corresponding to $\bar{h}_f^{(+)}(x, y, z)$
$H_r(u, v)$	Frequency response of averaging process on the circle of radius r
$H_{opt}(u, v)$	Frequency response of optimal Wiener filter
$H^{(n)}(u, v)$	Frequency response of vertical derivative of order n
$k_\alpha(\rho)$	General kernel
$\Lambda_{mn}(u, v)$	Two-dimensional Féjer kernel
$n(x, y)$	Noise component of surface field
$P(u, v)$	Power spectrum
$P_a(u, v)$	Power spectrum of sampled field
$\bar{P}_e(u, v), \hat{P}_e(u, v)$	Power spectrum estimates
$P_{mn}(u, v)$	Periodogram
$\tilde{\psi}$	Smoothed version of a function ψ
$\rho(\mathbf{r})$	Density function
ρ_N	Radial Nyquist frequency
$\tau_{\xi,\eta}$	Translation operator
$s(x, y)$	Signal component of surface field
$S(u, v)$	Fourier transform of $s(x, y)$
s_0	Grid spacing of square grid
σ	Variance
$U(\mathbf{r})$	Potential of the deviation masses of the earth
u, v	Frequencies corresponding to the x, y-directions
u_N, v_N	Nyquist frequencies
$V(\mathbf{r})$	Newton's potential of the earth
$w(x, y)$	General weighting function
$W(u, v)$	Fourier transform of $w(x, y)$

References

A. General works

Anderson, T. W. (1971). "The Statistical Analysis of Time Series." Wiley, New York.
Arsac, J. (1961). "Transformation de Fourier et théorie des distributions." Dunod, France.
Bartlett, M. S. (1962). "An Introduction to Stochastic Processes." Cambridge Univ. Press, London and New York.
Bendat, J. S. (1958). "Principles and Applications of Random Noise Theory." Wiley, New York.
Blackman, R. B., and Tukey, J. W. (1958). "The Measurement of Power Spectra." Dover, New York.
Courant, R., and Hilbert, D. (1962). "Methods of Mathematical Physics." Wiley, New York.
Dantzig, G. B. (1963). Linear Programming and Extensions." Princeton Univ. Press, Princeton, New Jersey.
Dobrin, M. B. (1960). "Introduction to Geophysical Prospecting." McGraw-Hill, New York.
Doob, J. L. (1958). "Stochastic Processes." Wiley, New York.
Garland, G. D. (1965). "The Earth's Shape and Gravity." Pergamon, Oxford.
Grant, F. S., and West, G. F. (1965). "Interpretation Theory in Applied Geophysics." McGraw-Hill, New York.
Grenander, U., and Rosenblatt, M. (1957). "Statistical Analysis of Stationary Time Series." Wiley, New York.
Hannan, E. J. (1960). "Time Series Analysis." Wiley, New York.
Heiskanen, W. A., and Meinesz, F. A. (1958). "The Earth and its Gravity Field." McGraw-Hill, New York.
Holloway, J. L. (1958). "Smoothing and filtering of time series and space fields. *Advan. Geophys.* 4, 351–389.
Kellog, O. D. (1960). "Foundations of Potential Theory." Dover, New York.
Kopal, Z. (1961). "Numerical Analysis." Chapman & Hall, London.
Lee, Y. W. (1960). "Statistical Theory of Communications " Wiley, New York.
Morse, P. M., and Feshbach, H. (1953). "Methods of Theoretical Physics." McGraw-Hill, New York.
Papoulis, A. (1965). "Probability, Random Variables and Stochastic Processes." McGraw-Hill, New York.
Rice, J. R. (1964). "The Approximation of Functions." Addison-Wesley, Reading, Massachusetts.
Sneddon, I. N. (1951). "Fourier Transforms." McGraw-Hill, New York.
Todd, J. (1962). "Survey of Numerical Analysis." McGraw-Hill, New York.
Van Trees, H. L. (1968). "Detection, Estimation and Modulation Theory." Wiley, New York.
Watson, G. N. (1958). "A Treatise on the Theory of Bessel Functions." Cambridge Univ. Press, London and New York.
Wiener, N. (1950). "Extrapolation, Interpolation and Smoothing of Stationary Time Series." Wiley, New York.

B. Upward continuation

Henderson, R. G. (1960). A comprehensive system of automatic computation in magnetic and gravity interpretation. *Geophysics* 25, 569–585.

Henderson, R. G., and Zietz, I. (1949). The upward continuation of anomalies in total magnetic intensity fields. *Geophysics* **14**, 517–534.

Nettleton, L. L., and Cannon, J. R. (1962). Investigation of upward continuation systems. *Geophysics* **27**, 796–806.

Peters, L. J. (1949). The direct approach to magnetic interpretation and its practical application. *Geophysics* **14**, 290–320.

C. *Downward continuation*

Bullard, E. C., and Cooper, R. I. B. (1948). Determination of the masses necessary to produce a given gravitational field. *Proc. Roy. Soc., Ser.* A **194**, 332–347.

Ku, C. C., Telford, W. M., and Lim, S. H. (1971). The use of linear filtering in gravity problems. *Geophysics* **36**, 1174–1203.

Nedelkov, I. P., and Burnev, P. H. (1962). Determination of gravitational fields in depth. *Geophys. Prospect.* **10**, 1–18.

Negi, J. G. (1967). Convergence and divergence in downward continuation. *Geophysics* **32**, 867–871.

Roy, A. (1966). Downward continuation and its application to electromagnetic data interpretation. *Geophysics* **31**, 167–184.

Roy, A. (1967). Convergence in downward continuation for some simple geometries. *Geophysics* **32**, 853–866.

Rudman, A. J., Mead, J., Whaley, J. F., and Blakely, R. F. (1971). Geophysical analysis in central Indiana using potential field continuation. *Geophysics* **36**, 878–890.

Trejo, C. A. (1954). A note on downward continuation of gravity. *Geophysics* **19**, 71–75.

D. *Convolution and frequency filtering*

Bhattacharyya, B. K. (1972). Design of spatial filters and their application to high-resolution aeromagnetic data. *Geophysics* **37**, 68–91.

Black, D. I., and Schollar, I. (1969). Spatial filtering in the wave-vector domain. *Geophysics* **34**, 916–923.

Byerly, P. E. (1965). Convolution filtering of gravity and magnetic maps. *Geophysics* **30**, 281–284.

Chan, S. H., and Leong, L. S. (1972). Analysis of least squares smoothing operators in the frequency domain. *Geophys. Prospect.* **20**, 892–900.

Clarke, G. K. C. (1969). Optimum second derivative and downward continuation filters. *Geophysics* **34**, 424–437.

Clarke, G. K. C. (1971). Linear filters to suppress terrain effects on geophysical maps. *Geophysics* **36**, 963–966.

Connes, J., and Nozal, V. (1971). Le filtrage mathématique dans la spectroscopie par transformation de Fourier. *J. Phys. Radium* **22**, 359–366.

Darby, E. K., and Davies, E. B. (1967). The analysis and design of two-dimensional filters for two-dimensional data. *Geophys. Prospect.* **15**, 383–406.

Dean, W. C. (1958). Frequency analysis for gravity and magnetic interpretation. *Geophysics* **23**, 97–127.

Fraser, D. C., Fuller, B. D., and Ward, S. H. (1960). Some numerical techniques for application in mining exploration. *Geophysics.* **31**, 1066–1077.

Gunn, P. J. (1972). Application of Wiener filters to transformations of gravity and magnetic fields. *Geophys. Prospect.* **20**, 860–871.

Lavin, P. M., and Devane, S. J. (1970). Direct design of two dimensional digital wave number filters. *Geophysics* **35**, 1073–1078.

Mesko, A. (1965). Some notes concerning the frequency analysis for gravity interpretation. *Geophys. Prospect.* **13**, 475–488.
Mufti, I. R. (1972). Design of small operators for the continuation of potential field data. *Geophysics* **37**, 488–506.
Naidu, P. S. (1966). Extraction of potential field signal from a background of random noise by Strakhov's method. *J. Geophys. Res.* **71**, 5987–5995.
Naidu, P. S. (1967). Two dimensional Strakhov's filter for extraction of potential field signal. *Geophys. Prospect.* **15**, 135–150.
Strakhov, V. N. (1964a). The smoothing of observed strengths of potential fields. Part I. *Bull. Acad. Sci. USSR, Geophys. Ser.* **10**, 897–904.
Strakhov, V. N. (1964b). The smoothing of observed strengths of potential fields. Part II. *Bull. Acad. Sci., USSR, Geophys. Ser.* **11**, 986–995.
Swartz, C. A. (1953). Some geometrical properties of residual maps. *Geophysics* **19**, 46–70.
Swartz, C. A. (1954). Filtering associated with selective sampling of geophysical data. *Geophysics* **20**, 402–419.
Treitel, S. (1967). Principles of digital Wiener filtering. *Geophys. Prospect.* **15**, 311–333.
Zurflueh, E. G. (1967). Application of two-dimensional linear wavelength filtering. *Geophysics* **32**, 1015–1035.

D. *Power spectrum estimation*

Akcasu, A. Z. (1961). Measurement of noise power spectra by Fourier analysis. *J. Appl. Phys.* **32**, 565–568.
Bartlett, M. S. (1960). Periodogram analysis and continuous spectra. *Biometrika* **37**, 1–16.
Cooley, J. W., Lewis, P. A. W., and Welch, P. D. (1967). "The Fast Fourier Transform and its Applications," Res. Pap. RC-1743. IBM Watson Res. Cent., New York.
Goodman, N. R. (1961). Some comments on spectral analysis of time series. *Technometrics* **3**, 221–228.
Grenander, U. (1958). Bandwidth and variance in estimation of the spectrum. *J. Roy. Statist. Soc., Ser. B* **20**, 152–157.
Hinich, M. J., and Clay, C. S. (1968). The application of the discrete Fourier transform in the estimation of power spectra coherence and bispectra of geophysical data. *Rev. Geophys.* **6**, No. 3, 347–362.
Horton, C. W., Hempkins, W. B., and Hoffman, A. J. (1963). Statistical analysis of some aeromagnetic maps from the North-western Canadian shield. *Geophysics* **29**, 582–601.
Jenkins, G. M. (1961). General considerations in the analysis of spectra. *Technometrics* **3**, 133–166.
Jenkins, G. M., and Priestley, M. B. (1957). The spectral analysis of time series. *J. Roy. Statist. Soc., Ser. B* **19**, 1–12.
Jones, R. H. (1965). A reappraisal of the periodogram in spectral analysis. *Technometrics* **7**, No. 4, 531–542.
Lomnicki, Z. A., and Zaremba, S. K. (1957). On estimating the spectral density function of a stochastic process. *J. Roy. Statist. Soc., Ser. B* **19**, 13–37.
Murthy, V. K. (1961). Estimation of the spectrum. *Ann. Math. Statist.* **32**, 730–738.
Naidu, P. S. (1968). Spectrum of the potential field due to randomly distributed sources. *Geophys* **33**, 337–345.
Naidu, P. S. (1969). Estimation of spectrum and cross-spectrum of aeromagnetic field using Fast Digital Fourier transform (FDFT) techniques. *Geophysics* **34**, 344–361.
Parzen, E. (1956). On consistent estimates of the spectral density of a stationary time series. *Proc. Nat. Acad. Sci. U.S.* **42**, 154–157.

Parzen, E. (1957a). On consistent estimates of the spectrum of a stationary time series. *Ann. Math. Statist.* **28**, 328–348.
Parzen, E. (1957b). On choosing an estimate of the spectral density function of a stationary time series. *Ann. Math. Statist.* **28**, 921–932.
Parzen, E. (1958). On asymptotically efficient consistent estimates of the spectral density function of a stationary time series. *J. Roy. Statist. Soc., Ser. B* **20**, 303–322.
Parzen, E. (1961). Mathematical considerations in the estimation of spectra. *Technometrics.* **3**, 167–190.
Sax, R. L. (1966). Application of the filter theory and information theory to the interpretation of gravity measurements. *Geophysics* **31**, 570–575.
Tukey, J. W. (1961). Discussion, emphasizing the connection between analysis of variance and spectrum analysis. *Technometrics* **3**, 191–219.
Whittle, P. (1957). Curve and periodogram smoothing. *J. Roy. Statist. Soc., Ser. B.* **19**, 38–47.
Wonnacot, T. H. (1961). Spectral analysis combining a Bartlett window with an associated inner window. *Technometrics* **3**, 235–243.

F. *Fast fourier transform algorithm*

Bergland, G. D. (1967). The Fast Fourier Transform recursive equations for arbitrary length records. *Math. Comput.* **21**, 236–238.
Bingham, C., Godfrey, M. D., and Tukey, J. W. (1967). Modern techniques of power spectrum estimation. *IEEE Trans. Audio Electroacoustics*, **15**, No. 2, 56–66.
Bogert, B. P. (1967). Informal comments on the uses of power spectrum analysis. *IEEE Trans. Audio Electroacoustics.* **15**, No. 2, 74–75.
Cochran, W. T. (1967). What is the Fast Fourier Transform. *Proc. IEEE* **55**, No. 10, 1664–1674.
Cooley, J. W., and Tukey, J. W. (1965). An algorithm for the machine calculation of complex Fourier series. *Math. Comput.* **19**, 297–301.
Cooley, J. W., Lewis, A. W., and Welch, P. D. (1967). Historical notes on the Fast Fourier Transform. *Proc. IEEE* **55**, 1675–1677.
Cooley, J. W., Lewis, A. W., and Welch, P. D. (1969). The Fast Fourier Transform and its applications. *IEEE Trans. Educ.* **12**, 27–34.
Cooley, J. W., Lewis, A. W., and Welch, P. D. (1969b). Application of the Fast Fourier Transform to computation of Fourier Integrals, Fourier series and convolution integrals. *IEEE Trans. Audio Electroacoustics* **15**, 79–84.
HARM. (1966). " Harmonic Analysis Subroutine for IBM 7090," SDA No. 3425. SHARE Distribution Agency, Program Inform. Dept., IBM Corp., 40 Saw Mill Road, Hawthorne, New York 10532.
Helms, H. D. (1967). Fast Fourier Transform method of computing difference equations and simulating filters. *IEEE Trans. Audio Electroacoustics* **15**, 85–90.
Singleton, R. C. (1967). A method of computing the Fast Fourier Transform with auxiliary memory and limited high-speed storage. *IEEE Trans. Audio Electroacoustics* **15**, 91–98.
Welch, P. D. (1967). The use of the Fast Fourier Transforms for the estimation of power spectra, a method based on time averaging over short, modified periodograms. *IEEE Trans. Audio Electroacoustics* **15**, 70–73.

G. *Second derivative method*

Agarwal, B. N. P., and Lal, T. (1972a). A generalized method of computing second derivative of gravity field. *Geophys. Prospect.* **20**, 385–394.

Agarwal, B. N. P., and Lal, T. (1972b). Calculation of the vertical gravity field using the Fourier transform. *Geophys. Prospect.* **20**, 448–457.

Danes, Z. F., and Oncley, I. A. (1962). An analysis of some second derivative methods. *Geophysics* **27**, 611–615.

Elkins, T. A. (1950). The second derivative method of gravity interpretation. *Geophysics* **16**, 29–50.

Henderson, R. G., and Zietz, I. (1949). The computations of second vertical derivatives of geomagnetic fields. *Geophysics* **14**, 508–616.

Meskö, C. A. (1966). Two-dimensional filtering and the second derivative method. *Geophysics* **31**, 606–617.

Rosenbach, O. (1953). A contribution to the computation of the second derivative from gravity data. *Geophysics* **18**, 894–912.

H. *Related topics*

Agarwal, R. G., and Kanasewich, E. R. (1971). Automatic trend analysis and interpretation of potential field data. *Geophysics* **36**, 339–348.

Agocs, V. B. (1951). Least squares residual anomaly determination. *Geophysics.* **16**, 686–696.

Al-Chalabi, M. (1971). Some studies relating to nonuniqueness in gravity and magnetic inverse problems. *Geophysics* **36**, 835–855.

Al-Chalabi, M. (1972). Interpretation of gravity anomalies by non-linear optimisation. *Geophys. Prospect.* **20**, 1–16.

Botezatu, R., Visarion, M., Scurtu, F., and Cucu, G. (1971). Approximation of the gravitational attraction of geological bodies. *Geophys. Prospect.* **19**, 218–227.

Bott, M. H. P., and Smith, R. A. (1958). The estimation of the limiting depth of gravitating bodies. *Geophys. Prospect.* **6**, 1–10.

Cordell, L., and Henderson, R. G. (1968). Iterative three-dimensional solution of gravity anomaly data using a digital computer. **33**, No. 4, 596–602.

Grant, F. S. (1951a). Three-dimensional interpretation of gravitational anomalies. Part I. *Geophysics* **17**, 344–364.

Grant, F. S. (1951b). Three-dimensional interpretation of gravitational anomalies. Part II. *Geophysics* **17**, 756–789.

Grant, F. S. (1953). A theory for the regional correction of potential field data. *Geophysics* **19**, 23–45.

Grant, F. S. (1957). A problem in the analysis of geophysical data. *Geophysics* **22**, 309–344.

Grant, F. S. (1972). Review of data processing and interpretation methods in gravity and magnetics, 1964–1971. *Geophysics* **37**, 647–661.

Griffin, W. R. (1948). Residual gravity in theory and practice. *Geophysics* **14**, 39–56.

Henderson, R. G., and Cordell, L. (1971). Reduction of unevenly spaced potential field data to a horizontal plane by means of finite harmonic series. *Geophysics* **36**, 856–866.

Kreisel, G. (1948). Some remarks on integral equations with kernels, $L(\xi_1 - x_1, \ldots, \xi_n - x_n, \alpha)$. *Proc. Roy. Soc. Ser. A* **107**, 160–178.

Kunaratnam, K. (1972). An iterative method for the solution of a non-linear inverse problem in magnetic interpretation. *Geophys. Prospect.* **20**, 439–447.

La Porte, M. (1972). Elaboration rapide des cartes gravimétriques déduites de l'anomalie de Bouguer à l'aide d'une calculatrice électronique. *Geophys. Prospect.* **10**, 238–257.

Nagy, D. (1966). The gravitational attraction of a right rectangular prism. *Geophysics* **31**, 362–371.

Naidu, P. S. (1967). Statistical properties of potential fields over a random medium. *Geophysics* **32**, 88–98.

Naidu, P. S. (1970). Statistical structure of aeromagnetic field. *Geophysics* **35**, 279–292.
Nettleton, L. L. (1953). Regionals, residuals and structures. *Geophysics* **19**, 1–22.
Paul, M. K. (1967). A method of computing residual anomalies from Bouguer gravity map by applying relaxation technique. *Geophysics* **32**, 708–719.
Polya, G. (1929). Untersuchung über Lücken und Singularitäten von Potentzreihen. *Math. Z.* **29**, 549–600.
Qureshi, I. R., and Idries, F. M. (1972). Two-dimensional mass distributions from gravity anomalies. *Geophys. Prospect.* **20**, 106–108.
Qureshi, I. R., and Mula, H. G. (1971). Two-dimensional mass distribution from gravity anomalies: A computer method. *Geophys. Prospect.* **19**, 180–191.
Simpson, S. M. (1953). Least squares polynomial fiitting to gravitational data and density plotting by digital computers. *Geophysics* **19**, 255–269.
Skeels, D. C. (1947). Ambiguity in gravity interpretation. *Geophysics* **12**, 43–56.
Spector, A., and Grant, F. S. (1970). Statistical models for interpreting aeromagnetic data. *Geophysics.* **35**, 293–302.
Syberg, F. J. R. (1972a). A Fourier method for the regional residual problem of potential fields. *Geophysical Prospect.* **20**, 47–75.
Syberg, F. J. R. (1972b). Potential field continuation between general surfaces. *Geophys. Prospect.* **20**, 267–282.
Talwani, M., and Euring, M. (1960). Rapid computation of gravitational attraction of three-dimensional bodies of arbitrary shape. *Geophysics* **25**, 103–225.

AERIAL METHODS IN GEOLOGICAL–GEOGRAPHICAL EXPLORATIONS*

B. V. Shilin and V. B. Komarov

Laboratory of Aeromethods,
Ministry of Geology, Leningrad, U.S.S.R.

1. Introduction 263
2. Radar Aerial Survey 265
   2.1. Procedures 265
   2.2. Interpretation of Rock Composition 272
   2.3. Interpretation of Structural-Tectonic Texture 277
   2.4. Interpretation of Geological Structure in Closed Regions 279
3. Infrared Aerial Survey 282
   3.1. General 282
   3.2. Infrared Aerial Survey of Active Volcanoes 282
   3.3. Study of Regions of Strong Geothermal Activity 286
4. Aerogeochemical Survey: Remote Sensing of Gases and Vapors 304
   4.1. Method 304
   4.2. Equipment and Procedure 305
   4.3. Aerial Measurements over Fumarole Fields of Caldera Uson 308
   4.4. Active Volcanoes of Kamchatka 313
5. Conclusions 321
   References 322

1. Introduction

When the first images of the earth's outer appearance were obtained from orbital altitudes, specialists in various fields of research became aware of a new, exceptionally effective means of studying the earth's natural resources and the processes taking place on its surface. Sciences dealing with the earth were offered the possibility, on the one hand, to rise to a new level of extensive generalizations regarding the geological and geographical structure of the planet and, on the other hand, to obtain regional information enabling more reliable regional assessments for the rational development and mastering of natural resources.

The use of information obtained from the orbital altitudes was preceded by the successful application of other aerial methods. The necessity for rapidly acquiring data on the natural resources of the enormous territory of the USSR for their efficient exploration and use required the development of methods of remote sensing from aircraft, based on the use of the electro-

---

* Published in cooperation with Advances in Space Science and Technology, edited by Frederick I. Ordway.

magnetic spectrum and other physical phenomena. In spite of the limited spectral range used for classical aerial photographic surveys (400–1000 nm), this method is the most universal and efficient of all aerial methods with respect to both the amount of information obtained and the range of its application in economic and scientific research.

Aerial photographic surveys are especially efficient for various kinds of mapping: topographic, geological, landscape, soil, geobotanical, forestry, etc. They have been used for inventory of forests, land surveying, land reclamation, planning of railways and highways, power lines and pipe lines, and various explorations.

Aerial photographs are also widely used for: (1) geographical investigations of all kinds, including a complex study of landscapes, their typology, rhythm, and dynamics of their development, the realization of measures to protect the environment against erosion, plant diseases, water and air pollution, etc.; (2) geological research connected with the study of the earth's crustal structure and tectonics (including recent movements), the search for mineral deposits, geological study of shallow seas, the hydrogeological and engineering-geological explorations; (3) the study of forests, soils, swamps, vegetation (including submerged plants), pastures, etc.; (4) the study of hydrological conditions of dry land, processes on shores and coasts of various reservoirs, the course of the rivers and processes in the river beds; (6) the study of sea streams and sea heaving, hydro-optical characteristics of the sea, and processes in the estuaries; (7) glaciological research; (8) archeological investigations; (9) the study of cloud cover; (10) the study of fishery resources (stock-taking of the number of animals and fish), their protection, and proper exploitation (wildlife management). They are also used for many other special tasks.

The utilization of aerial photography data in the above-mentioned fields increases the amount of information and improves the accuracy of the data on the objects under study. It also permits a lessening in the amount of ground exploration and a general reduction of the expense of solving the problems.

The increase of efficiency in aerial photographic methods is directed toward: (1) the use of small-scale aerial photographs which allow for sufficient optical and photographic magnification and permit, while keeping sufficient detail, the encompassing of large areas for obtaining in a number of cases new information; (2) stipulation of new, more profound connections between the objects and photo images of their indicators while obtaining quantitative characteristics; (3) the development of algorithms to identify various landscape objects for the automation of photoreading and interpretation processes; (4) the design of complex automated systems to handle the aerial photographs by use of computers and peripheral equipment.

It should be noted here that in spite of the achievements in the field of cybernetics and automated devices, the interpretation process remains one of the most complicated tasks. Therefore, the final analysis of the results obtained while using the automated systems for handling the photographs should still be performed by a specialist comparing them with the initial data.

For geological mapping and in the search for minerals the usual aerogeophysical methods are utilized. They are based on the recording of gamma radiation, electromagnetic and gravitational fields, as well as the radio wave portion of the electromagnetic spectrum. These methods allow the acquisition of data on deep structures of the earth's crust and on its types of rocks. When used in combination with aerial photography, they considerably increase the reliability of geological results.

The methods of aerial photography and the aerogeophysical methods, used successfully for a long time for the purpose of studying natural resources, are fairly well developed and described in the extensive special literature.

In recent years, ever increasing attention has been centered on the development of the techniques of remote sensing and methods for studying natural resources, especially their installation in space vehicles. However, the criteria for using some remote sensing receivers developed very recently (infrared, radar, and others) are not yet sufficiently established. The efficient use of these receivers for studying the natural resources from orbital platforms cannot be anticipated without carrying out this stage of investigation. The present review concerns some results on that stage of research using infrared, radar, and geochemical aerial surveys with emphasis on trends in the application of these methods.

2. Radar Aerial Survey

2.1. Procedures

The advent of side-looking radar systems brought with it a considerable increase in the resolution of the images obtained, enabling their use for studying wider areas of the country. This was also stimulated by the quest of specialists to use the small-scale images for study of large territories, an aim which cannot always be attained through aerial photography because of meteorological obstacles. In this case the small-scale images served as a kind of bridge in passing from the materials of traditional aerial photography to extra small-scale images obtained from space altitudes.

In the USSR the side-looking radar system TOROS was developed for the purpose of ice exploration, which in a very short time has found wide application for both the evaluation and the prognosis of ice conditions. This system functions under the complex meteorological conditions of the Arctic and rapidly obtains images of large territories. These images have further use for solving geological and other problems.

The TOROS system scans the earth's surface by means of two narrow vertical beams formed by parabolic antennas mounted on both sides along the aircraft fuselage and directed normally to their axes.

Scanning along the flight line is accomplished by the aircraft's forward movement. An image of the terrain is formed on the screens of cathode-ray tubes and is simultaneously recorded onto photographic film 19 cm wide. The speed of the film's motion is proportional to the aircraft's flight speed and the survey scale.

Thus, an image is obtained on the film. The density of this image at any point is a function of the reflecting properties of the irradiated terrain objects scanned in the given wave range.

The centimeter wavelength range, the narrow directivity pattern in a horizontal plane, and short pulses (fractions of a millisecond) enable the TOROS system to obtain an image of terrain with a fairly high resolution close to that of the small-scale aerial photographs.

The radar image tone determined by the reflected signal intensity depends on the operating frequency of the station, the angles of radiation, the texture of the surface, and physical properties of the irradiated objects. Depending on the reflection properties of the irradiated surface, the diffusive dispersion, the specular reflection, and radio-wave absorption will take place.

Most of the objects, because of their inherent surface texture and physical properties, diffusively disperse the radio waves, contributing to the generation of various reflected signals and, accordingly, the contours of different tone on the radar image.

When the texture of the surface has irregularities smaller than one-half wavelength, a radio signal reflection will appear close to specular reflection. This will result in its disappearance on the receiving device and the absence of an image (more precisely—in the appearance of black tone on the positive image). In this way the water surfaces, regions of salt marshes, takyrs, highways, and other even surfaces will be displayed. The maximum absorption of radio waves and the dark tone of the image will be observed also when objects or regions with great moisture content and, consequently, high electric conductivity, are irradiated. These objects may be: freshly tilled plow land, the regions of tectonic breaks with outlets of subterranean waters, and objects that absorb moisture readily and are surveyed after rain. The small angles of radiation especially emphasize the relief on the radar images, creating a pronounced "sculptural" picture.

Owing to these specific properties of image production, as well as a certain generalization at the expense of small scale, the radar aerial images of the

TOROS system present new information beyond that obtained by aerial photographs. They are widely used in the USSR for compiling ice charts and in geological exploration.

The TOROS system permits the compilation of maps of general ice cover distribution based on the results of ice surveys for carrying out navigational reconnaissance and permits the compilation of detailed large-scale maps of ice distribution and condition. These are of direct use for navigation and for conducting experiments concerned with hydraulic engineering, rescue operations, and other work at sea.

The intensity of the radar signal reflected from the ice surface depends on the relief and "roughness" of the latter as well as on the physical properties and condition of the ice fields. These characteristics may vary within wide limits. However, there exists a certain set of indications for each of the ice categories, which relate to the particular features of the surface relief and the configuration and size of the ice fields.

Contour details and distinctions in character of the micro-relief of the floe surfaces, which cannot be detected either by visual air reconnaissance or in aerial photographs, may be observed in radar images with enhanced contrast. They are the edges of sea ice floes frozen together and covered with snow, narrow frozen cracks, and the character of their surface relief.

The following characteristics of the sea ice cover may be determined by radar images: degree of packing, age (using enlarged gradations), relative amount of floes of different size, the degree of hummocking of ice cover, the amount, size, and orientation of free water areas and channels, the speed and direction of the ice drift (by repeated surveys), and in specific cases also the ice compression.

The determination of the ice packing is the simplest and most exact operation. The discrepancies between estimates of ice packing made with radar images and aerial photographs amount to only 3–4 %.

In contrast, the estimation of ice age with radar images is the most complicated process. However, as experience has shown, some of the ice categories can be determined with sufficient accuracy. Among these are: (a) initial ice such as slush ice, nilas, sludge; (b) young ice (gray and gray-white) differing from nilas by greater brightness of its image and by its tendency to break in motion; (c) first-year ice without classification by color, first-year ice of medium thickness, and thick first-year ice; (d) old ice (its estimation becomes considerably more complicated during the summer when melted snow accumulates on the ice surface). Figure 1 presents an example of a radar image for different ages of ice and the results of its interpretation; an aerial photograph of the same region is added for comparison.

The radar images transmitted on board an ice breaker provide the most objective information on the state of ice cover and permit a quick choice of

Fig. 1.(a) Aerial radar image of ice cover. (b) Aerial photograph of the same part of ice cover

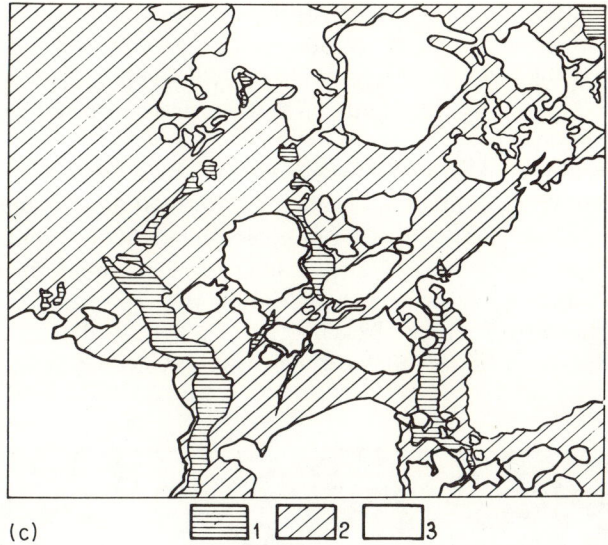

Fig. 1.(c) The scheme of aerial radar image interpretation results (composed by V. S. Loshtshilov). 1—nilas and open water; 2—gray and gray-white ice; 3—thick one-year ice.

route for a ship convoy. Figure 2 gives a radar image fixing the route of a ship convoy through gray ice to open water.

The interpretation of radar images permits compilation of operational ice charts along the whole route of a ship convoy. They are much more detailed and accurate than those compiled from visual observations.

The extensive application of radar images in geological exploration was prompted by the " all-weather " availability of the method and the possibility

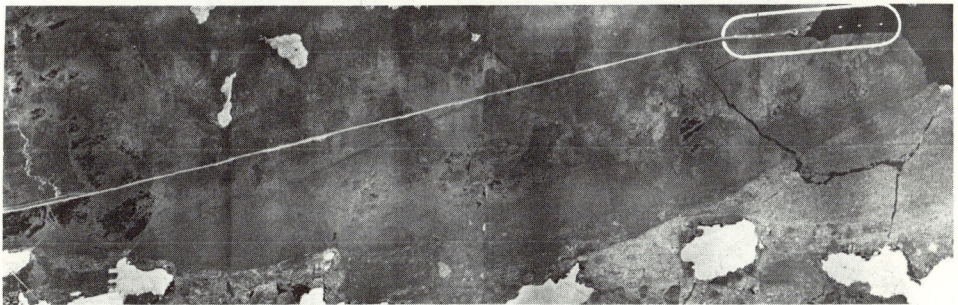

Fig. 2. Aerial radar image of gray ice floe. White line—route of the caravan; white bright points on open water and near the outlet from the ice—separate ships of the caravan (V. S. Loshtshilov's material).

of quickly obtaining small-scale images for vast areas (up to ten thousands of square kilometers), as well as adequate representation of the main indicators of geological structure—relief, hydrographic pattern, surface texture, humidity, vegetation, and top soil—as used in usual aerial photographic interpretation.

In some instances the radar survey has an advantage over the photographic survey as a result of a peculiar "sculpturedness" defining various elements of relief (the main indicator of geological structure) and the generalization of small immaterial details of landscape on radar images, which are often a hindrance in studying the aerial photograph (Fig. 3).

The morphological features of each type of relief are determined by the material composition of the rocks, their stability against weathering, the

Fig. 3. Aerial photograph (a) and aerial radar image (b) of a district of the Kamchatka Eastern volcanic belt. Due to lower resolution on radar image small landscape details are smooth, particularly the small structure of lava flows and covers. Accordingly, the regional break marked by the chain of small volcanic cones is shown somewhat more distinctly. In a number of cases the contrasts of some objects (e.g. bright bare sections on the aerial photograph) are changed.

character of clefts, and the amplitude of the most recent tectonic movements which are especially sharply defined on radar images.

Besides the morphological types of relief, its individual forms, such as ridges, mounds, washout cones, beds of waterways, terraces, as well as the elements of its forms (shelves, slopes, edges), are well shown on the radar images. Each specific morphological type and subtype of relief is distinguishable and permits one to obtain a general concept of geomorphological structure of the territory explored.

2.2. *Interpretation of Rock Composition*

The unequal response of rocks to radio wave diffusion which becomes apparent in radar images through the changes in tone, makes it possible in some instances to determine the rock type. Thus, the image of coarse-grained rocks will be of lighter tone than that of fine-grained rocks, all other things

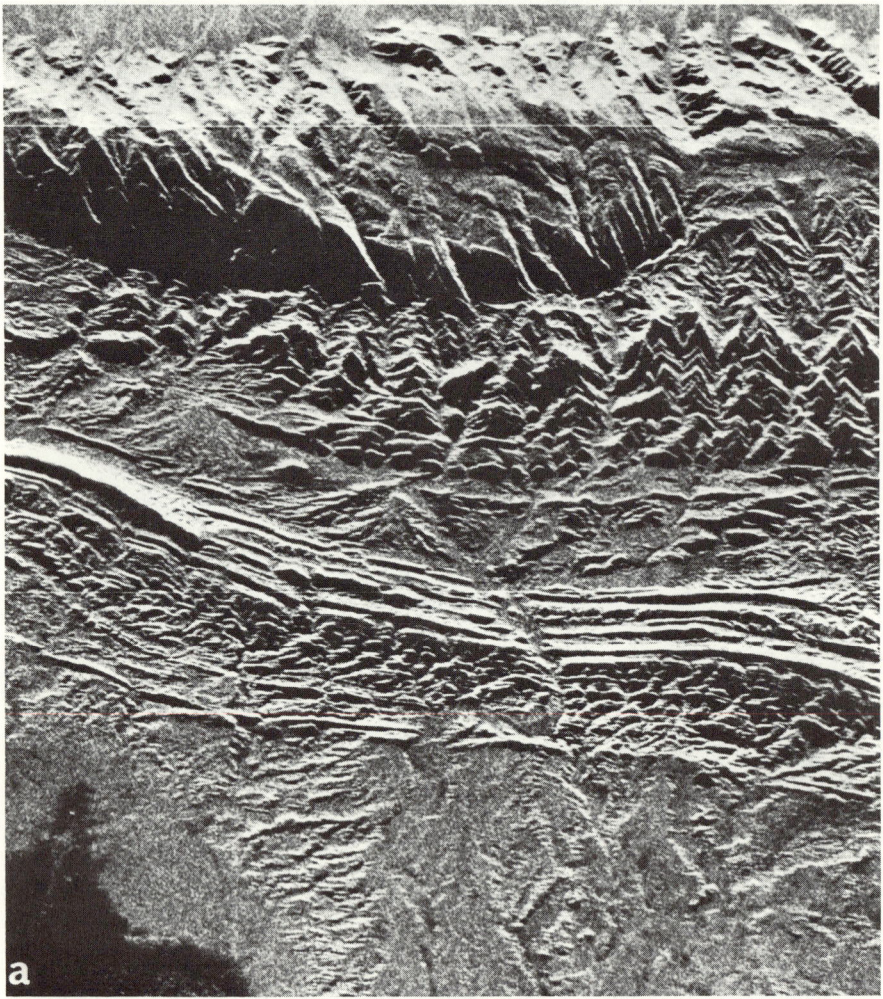

FIG 4a. See facing page for legend.

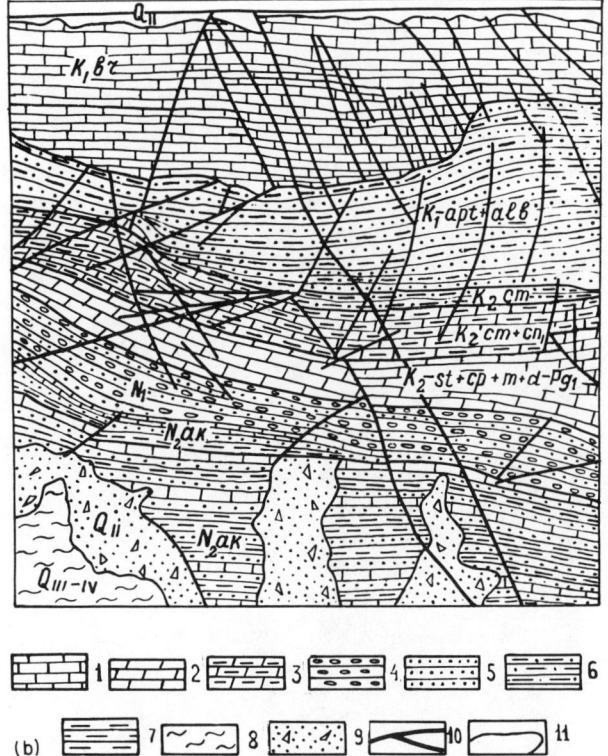

Fig. 4. Aerial radar image of Middle Asia mountain region (a) and scheme of its geological interpretation results (b) (composed by V. A. Starostin). 1—limestone, 2—marl, 3—clay marl, 4—conglomerate, 5—sandstone, 6—aleurolite, 7—clay, 8—loam, 9—sandy l am, 10—faults, 11—boundaries of lithologo-stratigraphic divisions.

being equal. In some cases a sharp difference is observed in the tone range of the objects on the aerial photographs and radar aerial images (Fig. 3). Occasionally, when rocks lie horizontally and irregularities of their texture are less than one-half wavelength, there will be a nearly specular reflection of the latter and the rocks will be of dark tone in the image.

In this way, with due consideration of the character of the surface texture, the relief features (slope angles), cleft characteristics, the character of the erosion pattern, the degree of humidity, and the development of vegetation, it is possible to distinguish between rocks of different composition. Clearly defined border lines between the lithologic-stratigraphic subdivisions emphasized by the above-mentioned features enable one to determine the geological structure of the region.

An example of interpreting a sedimentary series of rocks is presented in Fig. 4. On the radar image one can easily identify the carbonate and terrigene

Fig. 5a. Aerial radar image of Karymski volcano and its environment.

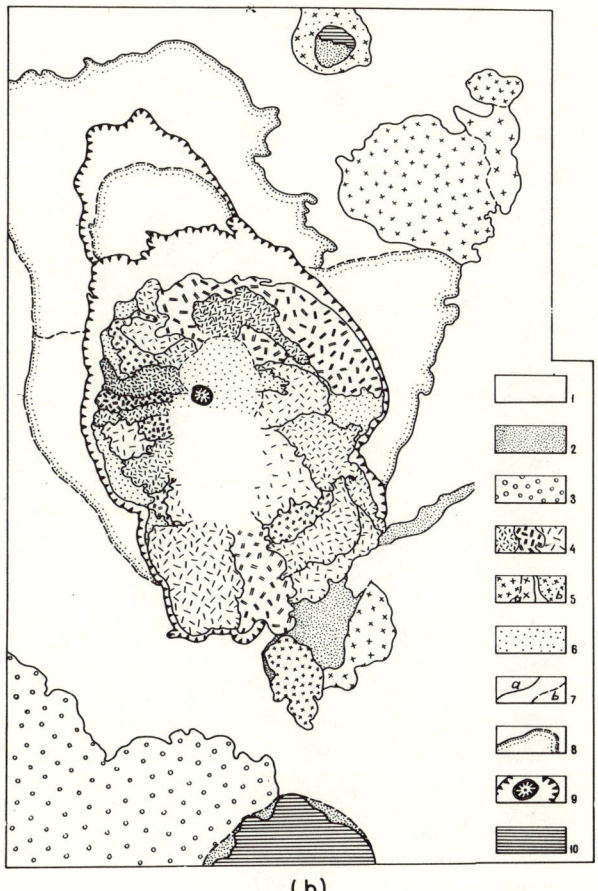

(b)

Fig. 5b. Various geological formations in the region of Karymski volcano (composed by N. A. Gussev). 1—volcanic foundation, plateau composed of pyroclastic and lava products of different thickness; 2—recent alluvial and lake deposits filling the most depressed parts of relief; 3—tuff deposits of the southern complex bearing no relation to the structure of Karymski volcano; 4—young lava and agglomerate flows of the recent Karymski volcano; 5—Volcanic deposits of other eruption centers: a—mainly lava, b—mainly pyroclastic; 6—recent ash–lava deposits of Karymski volcano; 7—geological border lines: a—interpreted with assurance, be—interpreted with incertitude; 8—border of ancient Karymski volcano and Dvor volcano; 9—caldera scarp of the main structure and active volcano; 10—Lakes.

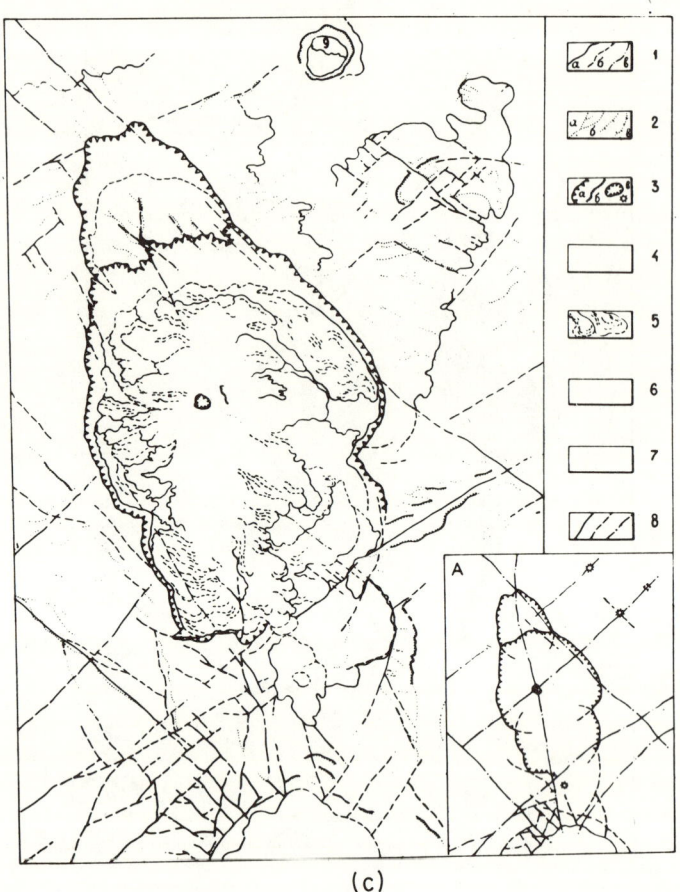

Fig. 5c. Structural-tectonic structure of the Karymski volcano (composed by N. A. Gussev). 1—geological borders of most considerable subdivisions: a—interpreted with assurance, b—interpreted with less assurance, c—supposed continuations of geological borders; 2—some morphological linear elements (erosion valleys, slope discontinuities, scarps); 3—caldera scarp of the main volcanic structure and the active crater; 4—centers of extinct volanic activity; 5—recent lava and agglomerate flows of contemporaneous Karymski volcano; 6—the most notable erosional and tectonic scarps; 7—lakes; 8—fractures and splits: a—more significant, b—less significant and interpreted with incertitude, c—supposed principal directions of tectonic breaks. A—Main tectonic directions of the Karymski volcano area.

rocks of chalk, paleogene, and neogen, while the border lines of individual formations are traced more distinctly as compared to aerial photographs. Friable Quarternary deposits, especially the takyr regions, and shores with different degrees of salinity, moisture, and surface texture are well distinguished as regards their material composition.

The volcanogenic and intrusive rocks, especially dikes and dike belts, stand out by the morphology of relief, the character of clefts, and other features. It should be noted that the image contrast of the dikes and other linear geological objects, including faults, depends upon their position relative to the scanning beam. The dikes are especially well displayed when in a perpendicular position relative to the direction of radiation and much worse when oriented in parallel.

The interpretation of volcanic formations is exemplified by Fig. 5, which presents the varying composition of eruptive products for the active volcano Karymski on Kamchatka [Fig. 5(b)]. Aerial radar surveys permit adequate interpretation of the lava streams of the volcano's main cone and their relationships, and the flat surfaces characteristic of the areas of young lake and river sediment deposits.

2.3. Interpretation of Structural-Tectonic Texture

Aerial radar surveys allow one, on the basis of different rock material composition and the morphology of folds, to distinguish between various structural-tectonic subdivisions: structural zones, specific folds of different orders, ring structures, and zones of tectonic fractures.

Owing to the differences in the surface texture resulting from the relief features, clefts, sheet jointings, and other indicators of geological structure, the folded structures are well distinguished on the aerial radar images.

The change of intensity in radio waves reflected from the surfaces, which differ in their physical properties, indicates, in principle, the possibility of using the rocks with high moisture content and, consequently, high electric conductivity as marker horizons.

The degree of interpretability on aerial radar images of structural elements (marker horizons, fractures) depends not only upon the material composition of the rocks, their distinctness in relief, or conditions of occurrence, but also upon their orientation with respect to the direction of radiation.

It might be well to point out the feasibility of interpretation of tectonic fractures of different orders on the aerial radar images and the possibility of distinguishing additional fractures that are imperceptible or poorly noted on aerial photographs.

In the territory of Kazakhstan and Uzbekistan, the interpretation of aerial radar surveys permitted the identification of a number of large-size systems, fractured zones, differing in age and significance, which might serve for ore discovery. On the basis of the geological analysis of aerial radar images schemes of tectonic fractures were plotted for these regions.

Comparison of the interpretation results for aerial radar images and geophysical explorations revealed good agreement. This suggests the possibility of using the aerial radar images for mapping the fractures which are not sufficiently visible on the surface and which are inadequately fixed by the usual geological methods, but at the same time are noted during geophysical observations.

A distinct interpretation of tectonic fractures on radar images suggests their use for the study of block tectonic structure (Fig. 6). An example of

FIG. 6a. See facing page for legend.

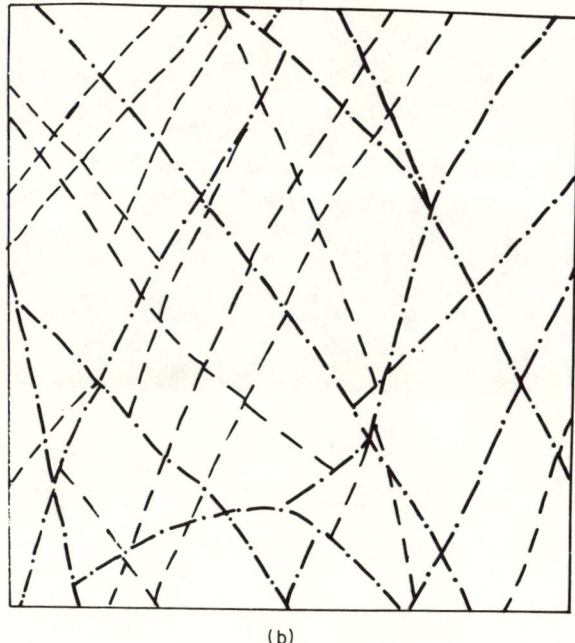

(b)

Fig. 6. Aerial radar image of block structure region of upper Paleozoic effusives in Kazakh fold country (a) and fractures limiting the rocky blocks (b). Fractures are interpreted with certainty according to the orientation of the erosional pattern large elements, zones of increased humidity, and zones of vegetation cover development.

compiling sufficiently detailed structural tectonic schemes is presented by the results obtained for the volcano Karymski and for the adjacent region [Fig. 5(c)] where the main directions of fractures whose intersection controls the position of extinct and active eruption centers are displayed. The analysis of data for a somewhat greater territory allowed the establishment of previously unknown relationships of volcanic structures and volcano-tectonics.

2.4. Interpretation of Geological Structure in Closed Regions

It is common knowledge that the main indicators of the closed geological structures on the surface of friable deposits are: (1) morphology of contemporaneous relief forms (their orientation, position in space), (2) changes of material composition of deposits and orientation of the contours of their facial differences, (3) change of soil-vegetation cover character, (4) changes of physical condition of friable deposits (in particular, humidity) and orientation of the contours of humidified lithological differences, and (5) orientation of elements of erosion pattern.

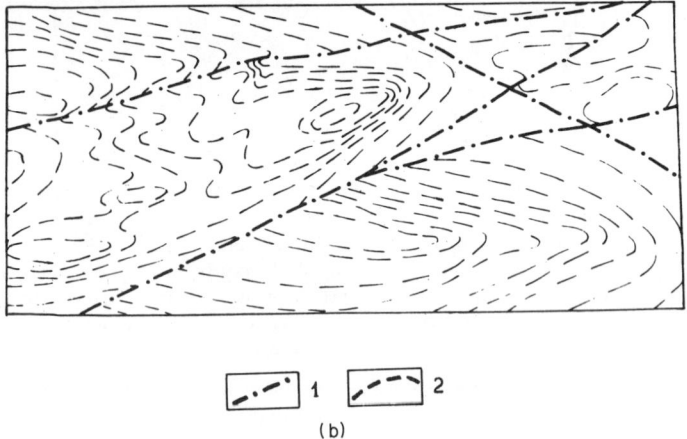

Fig. 7. Aerial radar image of the region of submountain proluvial plane (a) and its structural–tectonic structure (b) (composed by V. A. Starostin). Designations: 1—fractures, 2—structural lines.

As stated above, these indicators are easily identified on the radar aerial images and, using them, it is possible to interpret the hidden structures. In the regions of recent sinking and rising of foundation the tone differences are very important. In addition, the orientation of the relief and hydropattern forms, in particular the sharp changes in direction of elements of the erosion patterns, are also of vital importance.

The closed tectonic fractures along which the tectonic movements took place are interpreted on the radar aerial images by the oriented contours of the humidification zones, "straightened" portions of erosional pattern, contacts of the contour configuration and structure differing in tone, and pattern of the image.

s, science and research have gained a new highly efficient means for the gation of the earth's surface which, together with other available means taining information (aerial photography, aerogeophysics, etc.), will te the extension of our knowledge of the natural resources.

3. INFRARED AERIAL SURVEY

General

development of this method for compiling thermal charts of the earth's e is associated with the great success of infrared techniques in recent . The usefulness and extension of this application in studying natural rces are predicated on the fact that many natural and artificial objects phenomena are accompanied by temperature contrasts. The landscape nts with abnormally high temperatures are most readily studied by red aerial surveys. Consequently, at the present time infrared aerial y has found a fairly wide and efficient application in the study of regions cent volcanism (Gussev *et al.*, 1972; Shilin *et al.*, 1969, 1971) and for ing forest fires.

me results for the first of the above-mentioned applications are given w. The specifics of using the infrared aerial surveys in this case require a rate consideration of their use for studying the active volcanoes and the iated regions of strong geothermal activity.

ie infrared aerial survey was carried out from the aircraft Li-2 with the of an infrared scanner, whose detailed description has been given by in *et al.* (1971).

Infrared Aerial Survey of Active Volcanoes

n example of the detailed study of active volcanoes by means of infrared al survey is the Karymski volcano—one of the most active volcanoes in Kurilo Kamchatka volcanic zone. It is a regular cone situated in a large lera. The diameter of the crater is about 250 m.

o information has been found in the literature about the arrangement . intensity of this volcanic thermal field, although in recent years the cano was systematically investigated by volcanologists (Ivanov, 1970). s is explained by the inaccessibility of the crater for a direct land-based dy because of the contant high activity.

he infrared aerial surveys were carried out in 1967 and 1972. During the t period, the volcano was comparatively inactive—only a strong gas ission by the crater fumarole fields was observed visually. During the rs that followed, the volcanic activity sharply increased, attaining its

The closed structures of West Turkmenia pro⟨...⟩
preting buried structures on the radar images. T⟨...⟩
divided by the break of northeastern bearing ma⟨...⟩
on the radar image (Fig. 7). On the aerial photogr⟨...⟩
above-mentioned features of geological structure ca⟨...⟩

The conformity of the elements of structural-tect⟨...⟩
pretation of radar aerial images, to the data of g⟨...⟩
drilling corroborates the obvious potentialities o⟨...⟩
geological explorations in closed territories.

The examples described above demonstrate wi⟨...⟩
radar images for photogeological interpretation and⟨...⟩
tectonic and geomorphological maps, as well as for⟨...⟩
composition of rocks.

The possibility of obtaining quickly the informat⟨...⟩
regardless of the time and weather conditions, offers⟨...⟩
in a short time radar surveys in different areas of t⟨...⟩
their geological structure and geographical location)⟨...⟩
data to geologists for practical use.

Interesting results were also obtained by using rad⟨...⟩
field crop studies. One can differentiate plowed land⟨...⟩
various crops by the degree of protective cover, the s⟨...⟩
tion cover, and the moisture, which are manifested in⟨...⟩
change in its tone (Fig. 8).

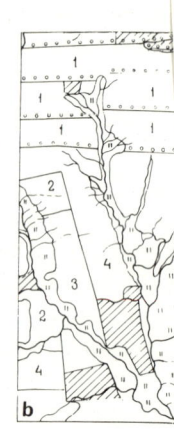

Fig. 8. Aerial radar image of agricultural region (a) and the res⟨...⟩
(composed by N. N. Semyonova). Designations: 1—fallow fields,
old arable lands. The numerals on the scheme indicate: 1—wheat⟨...⟩
4—maize.

maximum in 1971. Fresh lava streams descended the slopes of the volcano. The period of gas–ash outbursts was reduced to a few minutes. In 1971 the activity decreased; during the surveys in October 1972, the outbursts occurred every few hours. The crater configuration changed, compared to 1967: it had the shape of a slightly extended oval, but the direction of the long axis changed from NW to SW. The results of the survey in 1967 are given by Shilin *et al.* (1969).

In 1972 the infrared aerial survey was carried out from two altitudes (400 and 800 m) and at different levels of signal discrimination (Fig. 9), which allowed the obtaining of extensive information on the crater structure. The infrared image 9(b) was obtained in the spectral interval with a minimum width of 4.6–5.3 μ (enabling the complete elimination of the influence of the sun) and with a small level of signal discrimination. As a result, the general scheme of crater thermal fields and adjacent regions was revealed [Fig. 9(c)]. A similar, but slightly more detailed, scheme is provided by the interpretation of the infrared image 9(d) obtained at a somewhat higher level of discrimination. On that image the weakly heated peripheral areas of the anomalies have disappeared and the infrared image 9(d) acquired a somewhat greater distinctness as compared to the preceding one. The analysis of infrared images demonstrates that the thermal fields of the crater (slightly nonuniform in their structure) have the shape of a figure eight extended in the direction of the crater oval. Intense anomalies of arch-like form are observed on the NW rim of the crater. A large but somewhat weaker anomaly is noted on the southern slope, about 80–100 m below its edge. The visual observations and analysis of the aerial photograph indicate that all these anomalies are connected with exceptionally active fumarole fields and large ascent channels over which an intensive gas emission is observed. The chains of anomalies on the NW slope are due to the anomalous heating of a young lava stream.

The thermal structure of the crater shown in Fig. 9(c) is confirmed and shown with improved accuracy by the infrared images 9(e) and (f) obtained from a lower altitude under close measuring conditions—an average level of discrimination and narrow spectral limitations of 4.6–5.3 μ in (e) and 4.2–5.3 μ in (f). In accordance with the wider interval in the last case, the portions which are accentuated on the infrared image have a slightly greater area than the images of 9(e) due to the passing of the signal from the less heated parts of the crater. The analysis of these infrared images demonstrates that in the crater part of the volcano the most active and most heated is its SW area, whose thermal structure has almost not changed as compared to that shown in Figs. 9(b) and (d). The intensity of the anomaly is considerably reduced in the NE area, where some local parts are accentuated. The anomaly chain of arch-like shape is distinctly observed on the NW rim of the crater. The

Fig. 9.(a) Plan of aerial photograph of the Karymski volcano crater. Scale 1: 5,000. (b) Infrared image of the Karymski volcano crater. Scale 1: 20,000. Spectral interval 4.6–5.3 μ. Flight altitude 800 m. Time 11: 10 a.m. Low level of signal discrimination. (c) General scheme of the volcano crater and its environment's geothermal fields. Scale 1: 20,000. (d) Infrared image. Scale 1: 20,000. Spectral interval 4.6–5.3 μ. Flight altitude 800 m. Time 11: 12 a.m. Low level of signal discrimination. (e),(f) Infrared images. Scale 1: 10,000. Spectral interval 4.6–5.3 μ. Flight altitude 400 m. Time 11: 18 a.m., 11: 22 a.m. and 11: 28 a.m. Medium level of signal discrimination. (g) Geothermal fields of volcano crater from data of infrared images 9(e) and (f) interpretation. Scale 1: 10,000. Medium level of signal discrimination allows the revelation of the geothermal fields of high intensity. (h),(i) Infrared images. Scale 1: 10,000. Spectral interval 4.6–5.3 μ. Flight altitude 400 m. Time 11: 32 a.m. and 11: 38 a.m. High level of signal discrimination. (j) Volcanic crater ascent channels scale from data of infrared images 9(h) and (i) interpretation. Scale 1: 10,000. High level of signal discrimination allows the revelation of the crater ascent channels—the objects of the highest temperature and intensity of radiation.

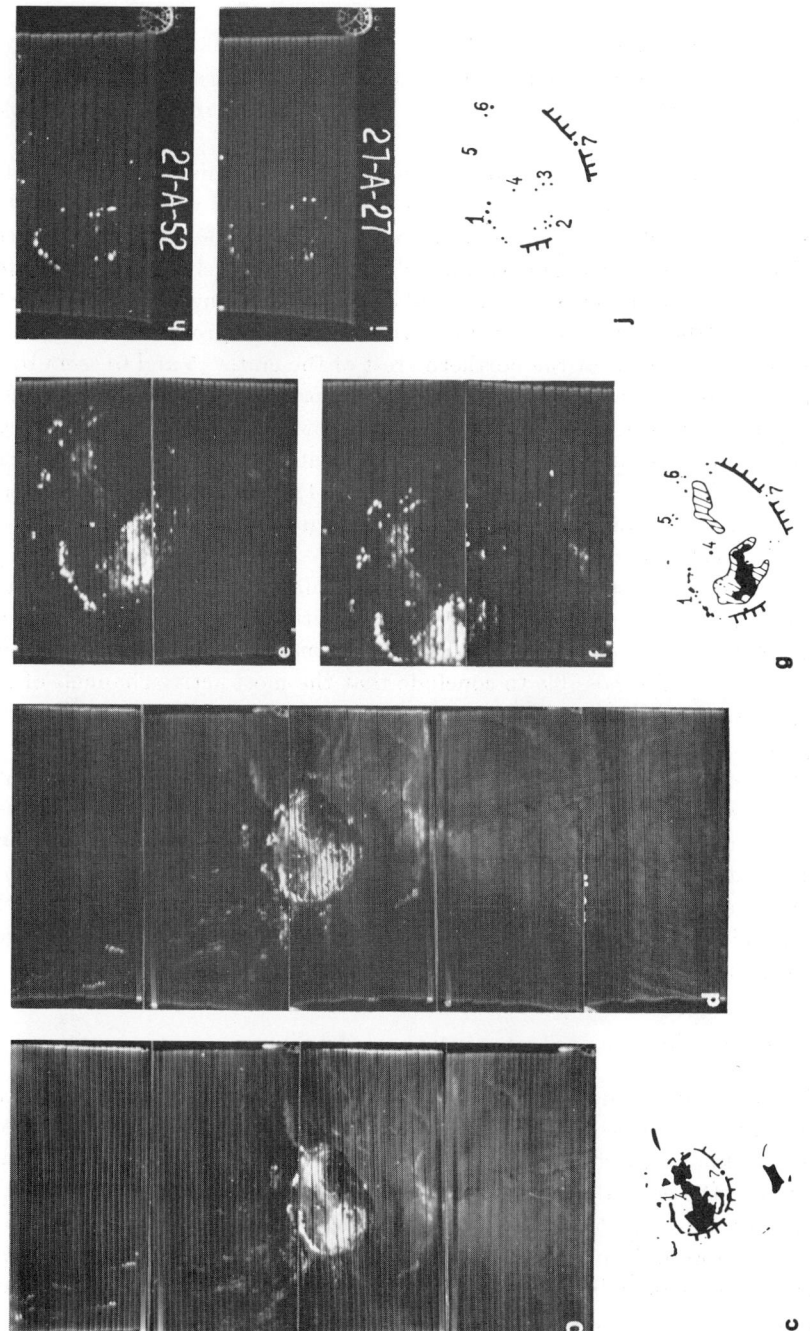

anomaly of the southern slope proved to be considerably weakened. Interpretation of the infrared images 9(e) and 8(f) enables one to plot the scheme of thermal sections of mean intensity for the crater and its environment [Fig. 9(g)].

The increase of the signal discrimination level [compared to the image 9(b) the radiation intensity is reduced about 25 times] while obtaining the infrared images 9(i) and (j) made it possible to compose a scheme of the volcano crater ascent channels 9(k). The most active channels are located on the NW rim of the crater (arch-like chain of channels—1) in the SSW area, where they are separated into two nearly equal groups (2 and 3). The two pairs of channels at the northern crest of the crater (5 and 6) seem to be equal in intensity and only a thorough comparison of infrared image 9(i) and (j) (the latter has been obtained in a narrower spectral interval) reveals a somewhat lesser activity of the channels located more to the east (6). The same may be applied to the central channel (4), which is very slightly perceptible on the infrared image 9(j). All the images display a single channel on the SSE crest of the crater (7).

In this way, the method of infrared aerial survey over the active volcano makes it possible to obtain a variety of information, ranging from the general scheme of thermal field location to the arrangement of the most active ascent channels. It is also possible to conclude that the most active channels of the volcano are drawn, essentially, to the peripheral parts of the crater, a matter of practical interest.

The comparison of the material obtained in the aerial surveys of 1967 (Shilin et al., 1969) and 1972 reveals that in this time interval sharply defined changes occurred in the structure of the thermal fields of the crater. This seems to be quite natural when one considers that the volcano passed through a period of activity in 1970–72. The scheme of thermal field arrangement in the NNE and SSE parts of the crater or its periphery has been preserved only in its most general form.

3.3. *Study of Regions of Strong Geothermal Activity*

The possible application of infrared aerial surveys to the caldera Uson on Kamchatka (Svyatlovski, 1959), where thermal phenomena of various types are known to exist over a vast territory was investigated. The caldera is located in the central part of the eastern volcanic belt to the south of Lake Kronotski and constitutes an extensive (diam 10 km) relatively flat depression cut 50 to 100 m deep into the surrounding plateau. To the central elevated part of the caldera belong the main groups of active and extinct thermal activities. The first are represented by numerous hot and warm springs, small lakes, baths, streams, gas and steam jets, mud pots, and small volcanoes, as well as by thermal platforms (Gussev et al., 1972). The areas closely adjacent to these objects are quite devoid of vegetation and are composed of hydrothermally altered bright opalized rocks. The edges of hydrothermally altered zones are brightly colored due to the presence of various oxides. Accordingly, the extinct and active thermal phenomena are often determined on the aerial photograph by the bright tone of hydrothermally altered rocks.

Near the active and partly extinct thermal activities sometimes arises thick vegetation of high grass which, apparently, may be explained by the presence of an anomalous heat flow favorable to vegetation. The temperature of the active thermal phenomena varies from a few degrees over the background values to 96–98°C.

In 1967 an infrared aerial survey was carried out from an aircraft in September in the daytime from an altitude of 800 m which produced infrared images with an average scale of 1:18,000. Simultaneously, an aerial photographic survey was carried out for a combined interpretation with the infrared images. The aerial photograph with a scale of about 1:18,000 for the central part of the caldera Uson and the mosaic of infrared images are shown in Figs. 10 and 11, respectively.

On comparing the infrared image with the aerial photograph their comparability in contour and tone becomes evident. However, their contrasts depend upon quite different qualities. In the aerial photographic survey these are the differences in spectral reflectance of the landscape elements, while in the infrared aerial survey the differences reside in thermal emissions or in radiative temperatures. The contour similarity between the infrared image and the aerial photograph is explained by the coincidence of certain landscape elements as regards reflectance and thermal differentiation (e.g. the contrasts of water and land, of sunlit and shaded slopes in conditions of dismembered relief, etc.). An appreciable tonal differentiation of infrared images results from a very wide range of temperatures, characteristic of various landscape elements in the daytime—from 5–7°C near the cold springs and small rivers up to nearly 100°C at the gas–steam jets and hot springs.

The interpretation of infrared images of the central part of the caldera enables one to distinguish a large group of bright, anomalous objects of irregular shape extending in an approximately northwestern direction and a group of extended parallel anomalies of greater size (located in the upper part of the infrared image). The limited dynamic range of infrared equipment and the comparatively broad spectral interval of measurements (3.2–5.3 μ) makes it difficult to eliminate false insolation anomalies, which greatly hamper the interpretation. The comparison of infrared images with the aerial photographs, and especially with the stereocouples, enables one to draw the conclusion that the second group is connected with facing the sun

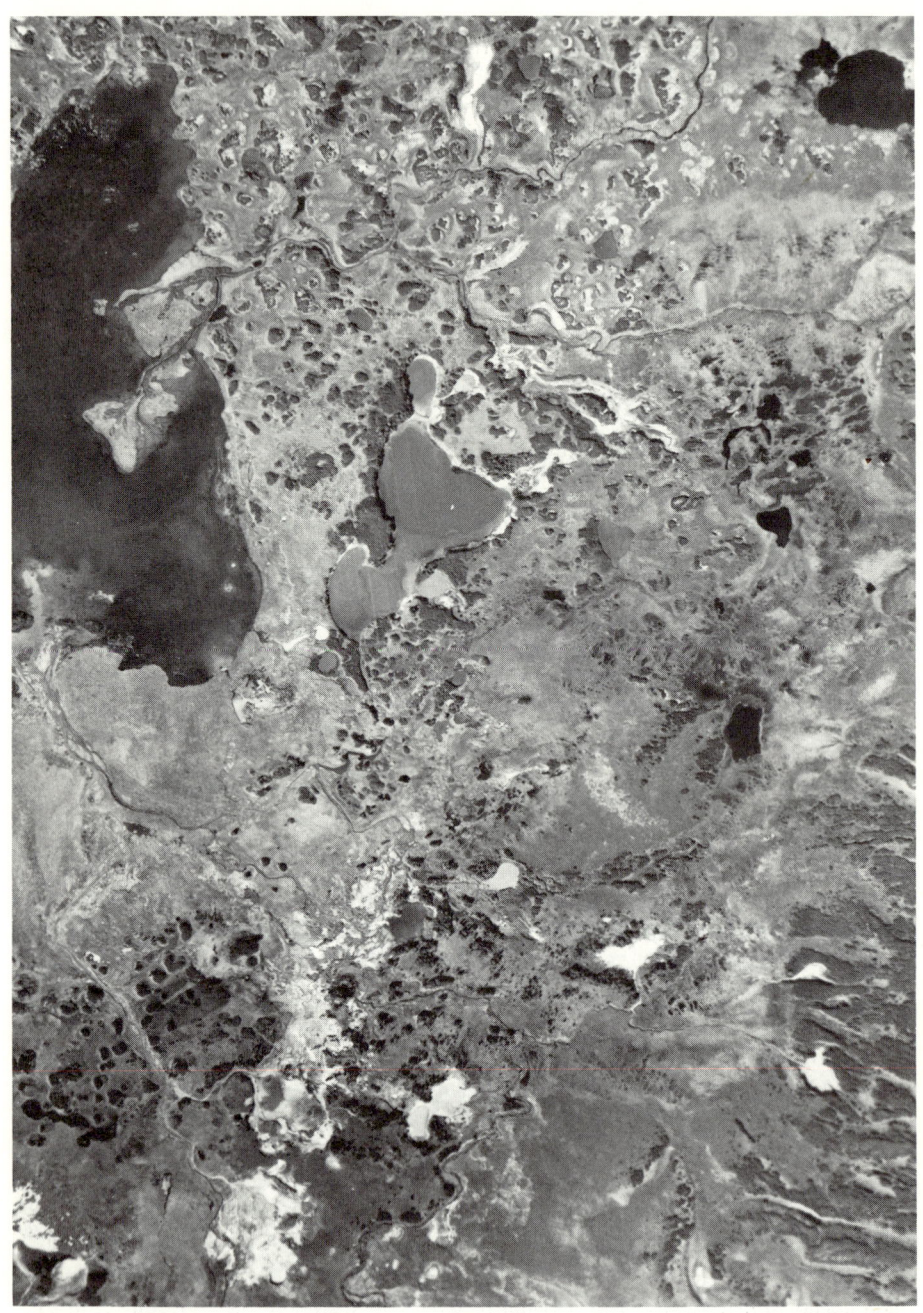

Fig. 10. Plan of aerial photograph of the central part of caldera Uson. Scale 1: 18,000.

FIG. 11. Mosaic of infrared images of the central part of caldera Uson. Spectral interval 3.2–5.3 μ. Flight altitude 800 m. Daytime, clear weather, separate clouds Survey of 1967.

and well heated elements of relief. It has been established that in September the contrast between shaded and sunlit slopes in this region may reach 15–18°C. However, the intensity of the apparent anomalies on the infrared images makes it possible to overlook the anomalies caused by geothermal phenomena, as further study has proved.

The first group of anomalies is interpreted more simply, as under the conditions of a practically flat relief the presence of apparent anomalies has been reduced to a minimum. An important role is played by the shape and location of these anomalies, their direct connection to the areas of bright (hydrothermally changed) rocks on the aerial photograph and—especially significant —with the water surfaces of lakes, rivers, and streams. The combined interpretation of large-scale aerial photographs (1: 5,000 1: 8,000) with infrared images allows the determination of some types of thermal phenomena according to their specific indications (Gussev et al., 1972). Many of the open thermal phenomena of the caldera (Fig. 14) may be determined by means of indirect interpretation, after performing some field work at the test sites.

However, in the initial stage of analyzing the data obtained by aeromethods in 1967, a scheme of thermal activities (Fig. 12) was compiled, and the anomalies were divided into two groups: the white color designated the anomalies which, with a great degree of certainty, may be connected with geothermal objects; the spotted part indicated the anomalies requiring ground exploration to determine their nature. The latter particularly concern the areas of weak heating of landscape elements near intensive thermal phenomena observed on infrared images.

Based on the results of 1967 and subsequent local checks, in 1972 the infrared aerial survey of the central part of caldera Uson was carried out from a lower altitude (600 m) and in the spectral interval of 4.2–5.3 μ, which allowed the infrared image of thermal phenomena to be obtained on a larger scale and almost without the interference of insolation (Fig. 13). This, in turn, allowed a scheme of caldera thermal phenomena (Fig. 14) to be plotted, differing in detail and reliability from the data of 1967. It suffices to say that in the survey of 1972 only two thermal areas in the lower part are doubtful (4 in Fig. 14).

In Fig. 13 the structure of lakes and thermal fields is interpreted in more detail; even small thermal objects, such as small single springs, baths, and mud pots are determined with assurance. Measurements of temperatures on the terrain at the moment of aerial survey provided the following results; a hot lake (1 on Fig. 13) $+38°C$, a cold stream (2) $+6°C$, a mud bath (3) $+52°C$, the lake Vosmyorka (4) $+18°C$, the spring near the lake shore on the central fumarole field (5) $45°C$, soil in the inactive area (6) $+12°C$.

For a detailed test of the representativeness of the infrared aerial survey

Fig. 12. Data of laboratory interpretation of aerial photosurvey and infrared aerial survey (composed by N. A. Gussev). The continuous tone white color denotes the anomalies connected with geothermal objects; the specks indicate the anomalies which need to be checked in the field to reveal their nature.

Fig. 13. Mosaic of infrared images of the caldera Uson central part. Scale 1:12,000. Spectral interval 4.2–5.3 μ. Flight altitude 600 m. Daytime, clear weather, separate clouds. Survey of 1972. Numerals and arrows indicate the sites of land temperature measurements during the aerial survey.

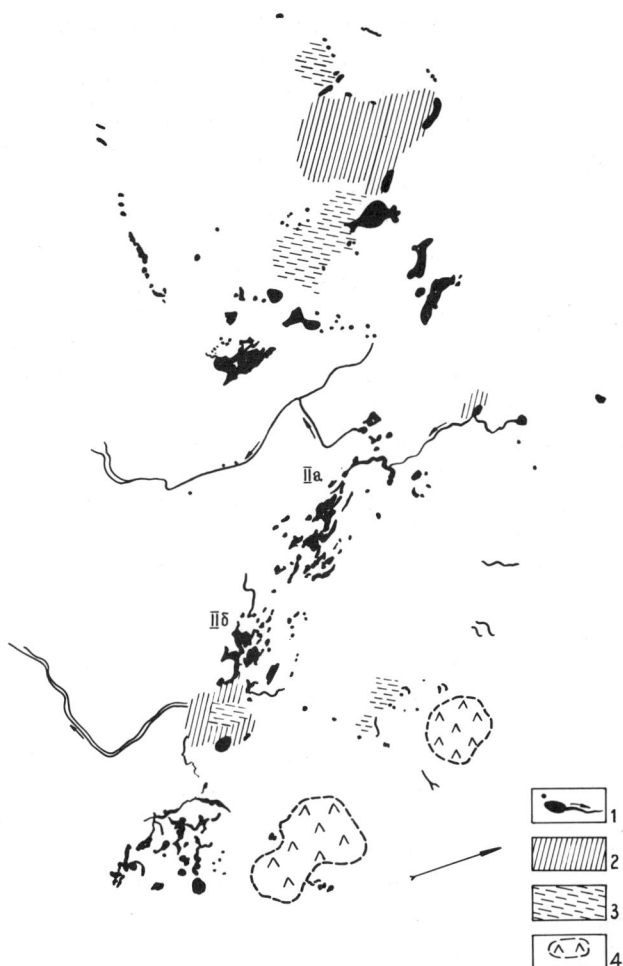

Fig. 14. Geothermal activities in the central part of caldera Uson from data of infrared aerial survey of 1972. Designations: 1—geothermal objects, warm rivers and the most heated parts of the lakes; 2—parts of lakes with medium geothermal heating; 3—parts of lakes with weak geothermal heating; 4—supposed anomalous regions requiring additional field checkup; IIa and IIb—two active parts of central fumarole field.

data while mapping the thermal activity of the area of anomalies revealed, surface thermometrical measurements with a grid of 50 × 50 m, as well as soil and geological observations, were conducted in 1968. The grid was transferred to a more accurate photomosaic with a scale of about 1:5,000. In the most interesting anomaly areas the grid was narrowed to 25 × 10 m. All the open thermal activities (springs, gas–steam jets, mud pots, etc.) were measured by contact mercury thermometers with a range of 0°C to 50°C subdivided into 0.5°C. For the open thermal activities maximum thermometers were also used. The observation point temperatures were determined, both on the surface and to depths of 20 and 50 cm.

The data at the 50 cm depths were used since the influence there of daily temperature fluctuations could be neglected, and they were therefore best for comparison with the infrared aerial survey data. The values of the measured temperatures were plotted on a photomosaic (scale 1:5,000) with the identification of the observation point on the aerial photograph. Isotherms were plotted separately for the ground and reservoirs in order to obtain a better description (Fig. 15).

Using the results of soil-geological observations at the points where temperature was measured, a map of open thermal activities was obtained (Fig. 16), where the areas of strongest geothermal activity are accentuated, namely the areas of the Eastern (Fig. 16, I) and Central (Fig. 16, II) fumarole fields as well as the area of Fumarole Lake–Central Lake.

A comparison of Figs. 12, 15, and 16 shows that all the anomalies revealed by the infrared aerial survey coincide with the areas of strong geothermal activity. As a rule, the anomalies of the first type (Fig. 12, white tone, and Figs. 11, 13) correspond to the open thermal activities or separate objects (Fig. 16)—gas steam jets, hot water and mud pots, thermal springs, etc., i.e. areas with the highest temperature (Fig. 15), whereas anomalies of the second type (Fig. 12, spotted part) correspond to the areas with lower anomalous temperature (20–25°C) that border the open thermal activities.

The comparative analysis of infrared images, photo interpretation data, thermometrical surveys, and the distribution of geothermal activities enables one to draw the following conclusions. All data of the infrared surveys are fully confirmed by the results of field observations, the zones of anomalies coinciding with the areas of strongest thermal activity (Fig. 16) and highest temperature (Fig. 15), i.e. areas with the convective type of anomalous heat flow (Shilin et al., 1969). The scheme of thermal activities in 1972 is much more detailed than the corresponding data for 1967. In this scheme it is possible to divide the anomalies of the Central fumarole field into two active areas (IIa and IIb, Fig. 14) and to note a whole series of thermal objects which have not been determined by the infrared survey of 1967.

On the thermometrical representation, the isolines of which result from

slopes the steam wells, mud pots, and small lakes are to be noted, mainly on the terrace in the middle part of the slope.

The aerial survey was carried out over the middle (and thermally most active) part of the Geyser River. The results are shown in Figs. 17–20.

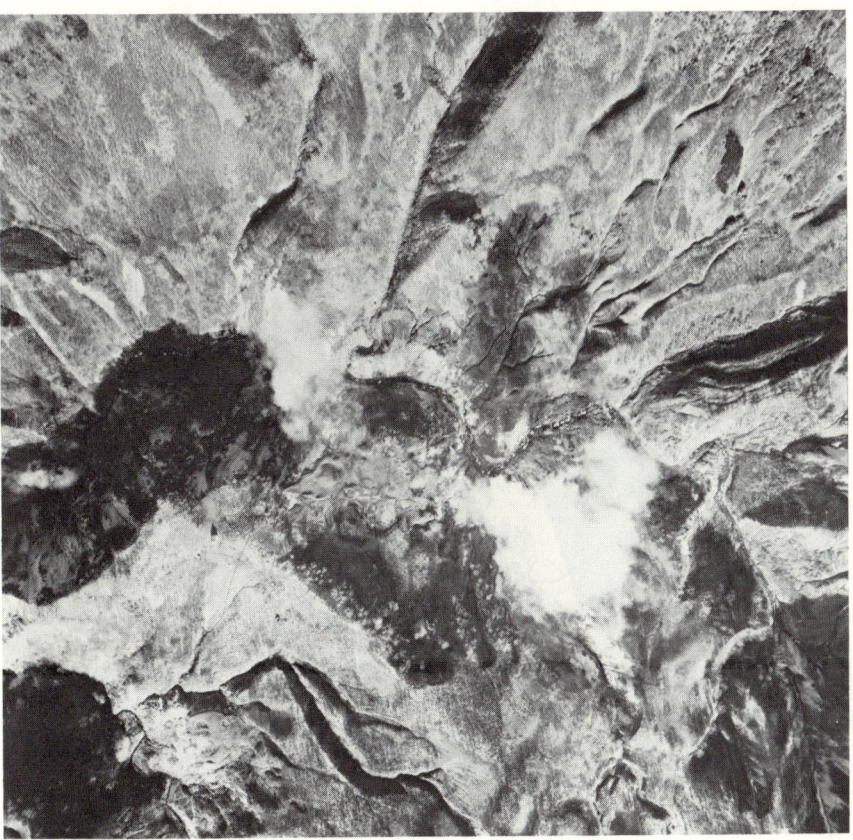

FIG. 17. Plan of aerial photograph of the most geothermally active part of Geyser Valley. Scale 1 : 10,000.

Figure 17 is an aerial photograph to a scale of about 1: 10,000; Fig. 18 is the mosaic of infrared images to a scale of about 1: 15,000 (1972); Fig. 19 is the arrangement of thermal activities of the same scale composed from the interpretation data; Fig. 20 is the mosaic of infrared images to a scale of about 1 : 20,000 (1967).

The rather favorable conditions of surveying in the middle of October 1972, the weakly heated slopes exposed to the sun, and the presence of large

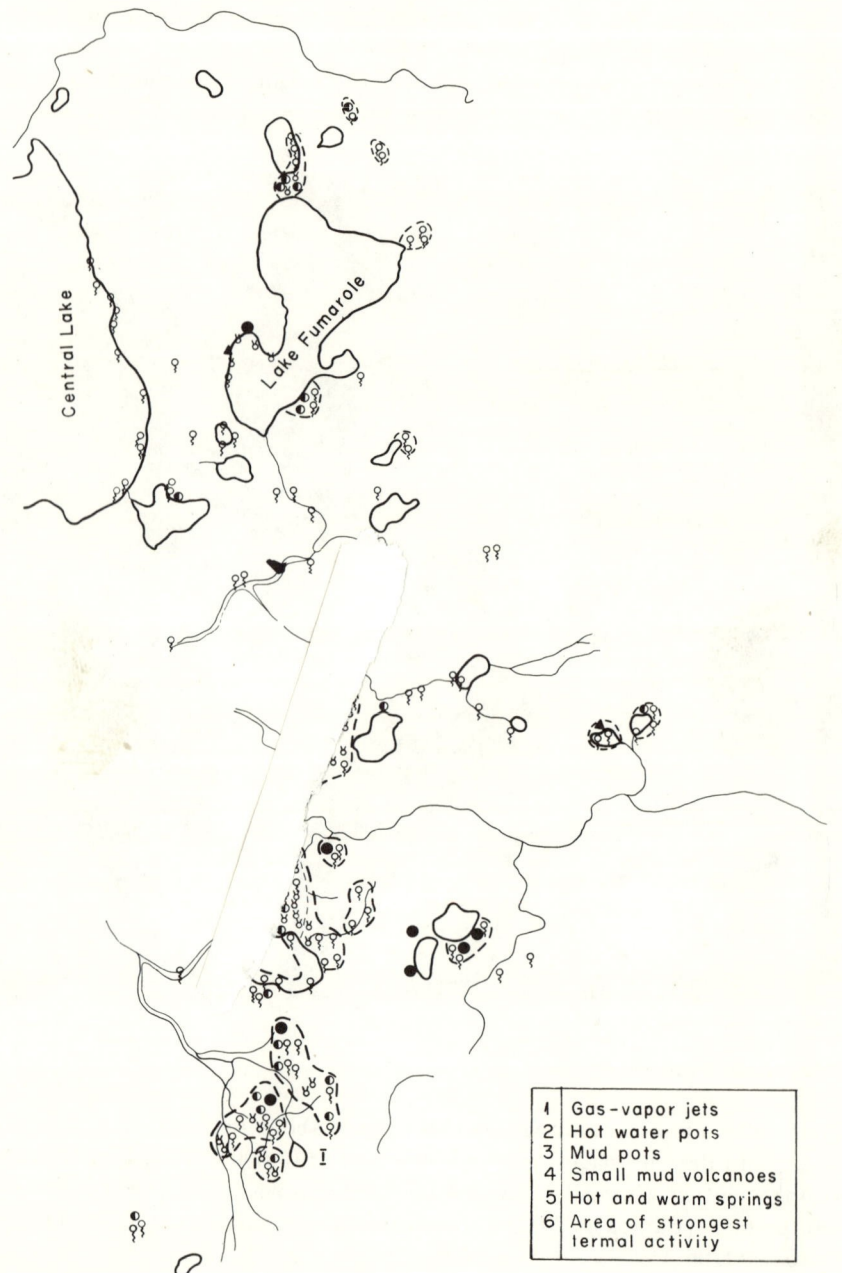

Fig. 16. Scheme of geothermal activities in the central part of caldera Uson (I—activities in the Eastern fumarole field, II—activities in the Central fumarole field). (Composed by E. I. Vavilov.)

considerable averaging of the true picture of the thermal field with respect to space and time (the measurements were conducted during two months), it is possible to identify only the general contours of the areas with a strong geothermal activity without revealing the specific objects, their shape or size. This information, which is of paramount importance for the further geological exploration, may be obtained in considerable detail only by means of infrared aerial surveys. As demonstrated in Shilin et al. (1969), the numerous surveys at different levels of registered signal discrimination in the receptor channel of the equipment enable one to reveal zones with various anomalous heating (as was done for the Karymski volcano and for hot streams on the slope of Kihpinych volcano), which greatly increases the scope of geological information.

The results of laboratory and field interpretation of the aerial photographic surveys and infrared aerial surveys allow one to discern the presence of two systems of breaks, in northwestern (principal) and northeastern directions, that cut the central part of the caldera, but these are characteristic for the whole caldera and the surrounding region as shown by the available geological data. These breaks in both directions and especially the weakened areas of their intersections are practically all associated with known geothermal activities.

On the extension of the main zone of the NW strike (Fig. 14), passing through the Eastern and Central fumarole fields and Lake Fumarole an exceptionally active Western fumarole field (studied by means of infrared aerial survey in 1972) is also known to exist.

Another example of mapping an area with various great thermal activities by means of aerial methods is presented in the results obtained for the well-known Geyser Valley (Ustinova, 1955; Vinogradov, 1964) located at the foot of the volcano Kihpinych. The Geyser River flowing from the volcano cuts into a powerful complex of volcanic rocks represented by lavas and tuffs of andesite and dacitic composition. In the area of the main thermal activities the river forms a narrow canyon cut to a depth of more than 300 m at the absolute depth of its surrounding plateau of about 800–1000 m. Because of the presence of the volcano in the vicinity of a high massif, the character of the relief seems to be very unfavorable for getting the data from the large-scale aerial surveys. However, in 1967 and in 1972 we succeeded in obtaining data enabling us to plot sufficiently detailed schemes of thermal activities.

The thermal activities in the form of hot springs with various discharges, steam jets, mud pots and small lakes, geysers, thermal platforms, etc. are known over the whole length of the Geyser River valley, with the following regularity being observed in their arrangement (Ustinova, 1955). Close to the river the massive-yield hot springs are situated, a little higher—the geysers and hot springs with small and varying discharge. In the upper part of the

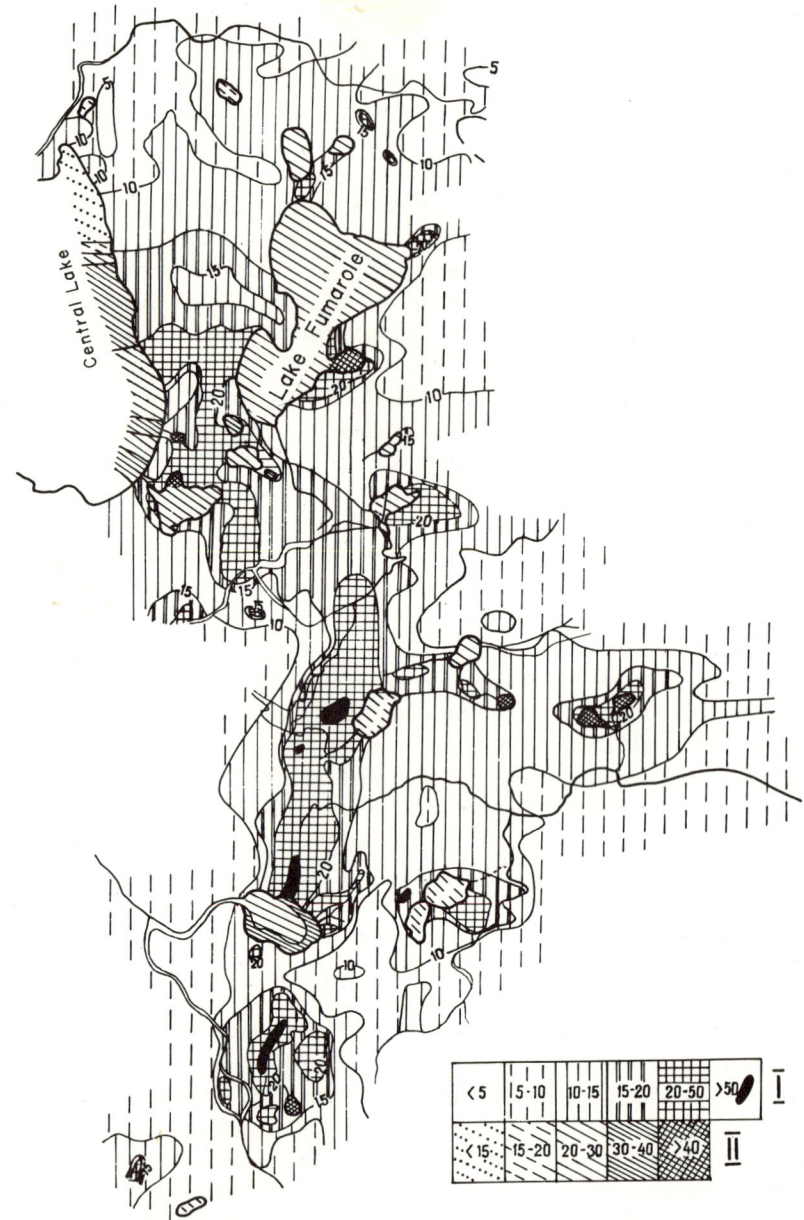

Fig. 15. Thermometric scheme of the central part of caldera Uson from data of field temperature measurements (in °C). Scale 1 : 12000. (Composed by E. I. Vavilov and B. V. Shilin.)

Fig. 18. Mosaic of infrared images of the same part of the valley. Scale 1:15,000. Spectral interval 4.2–5.3 μ. Flight altitude 600 m. Daytime, clear weather, separate clouds. Survey of 1972.

Fig. 19. Thermal activities of Geyser Valley from data of infrared aerial survey, 1972. The numerals show geothermal regions. The thermal field VIII is shown shaded due to the impossibility of detailed distinction between separate objects.

Fig. 20. Mosaic of infrared images. Scale 1: 20,000. Spectral interval 3.2–5.3μ. Flight latitude 900 m. Daytime. Considerable cloudiness. Survey of 1967.

shaded areas, as well as heavy cloudiness during the flight of 1967, enabled us to distinguish easily on the infrared images between the bright local anomalies of thermal activities with sharp contours and pale broad anomalies connected with solar heating.

Figure 21 shows the arrangement of thermal activities compiled by former investigators according to field observations (Vinogradov, 1964). The pattern includes the geysers (1), the main pulsing sources (2), and the so-called thermal areas (3) which imply the accumulation of active areas of various thermal activities and whose determination is very conventional.

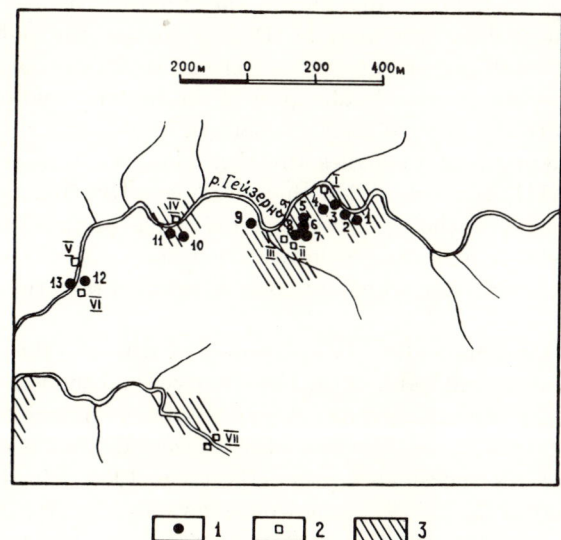

FIG. 21. Scheme of geothermal objects in Geyser Valley from the data of field observations of the Institute for Volcanology. Designations: 1—geysers (1—Rozovi Konus, 2—Horizontal, 3—Zhemchuzhni, 4—Velkian, 5—Nepostoyanni, 6—Dvoynoy, 7—Fontan, 8—Novi Fontan, 9—Shchel', 10—Bolshoy, 11—Malyi, 12—Konus, 13—Bolshaya Pechka); 2—main pulsating springs (1—Paryashchi, 2—Plachushchi, 3—Nepreryvni, 4—Grot, 5—Malahitovyi grot, 6—Malaya Pechka, 7—Kamenka, 8—Skalisti, 9—Vorota v Geyzernuju); 3—the most active geothermal regions.

The interpretation of aerial photographs, analyses of infrared images, and patterns of thermal activities prove that the data of aerial methods are fully confirmed by the data of field observations, but they immeasurably surpass the latter in the totality of information content.

In the easternmost part of the thermal activities, where the Geyser River forms two peculiar bends (I), anomalous heating of the river itself (due to the springs in its upper part) is observed as well as the areas where small warm springs situated in the vicinity discharge into the river. The latter are interpreted on the infrared image as slightly stretched anomalies suggesting "commas" by their shape. The heating of these objects essentially exceeds the heating of the river, which enables one to identify them as independent thermal activities. However, their direct connection with the river demonstrates that they are warm springs, whose discharge is not great and does not cause an appreciable increase in the temperature of the river water. Here again, near the river, a little higher along its left bank, a few small point-like thermal activities, apparently gas–steam jets, are revealed.

Further west occur anomalies of complex thermal structure (II), with the southwestern part undoubtedly due to the intensive heat of the geysers Velikan, Zhemchuzhni, Horizontalni, Rozovi Konus, the pulsating spring Paryashtshi of small geysers, and hot water and steam discharges connected with them. However, a considerable part of the thermal field of this district is connected with the slope of the right bank of Geyser River.

An intensive group of anomalies stretched along the following rectilinear river district (III) is connected with the geysers Dvoynoi, Nepostoyanni, Fontan, Novi Fontan, the high-discharge pulsating springs Grot, Malahitovi Grot, and others. A high, and comparatively constant, discharge of these thermal objects resulted in a high intensity of infrared radiation of this group of anomalies.

The ring of local anomalies (IV) is connected with the thermal activities of the terrace on the left bank of the Geyser River and the valley of its small warm effluent. The thermal activities are represented by numerous small warm lakes, mud pots, gas steam jets, heated platforms, and in the effluent valley also by hot springs and small geysers (Ustinova, 1955; Vinogradov, 1964). The temperature of the objects varies from a few tens of degrees in small warm lakes up to the boiling point in geysers and springs. The terrace consists almost completely of hydrothermally changed warm, claylike, opalized rocks covered with thick high grass and shrubbery. Only the areas of thermal activity are free from vegetation. The data of the infrared aerial survey and aerial photographic survey demonstrate that the western, more elevated part of the terrace has no active areas, whereas a ring structure (of great interest for further geological work) occupies its middle and eastern parts. In a similar region of Iceland this kind of structure, though much greater in scale, has been identified through infrared surveys.

About 150–200 m to the west of the ring structure, the infrared aerial survey reveals a group of three anomalies (V) connected with the hydrothermally changed and heated rocks in the upper part of the southern slope of the Geyser River valley. There, although only small discharges of steam jets are observed, the intensity of the anomaly indicates high intensity heat flow, which had not been noted in prior investigations.

The anomalous zone VI, which for some reason or other has not been mentioned before (Ustinova, 1955; Vinogradov, 1964), should be particularly identified, as the character of the heat anomalies proves that they are caused not only by the warm water of the Geyser River, but also by thermal objects located on the right bank. These, in all probability, are small springs; on the infrared image they are connected with the river.

Downstream of the Geyser River, where the valley is a narrow canyon with steep, up to 70°, slopes (Vinogradov, 1964), two thermal anomalies had been previously identified (Fig. 21): the region of geysers Bolshoi and Malyi at the beginning of the canyon and the region of the geysers Konus and Bolshaya Petchka. The analysis of the aerial photograph and the infrared images gives an essentially new picture of the thermal structure of this zone (VII): from the region of geysers Bolshoi and Malyi the zone is divided into two parallel subzones. The northern subzone is connected with the anomaly from the warm waters of the Geyser River flowing into it from the adjacent small springs, from geysers Konus and Bolshaya Petchka, and the pulsating springs Skalistyi and Kamenka. The southern subzone is connected with the thermal discharges of the right slope in the upper part of the river. Apparently, these discharges are not accompanied by outflow of warm water, as on the infrared image no connection with the river is observed.

The eighth anomalous group (VIII) coincides with the region of thermal activity located 500 m higher than the mouth of the stream flowing into the Geyser River on its right bank, where the active region rises up to a height of 50 m. Here many mud pots, gas steam jets, springs, etc. are situated, whose water appreciably increases the temperature of the stream, a fact well discernible on the infrared image.

In summary, the infrared aerial survey distinctly reveals all the anomalous zones of the main thermal region of the Geyser Valley and its inner structure, and the mass of information largely exceeds the data of many years of field observations. The data of infrared aerial surveys may serve as a basis for subsequent detailed geological field investigations of the structure and origin of thermal zones.

The comparison of the infrared aerial surveys of 1967 and 1972 proves that, when the flight altitude is reduced, the detail of geothermal objects is somewhat increased, and the reduction of the spectral interval width (from 3.2 μ to 4.2 μ) practically eliminates the losses resulting from solar heating and from reflected solar radiation. It should also be noted that no appreciable changes took place in the structure of the geothermal activities of the Geyser Valley during the last five years.

In conclusion, it may be said that the large-scale infrared aerial survey permitted the study with sufficient detail of the geothermal fields and the active volcanoes of the Eastern volcanic belt from the volcano Karymski in the south up to Lake Kronotski in the north. The study of the regional geological structure of this region undoubtedly indicates the presence of an interconnection between these objects. However, this important interconnection is revealed only on the infrared aerial surveys of sufficiently small scale (less than 1 : 100000) which were carried out many times under favorable conditions at night different levels of signal discrimination.

4. Aerogeochemical Survey: Remote Sensing of Gases and Vapors

4.1. Method

The method to be considered now is based upon recording from aircraft the volatile elements and compounds evaporated by surface objects into the atmosphere. By analogy with the widely used geochemical surveys studying the haloes of dispersion of chemical elements and compounds in soils and rocks (Ginzburg, 1957), one may label these "gas haloes" or "evaporation haloes" in the near-surface layer of the atmosphere and call such studies aerogeochemical surveys.

To the volatile elements and compounds, which are of interest from the geological-geochemical point of view, belong in the first place the halogens, sulfurous gas, hydrogen sulfide, a number of hydrocarbons, mercury (to a certain extent), etc. The work of Berringer discusses the great potential of using aerogeochemical surveys in geology for seeking sulfur-sulfide deposits by SO_2 haloes, oil deposits by SO_2 and iodine haloes, and polymetallic deposits by mercury haloes (Barringer, 1964; Barringer and Schork, 1966; Barringer et al., 1968).

Among recent work, the report (Rouse and Stevens, 1970) on the study of SO_2 haloes over the sulfide deposits merits consideration. According to the geological conditions (mainly the thickness of overlying sediments, the presence of faults, and weather factors), the SO_2 concentration varies from 0.03 to 1.0 mg/m^3, when the measurements are taken from an automobile and from a helicopter with flight speed up to 40 miles/hr.

The available general geological and geochemical research suggests a potential use for aerogeochemical survey. However, the actual data are evidently inadequate for organizing operations in the search for minerals. In particular, there are practically no data on field measurements of gas concentration in the air over mining fields and over oil fields. Accordingly, it seems more appropriate to conduct the experimental flights on objects with exceptionally pronounced haloes of evaporation, namely the active volcanoes and regions of strong geothermal activity.

4.2. Equipment and Procedure

The specific features of measuring in flight the haloes of evaporation from the objects on the surface impose two demands on the equipment—a continuity of measurement and rapidity of action. A great number of high precision methods for determining the gas concentration have been developed in recent years; yet in most cases they display an essential defect—the cycling of measurements or a considerable time needed for one measurement. Consequently, in the first stage of the research it seems to be an optimal solution to use the gas analyzers based on the coulomb-polarography principle with a sensitivity of 0.03 mg/m^3 (Alperin et al., 1968). Devices of this type enable one to take continuous measurements, although they display an appreciable inertia. The latter restricts their use to helicopters.

During the summers of 1970–72, the gas analyzer was used not only for SO_2, but also for H_2S measurements. The acquisition of data on the H_2S concentration is possible only in the presence of SO_2 absorption filters at the inlet of the gas analyzer. A cartridge with cotton was used as an SO_2 filter, which proved to be quite satisfactory (Manganell, 1970).

When the measurements are made from the helicopter, distortions resulting from the turbulent flow created by the rotor complicates matters. On the basis of theoretical and experimental data on helicopter aerodynamics, optimal conditions for measuring the gas haloes from a helicopter are provided when the sensors are installed at a distance of 1 m in front of the fuselage and speeds of 50–70 km/hr, so that the influence of turbulent flow from the rotor may be neglected.

However, in 1970–71, under the prevailing technical conditions, various methods of sampling the gas specimens were used. In 1970, the measurements were made by directly sampling the gases during the flights in extended gas plumes of an active volcano, or when the helicopter hovered at a low height over the fumarole fields. This permits one to neglect the inertness of the device and the air intake tubes were extended through the first left illuminator of the helicopter. The instantaneous intake of specimens into a specially designed system of samplers was accomplished through the same tubes. The system consisted of a few glass 1 liter ampoules with cocks at their ends, a vacuum pump, and a system of fluoroplastic tubes. When sampling was done in flight, the ampoules were connected to the pump and evacuated. On signal of the navigator, the cock of one of the ampoules connecting it with the air intake tube was opened. The air was taken into the ampoule and its subsequent analysis was performed in flight. After the measurement was completed, the ampoule was evacuated for the next sampling. The test results of this sampling system indicated that reliable results may be obtained only from the first measurements. During the subsequent use of the same ampoules, the gas concentration sharply decreased because of absorption by water which settled on the walls during the first measurements as a result of adiabatic expansion.

In 1971–72 the measurements were performed using an intake device in the form of a tube, 1 m long, installed in the front part of the helicopter. The sample was collected into polyethylene sacks (4–5 liters), which were filled in 1–2 sec.

All aerial surveys of 1970–71 were conducted over the active volcano Karymski and the thermal activities of the caldera Uson which covers a large area of thermal springs and gas–steam jets.

While working over the volcano the helicopter climbed to the level of the gas–ash plume (Fig. 22). The navigator determined visually the thickest

FIG. 22. Eruption of volcano Karymski. (Photo by N. A. Gussev.)

part of the plume and the flight proceeded normal to the plume, usually at a speed of 100 km/hr or less. On the signal of the navigator a recorder was switched on and the polyethylene sacks were filled. Usually the flight time through the plume lasted tens of seconds. During the turn of the helicopter preceding the next approach to the plume, the gas concentrations in the sacks were measured. In the plume of an explosion only one or two passes were made because the plume dispersed rapidly. Sometimes the sequence of volcanic explosions was so rapid that the helicopter did not land in between.

In only two years more than seventy measurements were made over the volcano Karymski. An example of the recording of the anomalous concentration of gas in the plume is shown in Fig. 23.

In flights over the thermal activities of the caldera Uson, the concentration of gases was measured under different flight regimes (altitude, speed) in various meteorological conditions. The altitude varied between 5 and 40 m, while the method of specimen sampling was nearly the same as that used over the volcano, although the navigator had to be very careful while giving the signals for sampling because of the extremely small size of the objects in this case. Measurements made directly by the apparatus were done only while hovering. Besides, all measurements, about 60 of them, were made only in flights against the wind.

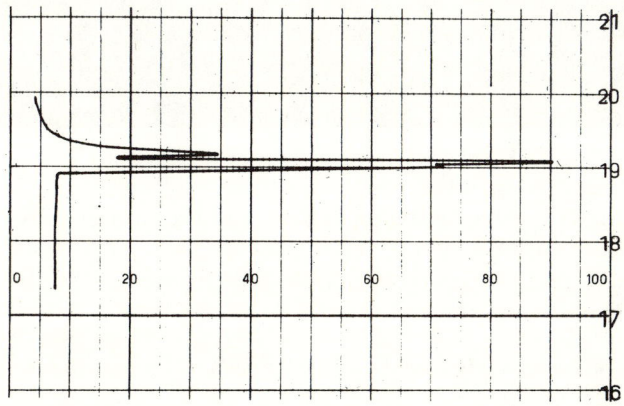

Fig. 23. Example of recording the anomalous concentration in the train of outbursts of volcano Karymski.

Over the fumarole fields of the caldera Uson some measurements were made in the regions of the most active exhalations of the Central and Eastern fumarole fields, represented by sectors that are very close to thermal springs, boiling baths, and gas–steam jets (Fig. 24) which noisily emit great quantities of gas and steam. The smell of hydrogen sulfide in calm weather is noted at a distance of many meters. During field measurements over these objects, the values of the total concentration of $SO_2 + H_2S$ varied from 1.5 mg/m³ (with the apparatus installed at a height of 3.5 m) up to 4.5 mg/m³ (for its position near the edge of the boiling baths). The maximum concentration (of more than 20 mg/m³) has been noticed when the intake tube was introduced directly into the mouth of the gas discharge channels.

Thus a few field measurements have shown a very rapid decrease of gas

FIG. 24. Fumarole fields of caldera Uson.

concentration on moving aside and upstream of the spring. These data are in good agreement with the results of the measurements in air, discussed below, which indicate that the gas content very quickly decreases with the altitude due to its dispersion in the air.

4.3. Aerial Measurements over Fumarole Fields of Caldera Uson

The results of aerial measurements over the thermal activities of the caldera Uson (Central and Eastern fumarole fields, Figs. 10 and 24) are summarized in Table I. In analyzing these data the following should be noted.

All the thermal activities under study are distinctly fixed by the aerogeochemical survey, while the results obtained are essentially dependent on the measuring conditions. An appreciable dependence of the measuring results on the flight altitude and speed as well as on the wind velocity is observed. Maximum concentrations were noticed at the minimum height of 5–10 m, small helicopter speeds, and in calm weather (cases shown in Table I—Serial Nos. 1, 3, 10, 11, 15, 16, 23, 24, 25, 28, 34, 40, 52, 53). When the flight altitude is increased, a rapid decrease of gas concentration down to the background values at a height of 25–40 meters is observed (Nos. 1–7, 15–18, 24–32, 40–45). This is the consequence of the decrease of the gas concentration by rapid dispersion in the atmosphere as the distance from the source increases.

TABLE I. Data on aerogeochemical survey over caldera Uson thermal activities

Serial No.	Region[a]	Time of measurement	Flight altitude (m)	Helicopter speed (km/hr)	Average wind speed (km/hr)	Course	Content of $SO_2 + H_2S$ (mg/m³)	Remarks
					1970			
1	CFF	12:50 p.m.	4–5	40	25	350°–360°	0.15	Sunny day, separate clouds, air temperature 3–5°C
2	CFF	12:53 p.m.	4–5	40	25	350°–360°	0.08	
3	CFF	12:55 p.m.	10	40	25	350°–360°	0.22	Wind gust, the train "floats"
4	CFF	12:56 p.m.	10	20	25	350°–360°	0.15	
5	CFF	12:58 p.m.	15	30		350°–360°	0.06	
6	CFF	13:02 p.m.	15	20–30	25	350°–360°	0.04	
7	CFF	13:04 p.m.	15	20–30	25	350°–360°	0.03	
8	EFF, C	13:06 p.m.	4–5	10	25	350°–360°	0.09	
9	EFF, C	13:08 p.m.	4–5	10	25	350°–360°	0.11	
10	EFF, C	13:10 p.m.	10	Hovering	25	350°–360°	0.2	
11	EFF, C	13:12 p.m.	15	Hovering	25	350°–360°	0.14	
12	EFF, C	13:14 p.m.	20	30	25	350°–360°	0.04	Wind gust, the train "floats"
13	EFF, C	13:16 p.m.	20	30	25	350°–360°	0.11	
14	EFF, C	13:18 p.m.	30	50	25	350°–360°	0.03	
15	EFF, W	13:22 p.m.	4–5	10	25	310°–320°	0.24	
16	EFF, W	13:26 p.m.	4–5	10	25	310°–320°	0.22	
17	EFF, W	13:26 p.m.	10	30	25	310°–320°	0.12	
18	EFF, W	13:28 p.m.	20	30–40	25	310°–320°	0.06	
19	EFF, W	13:30 p.m.	5–7	70–80	25	310°–320°	0.14	Wind gust
20	EFF, W	13:32 p.m.	15	100	25	310°–320°	0.05	
21	EFF, W	13:34 p.m.	15	100	25	310°–320°	0.04	
22	EFF, W	13:37 p.m.	15	100	25	310°–320°	0.04	
23	EFF, W	13:43 p.m.	5	10–20	25	310°–320°	0.3	

continued

TABLE I—*Continued*

Serial No.	Region[a]	Time of measurement	Flight altitude (m)	Helicopter speed (km/hr)	Average wind speed (km/hr)	Course	Content of SO_2 (mg/m³)	Content of H_2S (mg/m³)	Remarks
	10.15.71				1971				
24	EFF, C	11:30 a.m.	5	20	10–20	240°	0.00	0.13	Sunny day, separate clouds, air temperature −2°C
25	EFF, C	11:32 a.m.	5	40	10–20	240°	0.00	0.13	
26	EFF, C	11:33 a.m.	10	40	10–20	240°	0.00	0.03	
27	EFF, C	11:34 a.m.	15	40	10–20	240°	0.00	0.01	
28	EFF, C	11:36 a.m.	15	40	10–20	240°	0.04	0.11	
29	EFF, C	11:37 a.m.	20	80	10–20	240°	0.00	0.09	Wind gust
30	EFF, C	11:39 a.m.	25	80	10–20	240°	0.00	0.09	
31	EFF, C	11:40 a.m.	30	80	10–20	240°	0.00	0.01	
32	EFF, C	11:42 a.m.	40	80	10–20	240°	0.00	0.01	
33	EFF	11:44 a.m.	40	80	10–20	240°	0.00	0.00	
34	CFF	11:48 a.m.	5	40	10–20	280°	0.00	0.16	
35	EFF	11:51 a.m.	5	Hovering	10–20	280°	0.00	0.01	
36	CFF	13:10 p.m.	5	40	10–20	280°	0.00	0.06	The wind becomes somewhat stronger
37	CFF	13:10 p.m.	5	40	10–20	280°	0.00	0.06	
38	EFF	13:15 p.m.	10	40	10–20	280°	0.06	0.03	
39	CFF	13:16 p.m.	10	40	10–20	280°	0.00	0.01	

	10.26.1971							Sunny day, separate clouds, air temperature −5°C
40	EFF	12:37 p.m.	5	60	30	30°–40°	0.0	0.06
41	EFF	12:39 p.m.	10	60	30	30°–40°	0.0	0.03
42	EFF	12:41 p.m.	15	60	30	30°–40°	0.0	0.03
43	EFF	12:42 p.m.	20	60	30	30°–40°	0.0	0.03
44	EFF	12:43 p.m.	30	60	30	30°–40°	0.0	0.01
45	EFF	12:44 p.m.	40	70	30	30°–40°	0.0	0.01
46	EFF	12:45 p.m.	30	60	30	30°–40°	0.05	0.01
47	EFF	12:47 p.m.	20	60	30	30°–40°	0.0	0.02
48	EFF	12:48 p.m.	15	60	30	30°–40°	0.0	0.02
49	EFF	12:50 p.m.	10	60	30	30°–40°	0.03	0.03
50	EFF	12:52 p.m.	5	60	30	30°–40°	0.0	0.08
51	CFF	12:54 p.m.	5	40	20	30°–40°	0.0	0.06
52	CFF	12:55 p.m.	10	40	20	30°–40°	0.0	0.16
53	CFF	12:57 p.m.	15	60	20	30°–40°	0.0	0.01
54	CFF	12:58 p.m.	20	70	20	30°–40°	0.0	0.1
55	CFF	13:00 p.m.	30	80	20	30°–40°	0.0	0.06
56	CFF	13:01 p.m.	40	80	20	30°–40°	0.0	0.03
57	CFF	13:03 p.m.	30	70	20	30°–40°	0.0	0.07
58	CFF	13:04 p.m.	20	80	20	30°–40°	0.0	0.03
59	CFF	13:05 p.m.	15	80	20	30°–40°	0.0	0.06
60	CFF	13:06 p.m.	10	60	20	30°–40°	0.0	0.02
61	CFF	13:07 p.m.	5	40	20	30°–40°	0.0	0.07

[a] Key to abbreviations: CFF central fumarole field; EFF, eastern fumerole field, EFF, C eastern fumarole field, central part; EFF, W eastern fumarole field, western part.

The recorded value of the concentration decreases sharply when the speed of the flight is increased. This becomes evident when comparing, e.g., the values of Nos. 15, 16 and 19, 20, 9 and 10, 11, etc. In the last case, the influence of a small increase in speed is even more appreciable than the rise of the flight altitude which is probably explained by the inertia of the gas analyzer. This is confirmed by a good coincidence of the data obtained in measurements from the hovering helicopter (Nos. 10, 11) with the use of the sampler made of ampoules (Nos. 22, 23), where the intake of the gas for the subsequent analysis is nearly instantaneous, and excludes the influence of instrument lag. The disparity of the reading No. 21 (0.04 mg/m^3) results from the incorrect switching on of the sampler at the moment when the helicopter has nearly "slipped" over the gas halo. This was also the case when the sample was taken into a polyethylene sack (with No. 24). Here only some disparities of the values are noted in unsuccessful attempts to sample gas at the edges of the halo (Nos. 26, 27, 60).

The low value of the concentration when the helicopter was hovering at a low altitude (No. 35), was due to the unfavorable conditions of intake flow caused by the helicopter rotor. However, when the helicopter is hovering with a gas pickup through the intake tubes to the apparatus, high values are observed which decrease somewhat according to the altitude (Nos. 10, 11). At a low altitude (5 m) the helicopter practically hovers in the center of the halo mixing the gas with its rotor and causing an "averaging" of concentrations.

Various changes of the wind exert a great influence on the results of aerial gas surveys. When the wind is strong and gusty, the gas halo "creeps" along the earth, its upper edge drops, and measurements are hampered. Thus, for example, at a constant altitude and speed of flight (Nos. 12, 13, 1, and 2), but with sharp gusty wind, the gas concentration is reduced by one half as compared with calmer conditions. Evidently, this is also the reason for the general drop of gas concentration over the Eastern fumarole field on October 26, 1971 (Nos. 40–50), as compared to the data obtained ten days earlier (Nos. 24–33).

These conclusions indicate the changeability, of the gas halo and the variability of the measured values due to the numerous conditions for dispersal. This results in great difficulties for the development of such surveying methods and makes only very generalized rules in the search for objects of small dimensions possible:

1. The aerogeochemical survey should be carried out from a helicopter at minimum possible altitudes and optimal speeds.

2. The aerogeochemical survey should be performed under conditions of calm or steady wind with a speed of not over 10 km/hr. The direction of the flight should only be upwind, when the conditions for sampling are most favorable.

3. The lag of the equipment must be reduced and instantaneous sampling must be improved or other principles for measurements are required.

As follows from the analysis of Table I (Nos. 24–61), in the gas haloes of fumarole fields an appreciable predominance of hydrogen sulfide over sulfurous anhydride (90–100 % H_2S, seldom 50–60 %) is observed while in the gas outbursts of the volcanoes quite an opposite picture occurs. Such a difference in the final composition of gas haloes having a common or similar primary deep volcanic focus is explained by the influence of processes of SO_2 reduction while passing through thick deposits of various rocks. In the case of an active volcano the primary gases are ejected into the atmosphere without considerable changes in their initial composition.

In conclusion, we wish to emphasize the good agreement between the results of field and aerial observations of 1970 and 1971, regardless of some difference in the operational methods. Thus, for instance, the general character of the gas halo over the Eastern fumarole field is virtually not altered: the maximum concentrations amount to 0.24 and 0.25 mg/m³, respectively; the anomalous measurements of the value are recorded up to a height of 20–25 m. This suggests a satisfactory correlation of the aerogeochemical surveys of 1970 and 1971, as well as the absence of changes in the gas regime of geothermal fields.

4.4. Active Volcanoes of Kamchatka

The most interesting results were obtained over the volcano Karymski.

During October, 1970, the volcano was active, with a period of gas–ash explosions of 1–2 hr. The duration of an explosion usually did not exceed 1–2 min, the main outburst of ash and gas occurring in the first 5–10 sec. After the cloud broke away from the crater, only a weak gas escape was observed from the crater fumaroles. It was easy to notice visually the changing of color in the outbursts connected with the change of their composition and ash content. After the outburst the propagation of the gas cloud depended on the wind speed and direction, dispersing at a distance of 8–10 km from the volcano. During October 1971 the volcanic activity increased sharply, the interval between explosions was reduced to a few minutes, although the visual intensity of the strongest outbursts in our opinion hardly exceeded the intensity of the similar explosions in 1970. The stable gas plume extended from the volcano up to the Pacific Ocean coast, where near the shore an anomalous concentration of SO_2, 0.04 mg/m³ (No. 26), was observed. In October 1972, a considerable decrease of the volcanic activity was observed; the interval between outbursts increased up to 4–6 hr. In total, during four working days of 1970, eight outbursts were investigated, in 1971—twenty nine, in 1972—two. The results of the measurements are given in Table II.

A comparison of the results obtained in 1970 and 1971 indicates that, as compared to the previous season, the activity of the volcano increased considerably. This is confirmed by a sharp increase of total SO_2 and H_2S concentrations in the outbursts of 1971. Thus, for example, if in 1970 the maximum measured value amounted to 2.4 mg/m^3, in 1971 it exceeded 10 mg/m^3. In the small outbursts (12 : 50 p.m., 12 : 24 p.m., etc.) very high concentrations were observed which did not occur in 1970.

While analyzing the ratio of sulfurous gas and hydrogen sulfide in the composition of volcanic gases, a sharp predominance of the former may be observed. The content of hydrogen sulfide seldom amounts to 50 % of the total sum and usually varies within the limits of 0.2 to 2.0 mg/m^3, reaching its maximum sharply differing value of 5.7 mg/m^3. The gas relationship occurring in this case has already been explained above.

The idea of sulfurous anhydride being a sensitive indicator of growing volcanic activity (Masayo, 1961) is again confirmed. The increase of SO_2 concentration in volcanic gases attests to the rise of temperature and pressure in the volcanic focus, i.e. it is one of the first predictors of powerful eruptions.

A detailed examination of the data obtained in the aerial surveys of 1970 (Nos. 1–15) allows one to draw the conclusion that in the period of the survey (October 16–October 24, 1970) the activity of the volcano was characterized by the alternation of two types of explosions: gas–ash outbursts with a gas content of 0.8 to 2.4 mg/m^3 (10/16 at 11 : 08 a.m.; 10/21 at 11 : 07 a.m., 14 : 32 p.m.; 10/23 at 11 : 45 a.m., 12 : 53 p.m.; 10/24 at 11 : 09 a.m.) and gas outbursts with a gas content of less than 0.25 mg/m^3 (10/21 at 12 : 38 p.m., 11 : 27 a.m.). Although the data are obviously insufficient to give the exact sequence, they seem to be confirmed by visual observations.

In 1971, during the intensive activity of the volcano, there were many more measurements in gas outbursts, and appreciable fluctuations of concentration both within the limits of the data on one outburst and between different outbursts were noticed. In the first case these fluctuations were somewhat less and may be attributed either to the imperfection of the measuring methods, e.g. pilot errors (No. 48), or to the irregular gas distribution in the halo, the latter being very notable (see, for instance, Nos. 19–21, 22–25, 44–47, etc.).

As in the previous year, two types of explosions may be distinguished according to the data of aerogeochemical surveys: weak purely gaseous outbursts with a total gas content up to 2.5 mg/m^3 (the most frequently encountered is the value of about 1 mg/m^3) and powerful ash–gas outbursts with a total gas content up to 10 mg/m^3 and more (the most frequent value is 6–7 mg/m^3). Also, it is important to note a general weak tendency to a decrease of the quantity of sulfurous hydrogen in powerful explosions.

In 1972 the anomalous concentration of SO_2 sharply decreased and the

TABLE II. Data on aerochemical survey over active volcanoes

Serial No.	Date and time of explosions	Time of measurement	Flight altitude (m)	Distance to volcano from measurement point (km)	Wind speed (km/hr)	Helicopter speed (km/hr)	Course	Content of $SO_2 + H_2S$ (mg/m³)	Remarks
					1971				
				Volcano Karmski					
1	10.16.1970 11:00 a.m.	11:09 a.m.	1800	4.0	30	100	310	1.8	Clear weather, separate clouds
				Volcano Malyi Semlyachek, crater lake					
2	10.16.1970	11:18 a.m.	0	0.5–0.7	20	100	100	0.4	
				Volcano Karimski					
3	10.21.1970 11:00 a.m.	11:17 a.m.	2000	4	25	100	310	1.76	Clear weather, separate clouds
4	10.21.1970 11:00 a.m.	11:23 a.m.	2000	5	25	100	310	1.3	
5	10.21.1970 12:38 p.m.	12:45 p.m.	1800	5	25	100	310	0.25	
6	10.21.1970 12:38 p.m.	12:52 p.m.	1800	5	25	100	310	0.20	
7	10.21.1970 14:39 p.m.	14:46 p.m.	1700	5(?)	25	100	315	0.92	

(continued)

TABLE II—Continued

Serial No.	Date and time of explosions	Time of measurement	Flight altitude (m)	Distance to volcano from measuring point (km)	Wind speed (km/hr)	Helicopter speed (km/hr)	Course	Content of $SO_2 + H_2S$ by measuring device	Content of SO_2	Content of H_2S	Remarks
8	10.23.1970 11:27 a.m.	11:39 a.m.	1200	2.0	20	120	290	0.2			Clear weather, separate clouds
9	10.23.1970 11:45 a.m.	11:49 p.m.	1600	1.5	20	100	290	2.15			
10	10.23.1970 11:45 a.m.	11:53 a.m.	1600	3.0	20	100	300	0.5			
11	10.23.1970 12:53 p.m.	13:00 p.m.	1700	1.0–1.5	20	100	290	1.75			
12	10.23.1970 12:53 p.m.	13:04 p.m.	1900	5.0–6.0	20	100	290	2.4			
13	10.23.1970 12:53 p.m.	13:07 p.m.	2000	7.0–8.0	20	120	290	1.0			
14	10.24.1970 11:09 a.m.	11:17 a.m.	1600	5.0	30	100	320	0.8			
15	10.24.1970 11:09 a.m.	11:20 a.m.	2000	8.0	30	100	40	1.30			
						1971 Volcano Karymski					
16	10.12.1971 11:50 a.m.	12:05 p.m.	1800	4.0	40	100	230	2.0	0.98	0.25	Clear weather, separate clouds

316

#	Date	Time							Notes
17							0.31	0.1	
18							0.6	—	
19	10.12.1971	12:15 p.m.	2100	2.0	40	100	1.64	1.30	0.21
		12:12 p.m.							
20							2.15	0.25	
21							0.03	0.09	
22	10.12.1971	12:34 p.m.	2100	3.0	40	100	4.75	1.35	0.42
		12:30 p.m.							
23							4.75	0.7	
24							0.24	0.2	
25							0.14	0.08	
26	10.15.1971		1300	25.0	40	120	340	0.04	0.04	—
	10:25 a.m.								

Volcano Malyi Semlyachek, crater lake

| 27 | 10.15.1971 | | 1600 | 0 | 40 | 100 | 320 | 0.1 | 0.05 | — |
| | 10:35 a.m. | | | | | | | | |

Volcano Karymski

	10.20.1971									
28		11:43 a.m.	1400	2	20	120	360	—	0.32	0.01
29		11:47 a.m.	1600	1.5	20	80	180	—	0.05	0.03
30		11:47 a.m.	1600	1.5	20	80	360	1.16	1.00	0.11
31		11:56 a.m.	1500	1.0	20	100	180	0.48	1.60	0.21
32		11:56 a.m.	1500	1.5	20	100	360	3.0	1.80	0.21
33		12:08 p.m.	1700	1.5	20	100	260	0.9	0.26	0.06
34		12:09 p.m.	1700	1.0	20	100	180	0.08	0.42	0.06

The train of the volcano floats to the shore

(*continued*)

TABLE II—*Continued*

Serial No.	Date and explosions	Time of measurement	Flight altitude (m)	Distance to volcano from measuring point (km)	Wind speed (km/hr)	Helicopter speed (km/hr)	Course	Content of $SO_2 + HS_2$ by measuring device	Content of SO_2	Content of H_2S	Remarks
35	12:09 p.m.	12:14 p.m.	1700	1.5	20	100	360	—	6.2	2.0	
36	12:24 p.m.	12:26 p.m.	1700	1.5	20	100	190	—	5.4	0.6	
37	12:24 p.m.	12:27 p.m.	1800	1.5	20	100	250	—	3.25	0.38	
38	12:35 p.m.	12:42 p.m.	2300	2.5–3	20	100	360	—	0.18	0.06	
39	12:35 p.m.	12:43 p.m.	2300	3.5	20	100	180	—	0.56	0.11	
40	12:35 p.m.	12:44 p.m.	2000	4.0	20	100	360	—	0.20	0.07	
41	12:45 pm..	12:48 p.m.	2000	2.5	20	100	260	—	2.80	0.3	
42	12:45 p.m.	12:49 p.m.	2100	3	20	100	300	—	4.1	1.2	
43	12:51 p.m.	12:53 p.m.	1700	2	20	100	300	—	3.9	2.0	
44	12:56 p.m.	13:00 p.m.	2000	1.5	20	400	330	—	4.3	2.0	
45	12:56 p.m.	13:01 p.m.	2200	3	30	400	360	—	2.5	2.0	
46	12:56 p.m.	13:03 p.m.	2500	5	30	100	200	—	4.5	1.4	
10.22.1971											
47		13:05 p.m.	2600	6	30	100	330	—	5.3	0.8	
48	11:55 a.m.	11:59 a.m.	2100	1	40	100	360	—	0.04	—	
49	11:55 a.m.	11:59 a.m.	2100	1	30	100	360	—	0.71	0.13	
50	11:55 a.m.	12:00	2000	2	30	100	180	0.45	0.92	0.22	
51	11:55 a.m.	12:01 p.m.	2000	2	30	100	360	0.5	0.15	0.07	
52	12:06 p.m.	12:09 p.m.	2100	1	30	100	180	—	3.39	0.5	

53	12:06 p.m.	12:10 p.m.	2100	1.5	30	100	360	—	3.01	0.2	
54	12:06 p.m.	12:11 p.m.	2100	2	30	100	180	—	0.28	0.28	
55	12:15 p.m.	12:18 p.m.	2100	1	30	100	180	—	2.0	0.40	Train remote
56	12:15 p.m.	12:19 p.m.	2100	1.5	30	100	360	—	0.58	0.18	from volcano,
57	12:15 p.m.	12:20 p.m.	2100	1.5	30	100	180	—	1.39	0.07	human error
58	12:15 p.m.	12:25 p.m.	2100	8	30	100	360	—	0.07	0.07	
59	12:28 p.m.	12:31 p.m.	2000	1.5	30	100	180	—	0.06	0.05	
60	12:28 p.m.	12:31 p.m.	2000	1.5	30	100	180	—	1.08	0.26	
61	12:28 p.m.	12:33 p.m.	2400	3	30	100	360	—	3.7	0.76	
62	12:39 p.m.	12:41 p.m.	1800	1.5	30	100	180	—	2.1	0.26	

10.29.1971

63	12:15 p.m.	12:20 p.m.	1700	0.5	8–10	30	270	0.75	0.5	?
64	12:25 p.m.	12:29 p.m.	1700	0.5	8–10	80	90	1.0	—	—
65	12:36 p.m.	12:40 p.m.	1800	1.5	8–10	100	270	1.0	—	—
66	12:41 p.m.	12:43 p.m.	1800	1.0	8–10	100	90	7.5	3.7	2
67	12:45 p.m.	12:50 p.m.	1900	1.0	8–10	100	270	0.5	0.03	0.04
68	12:05 p.m.	12:54 p.m.	1900	2.0	8–10	100	90	0.1?	0.27	0.15
69	12:51 p.m.	12:55 p.m.	1900	0.5	8–10	100	270	0.1?	0.49	0.02
70	12:51 p.m.	12:57 p.m.	1900	1.0	8–10	100	90	0.1	0.05	0.03
71	13:05 p.m.	13:08 p.m.	1800	1.0	8–10	100	270	1	0.25	0.1
72	13:09 p.m.	13:10 p.m.	1800	2.0	8–10	100	90	0.35	0.48	0.11
73	13:09 p.m.	13:12 p.m.	1800	2.0	15	100	270	0.6	7.23	0.09
74	13:12 p.m.	13:13 p.m.	1800	1.0	15	100	90	7.5		

maximum value amounted to 0.2 mg/m^3. However, it should be noted that the long period between adjacent outbursts did not permit the acquisition of a sufficient number of statistical data.

The method of quantitative evaluation of the volcanic gas outburst effluent (Shilin and Beryland, 1971) reveals, in conformance with the data of 1971, that during the outburst the volcanic effluent amounts on the average to 2 kg/sec of sulfurous anhydride and 300 g/sec of hydrogen sulfide.

Thus, the following conclusions may be drawn:

1. In the activity of the volcano during the period 1970 to 1971 a considerable intensification not only in the increase of gas outburst frequency, but also in a sharp increase of gas content, is observed.

2. Both in 1970 and 1971, two types of explosion with low and high gas content can be distinguished, the latter being connected with a great quantity of ash in the outburst. The presence of short-period changes in the composition of volcanic gas outbursts are due to the specific processes in its depth focus.

3. In the composition of volcanic gases an appreciable prevalence of sulfurous anhydride over the hydrogen sulfide is observed.

It is well known that the study of the volcanic gas composition enables one to draw valuable conclusions on the composition of the mantle smelts, specifics of the magma differentiation, etc., which will contribute to the solution of the most important problems in geology and volcanology.

In the case of fumarole fields of the caldera Uson type or volcanoes in a static condition the study of the escaping gas can be performed by analysis of samples collected on the ground. But during the eruption of a volcano such measurements can only be made by aerogeochemical methods. Many investigators have stressed the exceptional importance and at the same time the practical difficulty of direct gas sampling from a volcano during eruption (Ivanov, 1970; Masayo, 1961).

Investigations of the SO_2 concentrations are of especial interest because this gas, as compared to the other components of volcanic outbursts, is sensitive to variations of temperature and pressure. It can be surmised (Masayo, 1961) that the character of SO_2 concentration change with time may become one of the indices of growing volcano activity. The aerogeochemical survey presents exceptional possibilities for speedy periodical measurements over active volcanoes. It should be added to the method of surveillance.

Further development of the aerogeochemical survey should be aimed at improving the method, the accumulation of data in single-component measurements (of SO_2 in the first place), and the measurements of some gas concentrations, as well as the development of the theory of the generation of evaporation haloes in the atmosphere and design of optical gas analyzers. The objective experimental studies, aside from active volcanism, should also be concerned with the discovery of mineral deposits.

5. Conclusions

The material presented above demonstrates the following facts:

1. New means of remote sensing of the environment, e.g. radar, infrared, and geochemical methods, used in surveys from aircraft allow one to obtain information which greatly improves the accuracy of our concepts of various objects and phenomena under study. In some cases radically new ideas are obtained. The experience in mounting the equipment which operates in the visible range of the spectrum, as well as the infrared receivers, on orbital platforms permits the conclusion that information can be obtaingd from space carriers and by means of other receivers. The method of studyine the objects in this case is different, as on the one hand information can be obtained periodically while on the other hand, the studies encompass simultaneously enormous areas with a considerable generalization of information due to high altitudes and small scale. In this case, the general approach in processing the information and its interpretation is changed. The generalization of information on the objects under study eliminates small, particular details and encourages the establishment of general principles in the structure of large regions. These distinctive features of information, obtained from satellites, require the development of new methods for interpretation with due regard for natural connections existing between the objects on a global scale.

2. The most comprehensive study of an object or the earth's surface is obtained when several types of remote sensing receivers are used in combination. An example of this process is the study of the contemporaneous volcanism (Kamchatka) by means of radar (regional geological structure), infrared (the structure of geothermal zones), and geochemical (gas haloes of volcanoes and geothermal zones) surveys.

It should be noted that in all investigations using new types of remote sensing devices they usually supplement the principal and the most informative kind—the photographic survey.

3. For the future, the application of new types of the environmental remote sensing studies should concentrate on development of the methodologies of surveying and of simultaneous studying the corresponding physical parameters on the ground. This will permit more reliable interpretation of data obtained by remote sensing for solving the specific problems dealing with natural resources by use of aerial and satellite carriers, and will provide a basis for developing new methods of remote sensing.

Acknowledgments

The authors wish to express their gratitude to N. A. Gussev, V. S. Loshtshilov, N. N. Semyonova, and V. A. Starostin for the courtesy of placing valuable data at their disposal.

References

Alperin, V. Z. et al. (1968). Automatic gas analyzer for continuous determination of sulphurous gas in atmospheric air. *Tr. Gl. Geofiz. Observ.* No. 234.

Barringer, A. R. (1964). *Proc. Symp. Remote Sensing Environ., 3rd,* **1964**, pp. 279–292.

Barringer, A. R., and Schork, J. R. (1966). *Proc. Symp. Remote Sensing Environ., 4th, 1966,* pp. 779–792.

Barringer, A. R., Newbury, B. C. and Moffat, A. J., (1968). *Proc. Symp. Remote Sensing Environ., 5th, 1968,* pp. 123–156.

Ginzburg, N. I. (1957). "An Experiment in Developing the Theoretical Principles of Geophysical Search Methods." Gosgeoltehizdat, Moscow.

Glushkov, V. M. et al. (1970). A new means for obtaining glaceological information. *Morsk. Flot.* **99**, No. 9.

Glushkov, V. M., Komarov, V. B., and Loshtshilov, V. S. (1972). *Rep. Congr. ISP, 12th, 1972,* pp. 1–7.

Gussev, N. A., Karizhenski, E. Ya., and Shilin, B. V. (1972). Infrared aerial survey in studying the regions of active geothermal activities. *Sov. Geol.* 1972, No. 1.

Ivanov, B. V. (1970). "Eruption of Volcano Karymski in 1962–65 and the Volcanoes of Karymski Group." *Nauka,* Moscow.

Komarov, V. B. (1971). *Proc. Symp. Remote Sensing Environ., 7th, 1971,* pp. 2275–2280.

Komarov, V. B., Starostin, V. A., and Nyavro, B. P. (1973). Radar aerial survey and its significance in the complex of aerial and space methods of geological investigations. Investigation of the natural environment by space means. Geologe and geomorphology. *Izv. Akad. Nauk SSSR,* Moscow pp. 111-119.

Manganell, R. M. (1970). *Air Pollut. Contr.* **20**, No. 7.

Masayo, M. (1961). *Bull. Geol. Surv. Jap.* **12**, No. 8.

Palmason, J., and others. (1970). *US. Symp. Develop. Util. Geotherm. Resour., 1970* Vol. 2, Part 1.

Rouse, G. C., and Stevens, D. (1970). *Mining Eng. (New York)* **22**, No. 12.

Shilin, B. V., and Berlyand M. E. (1971). *Proc. Symp. Remote Sensing Environ., 7th, 1971,* pp. 1733–1738.

Shilin, B. V., and others. (1969). *Proc. Symp. Remote Sensing Environ., 6th, 1969* pp. 175–188.

Shilin, B. V., and others (1971). *Proc. Symp. Remote Sensing Environ., 7th, 1971,* pp. 133–146.

Svyatlovski, A. E. (1959) "Atlas of Volcanoes of the U.S.S.R." Izv. Akad. Nauk, SSSR, Moscow.

Ustinova, T. I. (1955). "The Geysers of Kamchatka." Geografgiz, Moscow.

Vinogradov, V. N. (1964). On the regime of Kamchatka geysers. *Vop. Geogra. Kamchatki* No. 1.

SUBJECT INDEX

A

Aerial geological-geographic surveys, 263–322
  by radar, 265–282
  using infrared, 282–304
Aerogeochemical surveys, 304–320
  equipment and procedure for, 305–308
  method for, 304
Africa sandstorms in, as tropical cyclone sources, 7
Atmosphere, vacillation in, 178–182
Atmospheric flow, laboratory simulation of, 102–113

B

Baguio, definition of, 3
Bowen ratio, 17
Buoys, use in hurricane detection, 5

C

Carrier model of hurricane intensification, 76–77
  critique of, 78–79
Carrier model of tropical cyclone, 29–50
  critique of, 33–35
  maximum swirl speed estimate, 35–45
Christoffel symbol, 118
CISK, description and role in hurricane formation, 26–28
Computers, hurricane simulation on, 66–70
Convolution filtering
  definition of, 189–191
  frequency filtering comparison to, 244–247
  frequency response in, 191–195
  in gravity studies, 189–203

linear, discrete filters for, 192–195
  construction, 195–203
Coriolis parameter, 119
Cumulonimbi
  distribution of, in hurricane intensification, 81–82
  in tropical meteorology, 22–26
Cyclostrophic balance, in hurricane dynamics, 39
Cyclones (tropical), 1–100 (*see also* Hurricanes)
  definition of, 3
  energetics of, 50–55
  generation of, 5–8
  importance in global circulation, 13–17
  intensification of
    time-dependent flowfield in, 82–86
    theory, 75–86
  intensity of, relation to sea temperature, 55–59
  models of, 29–75
    Carrier, 29–50
    moist adiabatic process theory and, 89–91
  path prediction for, 12–13
  speed estimate of, 35–45
  swirl-divergence relation and, 45–50

D

Dirichlet's problem, 205
Dust devils, definition of, 3

E

Eye, of hurricane, 8

F

Filtering techniques
  convolution method, 189–203
  derivatives calculation of higher order, 248–250
  digitization of continuous field in, 237–240
  downward continuation operation, 213–219
    construction, 219–226
    frequency type, 226–247
  in gravity interpretation, 187–162
  power-spectrum estimation, 240–244
  in surface gravity studies, 203–226
  upward continuation type, 210–213
Frequency filtering
  in gravity studies, 226–247
    convolution filtering compared to, 244–247
    optimal Wiener type, 228–233
    procedure, 226–227
Friction in vacillation, 129

G

Geological-geographical surveys (aerial), 263–322
Geophysical flows, simulated, 102–105
Geothermal activity, infrared aerial surveys of, 286–304
Geyser Valley, infrared aerial surveys of, 295, 298–304
Gravity, filter techniques in interpretation of, 187–262
  surface type
    definition, 203–205
    upward and downward continuation of, 203–226

H

Hadley cell mechanism, hurricanes and, 15–16
Hadley flow, Rossby flow compared to, 106–108
Heat transport, northward sensible, 140–141

Heating effects, in vacillation, 129–131
Hurricane warnings, path predictions in, 3
Hurricane(s)
  computer simulation of, 66–70
  dangers and benefits of, 2
  definition of, 3
  facts about, 3–5
  kinetic energy and water content of, 87–88
  properties of, 8–12
  seeding of, implications of, 70–75
  temperature measurements in, 63–66
Hydrostatic approximation, 118

I

Ice, TOROS exploration of, 265–272
Infrared aerial surveys, 282–304
  of geothermal areas, 286–304
  of volcanoes, 282–286
Intertropical Convergence Zone (ITCZ), hurricane generation and, 4–6

K

Karymski volcano
  aerial radar survey of, 274–277
  aerogeochemical surveys of, 313–321
  infrared aerial survey of, 282–286
Kamchatka volcanoes, aerogeochemical surveys of, 313–321

M

Meteorology (tropical), 17–28
  CISK, 26–28
  cumulonimbi, 22–26
  stability, 17–22
  tensor calculus in, 117–118
  vacillation importance in, 112–113
Moist adiabatic process, theory of, 88–91
Motion field, dynamical properties of, 138–141

SUBJECT INDEX 325

P

Pacific, tropical storms in, Atlantic source of, 7
Papagallos, definition of, 3

R

Radar aerial surveys, 265–282
  of closed regions, 279–282
  of ice, 265–272
  of rock composition, 272–277
  of structural-tectonic texture, 277–279
Rainfall, hurricane effects on, 2–3
Riehl-Malkus extra oceanic heat source postulate, 26, 43–44, 53–55
  critique of, 59–63
Rock, radar aerial survey of composition of, 272–277
Rossby flow, Hadley flow compared to, 106–108

S

Sandstorms, hurricane genesis and, 7
Schuster periodogram, 242
Seeding of hurricanes, implications of, 70–75
Spectral dynamics, of vacillation, 131–138
Stability, in tropical meteorology, 17–22
Storm surges, from hurricanes, 2
Strakhov's method, for extraction of potential field signal, 233–237

T

Temperature, of hurricanes, measurement of, 63–65
Tensor calculus, general characteristics of, 114–117
Tornadoes, from hurricanes, 10
TOROS, in ice exploration, 265–272
Tropical depression, definition of, 3

Tropical disturbance, definition of, 3
Tropical storm, definition of, 3
Tropics meteorology of, 17–18
Trovado, definition of, 3
Typhoon, definition of, 3

U

Uson caldera
  aerogeochemical studies on, 308–313
  infrared aerial surveys of, 286–292, 296–297

V

Vacillation, 101–186
  atmospheric-flow simulation, 102–113
  characteristics of, 108–113
  cycles of, 175–178
  double symmetric, 178
  equations of quasi-geostrophic flow in, 120–123
  friction and heating effects in, 129–131
  Hadley and Rossby flows compared, 106–108
  kinetic energy type, 162–175
    in atmosphere, 180–182
    dynamics, 162–169
    energetics, 169–174
    physical interpretation, 174–175
  in laboratory flows, 108–111
  meteorological interest in, 112–113
  model atmosphere for study of, 113–131
  numerical study of, 101–186
  physical processes of model in, 128–129
  potential energy type, 113, 151–162
    in atmosphere, 178–179
    dynamics, 150–157
    energetics, 157–161
    physical interpretation, 162
  spectral dynamics of, 131–138
    baroclinic transfers, 145–148
    barotropic transfers, 144–145
    global forms, 141–144
    individual forms, 144

tensor formulation of quasi-hydrostatic flows in, 114–120
theoretical aspects of, 113–114
two-layer model of
  energetics, 125–128
  equations, 123–125
  unsymmetric, 175–176
Volcanoes, infrared aerial surveys of, 282–286

W

Waterspouts, from hurricanes, 10
Waves, from hurricanes, 2
Wiener filter theory, 232–233
Wiener-Hopf integral equation, 229
Willy-nilly, definition of, 3
Winds, from hurricanes, 2

QC
806
A3
v.17
1974

JAN 6 1974